WITHDRAWN FROM:
Nottingham Trent
University

Bio-Applications of Nanoparticles

ADVANCES IN EXPERIMENTAL MEDICINE AND BIOLOGY

Recent Volumes in this Series

A Continuation Order Plan is available for this series. A continuation order will bring delivery of each new volume immediately upon publication. Volumes are billed only upon actual shipment. For further information please contact the publisher.

Bio-Applications of Nanoparticles

Edited by

Warren C.W. Chan, Ph.D.

Institute of Biomaterials and Biomedical Engineering, The Terrence Donnelly Centre for Cellular and Biomolecular Research, Materials Science and Engineering, Toronto, Ontario, Canada

Springer Science+Business Media, LLC

Landes Bioscience

Springer Science+Business Media, LLC
Landes Bioscience

Printed in the U.S.A.

Please address all inquiries to the Publishers:
Landes Bioscience, 1002 West Avenue, 2nd Floor, Austin, Texas 78701, U.S.A.
Phone: 512/ 637 6050; FAX: 512/ 637 6079
http://www.landesbioscience.com

Bio-Applications of Nanoparticles, edited by Warren C.W. Chan, Landes Bioscience / Springer Science+Business Media, LLC dual imprint / Springer series: Advances in Experimental Medicine and Biology

ISBN: 978-0-387-76712-3

While the authors, editors and publisher believe that drug selection and dosage and the specifications and usage of equipment and devices, as set forth in this book, are in accord with current recommendations and practice at the time of publication, they make no warranty, expressed or implied, with respect to material described in this book. In view of the ongoing research, equipment development, changes in governmental regulations and the rapid accumulation of information relating to the biomedical sciences, the reader is urged to carefully review and evaluate the information provided herein.

Library of Congress Cataloging-in-Publication Data

Bio-applications of nanoparticles / edited by Warren C.W. Chan.
    p. ; cm. -- (Advances in experimental medicine and biology ; v. 620)
  Includes bibliographical references and index.
  ISBN 978-0-387-76712-3
  1. Nanoparticles. 2. Nanostructured materials. 3. Biomedical materials. I.
Chan, Warren C. W. II. Series.
  [DNLM: 1. Nanostructures. 2. Diagnostic Imaging--methods. 3.
Nanostructures--toxicity. 4. Nanotechnology. W1 AD559 v.620 2007 / QT 36.5
B6135 2007]
  RS201.N35B36 2007
  616.07'54--dc22

                                      2007042193

About the Editor...

WARREN C. W. CHAN received his B.S. degree in 1996 from the University of Illinois (Urbana-Champaign, IL, USA) and Ph.D. degree in 2001 from Indiana University (Bloomington, IN, USA). He conducted post-doctoral training at the University of California (San Diego, CA, USA). Currently, he is an assistant professor at the University of Toronto in the Institute of Biomaterials and Biomedical Engineering. He is also affiliated with the Department of Materials Science and Engineering and Chemical Engineering as well as the Terrence Donnelly Center for Cellular and Biomolecular Research. His research interest is in the development of nano- and microtechnology for cancer and infectious-disease diagnosis. He has received the BF Goodrich Young Inventors Award, Lord Rank Prize Fund Award in Optoelectronics, Canadian Research Chair in Bionanotechnology, and Early Career Award. Further, he is currently an Associate Editor of the "Journal of Biomedical Nanotechnology" and "Nanomedicine" and a panel member on the Council of Canadian Academies on Nanotechnology.

DEDICATION

I would like to dedicate this book to my grandmother Mrs. Yu Ye Yee (余玉叶), who has taught me many important values of life. One of those is "how to enjoy the little things in life."

PREFACE

There has been tremendous excitement in the hybrid research field of Bionanotechnology. The importance of nanotechnology stems from the ability to custom-design the properties of a material by changing the material's composition, size, and shape. For biological and medical researchers, novel detection systems and tools can be systematically designed to address research or application needs. The amount of grant dollars invested and number of published reports in this field have been growing at an exponential rate in the last ten years.

In this edited book, we highlight this exciting hybrid research field. We highlight the central players in the Bionanotechnology field, which are the nanostructures and biomolecules. We start the book by describing how nanostructures (Chapter 1) are synthesized and by describing the wide variety of nanostructures available for biological research and applications. We also show the techniques used to synthesize a wide variety of biological molecules (Chapter 2). Hopefully, the placement of these two chapters end-to-end shows similarities between how we make useful nanostructures and how nature makes useful molecules. Next, we focus on the assembly of nanostructures with biological molecules, which could lead to the design of multi-functional nanosystems. In Chapters 3 to 11, examples of the unique properties of nanostructures are provided along with the current applications of these nanostructures in biology and medicine. Some applications include the use of gold nanoparticles in diagnostic applications, quantum dots and silica nanoparticles for imaging, and liposomes for drug delivery. In Chapters 12 and 13, the toxicity of nanostructures are described. This book provides broad examples of current developments in Bionanotechnology research and is an excellent introduction to the field. Enjoy reading!

Warren C.W. Chan, Ph.D.

ACKNOWLEDGEMENTS

Dr. Warren C. W. Chan would like to acknowledge the various funding agencies for supporting his research efforts: Canadian Institute of Health Research, Natural Sciences and Engineering Research Council of Canada, Canadian Foundation for Innovation, Ontario Innovation Trust, Connaught Foundation, and Genome Canada through Ontario Genomic Institute. Dr. Chan would also like to thank all of the wonderful students, post-docs, technicians, and administrators in his research lab.

PARTICIPANTS

Lajos P. Balogh
Department of Radiation Medicine
and
Roswell Park Cancer Institute
Department of Oncology
University at Buffalo SUNY
Buffalo, New York
U.S.A.

Warren C.W. Chan
Institute of Biomaterials and Biomedical
    Engineering
The Terrence Donnelly Centre for
    Cellular and Biomolecular Research
 Materials Science and Engineering
Toronto, Ontario
Canada

Jinwoo Cheon
Department of Chemistry
    and
Nano-Medical National Core
    Research Center
Yonsei University
Seoul
Korea

Rahul Chhabra
Department of Chemistry
    and Biochemistry
Arizona State University
Tempe, Arizona
U.S.A.

P. Davide Cozzoli
National Nanotechnology Laboratory
    of NCR-INFM
Unita di Ricerca IIT
Distretto Tecnologico ISUFI
Lecce
Italy

Shivang R. Dave
Department of Bioengineering
University of Washington
Seattle, Washington
U.S.A.

Xiaohu Gao
Department of Bioengineering
University of Washington
Seattle, Washington
U.S.A.

Partha Ghosh
Department of Chemistry
University of Massachusetts
Amherst, Massachusetts
U.S.A.

Gang Han
Department of Chemistry
University of Massachusetts
Amherst, Massachusetts
U.S.A.

Keith B. Hartman
Department of Chemistry
Rice University
Houston, Texas
U.S.A.

Jung-tak Jang
Department of Chemistry
    and
Nano-Medical National Core
    Research Center
Yonsei University
Seoul
Korea

Travis Jennings
Institute of Biomaterials and Biomedical
  Engineering
University of Toronto
Toronto, Ontario
Canada

Young-wook Jun
Department of Chemistry
  and
Nano-Medical National Core
  Research Center
Yonsei University
Seoul
Korea

Jelena Kolosnjaj
UMR CNRS 8612, Faculté
  de Pharmacie
Université Paris-Sud 11
Châtenay-Malbry
France
and
Pharmacy Department
University of Ljubljana
Ljubljana
Slovenia

Isaac T.S. Li
Institute of Biomaterials
  and Biomedical Engineering
and
Edward S. Rogers Sr. Department of
  Electrical and Computer Engineering
University of Toronto
Toronto, Ontario
Canada

Yan Liu
Department of Chemistry
  and Biochemistry
Arizona State University
Tempe, Arizona
U.S.A.

Jasmina Lovrić
Department of Pharmacology
  and Therapeutics
McGill University
Montreal
Canada
and
Department of Pharmaceutical
  Technology
University of Zagreb
Zagreb
Croatia

Liberato Manna
National Nanotechnology Laboratory
  of NCR-INFM
Unita di Ricerca IIT
Distretto Tecnologico ISUFI
Lecce
Italy

Dusica Maysinger
Department of Pharmacology
  and Therapeutics
McGill University
Montreal
Canada

Fathi Moussa
UMR CNRS 8612
Université Paris-Sud 11
Châtenay-Malbry
France

Elizabeth Pham
Institute of Biomaterials
  and Biomedical Engineering
and
Edward S. Rogers Sr. Department of
  Electrical and Computer Engineering
University of Toronto
Toronto, Ontario
Canada

Vincent M. Rotello
Department of Chemistry
University of Massachusetts
Amherst, Massachusetts
U.S.A.

Reto A. Schwendener
Institute of Molecular Cancer Research
University of Zurich
Zurich
Switzerland

Jaswinder Sharma
Department of Chemistry
  and Biochemistry
Arizona State University
Tempe, Arizona
U.S.A.
Geoffrey Strouse
Department of Chemistry
Florida State University
Tallahassee, Florida
U.S.A.

Henri Szwarc
UMR CNRS 8612, Faculté
  de Pharmacie
Université Paris-Sud 11
Châtenay-Malbry
France

Weihong Tan
Department of Chemistry
Center for Research at the Bio/Nano
  Interface
Shands Cancer Center
  and UF Genetics Institute
University of Florida
Gainesville, Florida
U.S.A.

Kevin Truong
Institute of Biomaterials
  and Biomedical Engineering
and
Edward S. Rogers Sr. Department of
  Electrical and Computer Engineering
University of Toronto
Toronto, Ontario
Canada

Lin Wang
Department of Chemistry
Center for Research at the Bio/Nano
  Interface
Shands Cancer Center
  and UF Genetics Institute
University of Florida
Gainesville, Florida
U.S.A.

Lon J. Wilson
Department of Chemistry
Rice University
Houston, Texas
U.S.A.

Hao Yan
Department of Chemistry
  and Biochemistry
Center for Single Molecular Biophysics,
  The Biodesign Institute
Arizona State University
Tempe, Arizona
U.S.A.

Wenjun Zhao
Department of Chemistry
Center for Research at the Bio/Nano
  Interface
Shands Cancer Center
  and UF Genetics Institute
University of Florida
Gainesville, Florida
U.S.A.

CONTENTS

Section I. Nanostructure and Biomolecule Synthesis

Section II. Applications of Nanostructures

11. DENDRIMER 101 136

Lajos P. Balogh

Section III. Toxicity of Nanostructures

12. QUANTUM DOTS AND OTHER FLUORESCENT
NANOPARTICLES: QUO VADIS IN THE CELL? 156

Dusica Maysinger and Jasmina Lovrić

13. TOXICITY STUDIES OF FULLERENES AND DERIVATIVES 168

Jelena Kolosnjaj, Henri Szwarc and Fathi Moussa

14. TOXICITY STUDIES OF CARBON NANOTUBES 181

Jelena Kolosnjaj, Henri Szwarc and Fathi Moussa

Synthetic Strategies to Size and Shape Controlled Nanocrystals and Nanocrystal Heterostructures

P. Davide Cozzoli and Liberato Manna*

Abstract

The recognition of the strongly dimensionality-dependent physical-chemical properties of inorganic matter at the nanoscale has stimulated efforts toward the fabrication of nanostructured materials in a systematic and controlled manner. Surfactant-assisted chemical approaches have now advanced to the point of allowing facile access to a variety of finely size- and shape-tailored semiconductor, oxide and metal nanocrystals (NCs) by balancing thermodynamic parameters and kinetically-limited growth processes in liquid media. While refinement of this synthetic ability is far from being exhausted, further efforts are currently made to provide NCs with higher structural complexity as means to increase their functionality. By controlling crystal miscibility, interfacial strain, and facet-selective reactivity at the nanoscale, hybrid NCs are currently engineered, which consist of two or more chemically different domains assembled together in a single particle through a permanent inorganic junctions. In this chapter, we will review the strategies that have been so far developed for the synthesis of colloidal nanostructures, ranging from mono-material NCs with tailored dimensions and morphology to multi-material NC heterostructures with a topologically controlled composition.

Introduction

In the last decade, chemically synthesized colloidal nanocrystals (NCs) have become model systems to verify the dimensionality-dependent laws of nanosized materials, as advances in their synthetic protocols have made them available at the milligram-to-gram scale in a wide range of monodisperse sizes and morphologies.[1-4]

Two major reasons for the uniqueness of NCs: the significant fraction of atoms residing at the surface, as compared to that found in the corresponding bulk counterparts, and the restriction of charge carrier motion to a small material volume. Owing to these contributions, the chemical and physical properties systematically correlates with the NC size and shape.[5-12] For instance, NCs melt at much lower temperatures than those required for extended solids and can be easily trapped in crystalline phases that are unstable in the bulk.[13-16] In semiconductors, quantum confinement sets in once a critical threshold size is reached, leading to a widening of the band gap and of the level spacing at the band edges.[14,17] In noble metals, the decrease in

*Corresponding Author: Liberato Manna—National Nanotechnology Laboratory of CNR-INFM, Unità di Ricerca IIT, Distretto Tecnologico ISUFI, Via per Arnesano, Km 5, I-73100 Lecce, Italy. Email: liberato.manna@unile.it

Bio-Applications of Nanoparticles, edited by Warren C.W. Chan. ©2007 Landes Bioscience and Springer Science+Business Media.

size below the electron mean free path leads to intense surface plasmon absorption bands arising from the collective oscillations of the itinerant conduction electron gas on the particle surface.[18] In general, the size-dependence of the NC electronic structure provides a tool for tuning the red-ox potentials of the charge carriers and for modulating the dynamics of red-ox processes, while the dominance of electronic surface states in NCs confers them improved or unprecedented catalytic properties.[19-22] Finally, nanostructured magnetic materials behave as single magnetic domains whose magnetization can be easily influenced by thermal fluctuations of the local environment, depending on the particle size and on a variety of surface effects.[23,24]

Colloidal NCs are synthesized in a liquid solution containing some stabilizing organic molecules, broadly termed as surfactants. As extracted from their growing medium, they are typically made of a crystalline core with the desired chemical composition and a monolayer surface shell of tightly coordinated surfactants that provide them with solubility and, hence, with colloidal stability. These particles are chemically robust and thus amenable to be further processed after the synthesis to achieve disparate purposes, such as their integration with existing devices, or their incorporation into biological environments without substantial loss of the original properties.[25-29] Additionally, organization of NCs into ordered superstructures holds promise for the fabrication of new solid state materials whose chemical-physical behaviour would result from the collective interactions among proximal particles with individually relevant size-related properties.[30-33]

In colloidal solutions, NC growth evolution is driven by the interplay between thermodynamic factors (e.g., relative stability of crystal polymorphs) and kinetically-limited processes (e.g., diffusion of reactants, surface adhesion of surfactants). However, control over most of these parameters can be achieved by judicious adjustment of just a few experimental conditions, such as the type and the relative concentration of molecular precursors, catalysts, and organic stabilizers, in combination with a suitable growth temperature.[34-40] Indeed, NCs have been accessed with nanometer level precision in a variety of dimensional regimes, as well as in shapes as diverse as spheres, cubes,[41-43] rods,[12,44-49] wires,[43,50-53] tubes, stars,[54,55] disks,[56-58] and polypods,[43,53,59-64] by employing relatively inexpensive equipment. The substantial piece of knowledge gained from such synthetic success has contributed to trace useful size and shape guiding criteria, although these are yet far from being applicable in general to any material. A powerful extension of these concepts has been more recently devised with the synthesis of elaborate hybrid NCs possessing a topologically controlled composition, i.e., particles consisting of two or more chemically different material sections grouped together through an inorganic junction. The formation of such NC heterostructures depends on the ability to control additional parameters at the nanoscale, such as crystal miscibility, interfacial strain, and face-selective reactivity, by applying the same techniques as those used to fabricate single material NCs. Examples of these elaborate nanostructures include: (i) core/shell NCs, in which one or more layers of additional inorganic materials are uniformly grown around a NC core, forming an onion-like structure; (ii) hetero-dimers or hetero-oligomers, in which two or more spherical domains of different materials are joined together sharing specific couples of facets; (iii) matchstick- and dumbbell-shaped NCs, made of a rod-like section with one or two spherical particles of another material grown either on one or on both of its terminations, respectively;[65-67] (iv) multi-armed hybrid NCs comprising linear sections made of consecutive segments of different materials.[68] As a result of the combination of the properties that distinguish each crystalline domain, hybrid NCs hold promise as first prototypes of "smart" nanosized objects, potentially able to perform multiple technological tasks, such as in biomedical engineering, diagnostics, sensing, and catalysis.

This chapter will describe the general strategies that have been so far developed for the synthesis of size and shape controlled colloidal nanocrystals, and how such methods are currently being implemented to fabricate more elaborate hybrid nanocrystal heterostructures.

Colloidal Approaches: A Few General Concepts

In a colloidal synthesis, molecular precursors containing the atomic species that will form the NCs are introduced into a flask and allowed to react or decompose at a suitable temperature. Reactive species, commonly referred to as the NC "monomers", are consequently generated, inducing the nucleation of the NCs and sustaining their progressive enlargement. The key condition to achieve size- and shape-controlled growth of NCs is the presence of one or more surfactant species in the reaction environment. Surfactants are amphiphilic molecules composed of a polar head group and of one or more hydrocarbon chains with hydrophobic character. The most commonly used ones in colloidal syntheses include alkyl thiols, amines, carboxylic and phosphonic acids, phosphines, phosphine oxides, phosphates, phosphonates, and various coordinating (e.g., ethers) or noncoordinating solvents (e.g., alkanes, alkenes). Surfactants can act in two different ways, i.e., either by forming dynamic templates which pose physical constraints to the uncontrolled NC enlargement, or by behaving as surface complexing agents. Both mechanisms can ultimately guarantee slow growth rate, impart stability inhibiting inter-particle agglomeration, and ensure post-synthesis processability of the resulting NCs.

Dynamic Templates

When water and a nonpolar solvent are mixed in the presence of amphiphilic molecules (such as surfactants, lipids, some polymers and also some proteins), thermodynamically stable, either hydrophilic or hydrophobic micelles can spontaneously form, having different sizes and shapes depending on parameters, such as the relative 'water' to 'oil' phase abundance, solvent polarity, ionic strength, concentration of surfactants and additives. Therefore, micelles represent preformed compartments of nanometer size which can serve to confine the NC growth.[38]

Terminating Agents

In monophasic liquid systems, surfactants behave as terminating agents and perform two important tasks. They modulate the monomer reactivity by forming complexes with the precursors and assist particle growth by dynamically coordinating to the surface of the NCs, preventing them from aggregating irreversibly. The choice of surfactants varies from case to case: a molecule that binds too strongly to the surface of the particles is not useful, as it would not allow them to grow. On the other hand, a weakly coordinating molecule could promote the production of comparatively larger particles, or even insoluble aggregates.[20]

When the growth is stopped, a monolayer of surfactants remains tightly bound to the surface of the particles. Depending on its chemical nature, this capping shell can make NCs soluble in a wide range of liquid media. If desired, the coating can be deliberately exchanged with other organic molecules having different functional groups or polarity to purposely functionalize the NC surface. Based on such processing flexibility, after their synthesis NCs can be embedded into polymeric matrixes, conjugated with biological molecules, linked to other types of particles, immobilized onto substrates, or integrated with existing devices for creating a variety of electronic and sensor components.

Size Control

To obtain satisfactory control over the NC size, two basic conditions must be ideally fulfilled during the synthesis: a discrete, i.e., limited in time, nucleation event should occur first, that should be followed by a much slower growth process of the initially formed nuclei. As predicted by theoretical models, if the growth step occurs under a relatively high and nearly constant flux of monomers, the size distribution tends to narrow over time, as the smallest NCs in the population grow faster than the largest ones. In such "size focusing" regime, detrimental Ostwald ripening is inhibited until the monomer concentration does not become exceedingly consumed. The relative depletion of monomers between the nucleation and the growth stages will ultimately dictate the final NC size. An example is reported in Figure 1 for the case of $CoPt_3$.[69]

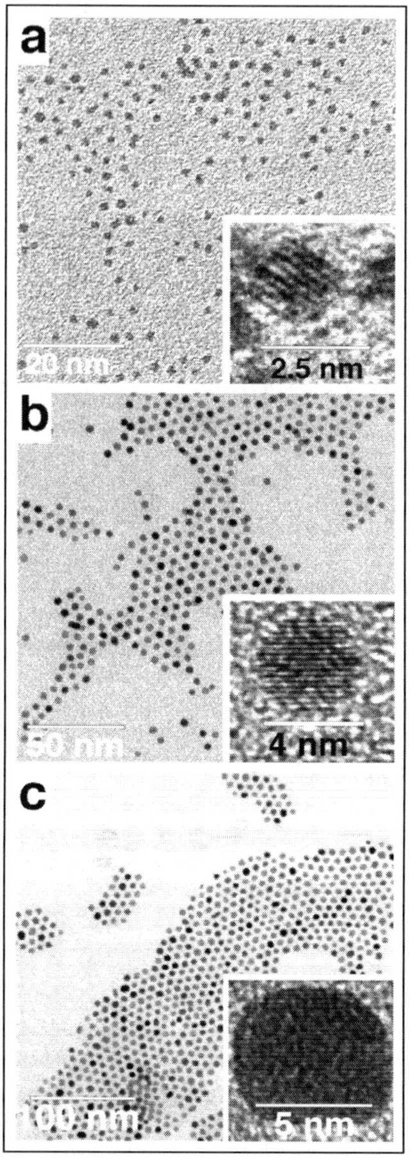

Figure 1. Example of size control achievable for colloidal NCs: Transmission Electron Microscopy (TEM) pictures of spherical $CoPt_3$ magnetic NCs with mean size of 1.5 nm (a), 3.8nm (b), 6.3 nm (c), respectively (reprinted with permission from Fig. 1; J Am Chem Soc 2002; 124:11480. Copyright 2007 American Chemical Society).

The above requirements can be often practically fulfilled by employing the so-called "hot injection technique", which relies on the fast addition of a comparatively "cold" (e.g., at room temperature) precursor solution into a high temperature (100–350°C) bath of surfactants. The rapid injection raises the monomer concentration above the nucleation threshold, resulting in a burst of nucleation that partially relieves the solution supersaturation. The simultaneous sudden drop in temperature that follows the injection greatly arrests nucleation, keeping it temporally seperatedfrom the subsequent growth stages. These latter are allowed to proceed at a temperature lower than that used for the injection, which allows monomers to be slowly released into the reaction mixture and the system to be maintained in a "size-focusing" regime for a long time.

In some cases, however, it is more advantageous to simply mix the reagents at low temperatures and then slowly heat the resulting mixture up to a target temperature. This syntetic scheme is especially suitable for those systems in which the rate limiting step is the nucleation stage, which is indeed naturally delayed as it is depends on the accumulation of a sufficiently high monomer concentration upon slow decomposition of surfactant-precursor complexes.

Relatively monodisperse colloidal NCs of a wide range of materials can be easily synthesized by means of variously modified approaches, all based on the above general synthetic schemes. Frequently, NCs synthesized as described above are roughly spherical or only slightly faceted, as high reaction temperatures and moderate monomer concentrations favour the formation of more thermodynamically stable isotropic shapes.

Shape Control

The production of NCs with anisotropic shapes (wires, rods, discs, branched shapes and so on) can be achieved in solution by several strategies. The most effective approaches exploit: (i) reactions confined in micelles;[36,38,70,71] (ii) catalysis with a metal particle;[47,53,72-74] (iii) seeded growth;[12,62,75-78] (iv) oriented attachment mechanisms;[63,79-82] (v) surfactant or solvent induced anisotropy;[34,35,41,48,50,54-56,83] and (vi) application of external electric or magnetic fields.[84-86] Figure 2 shows a few selected examples of variously shaped colloidal NCs of different materials. We will briefly review here each of these approaches.

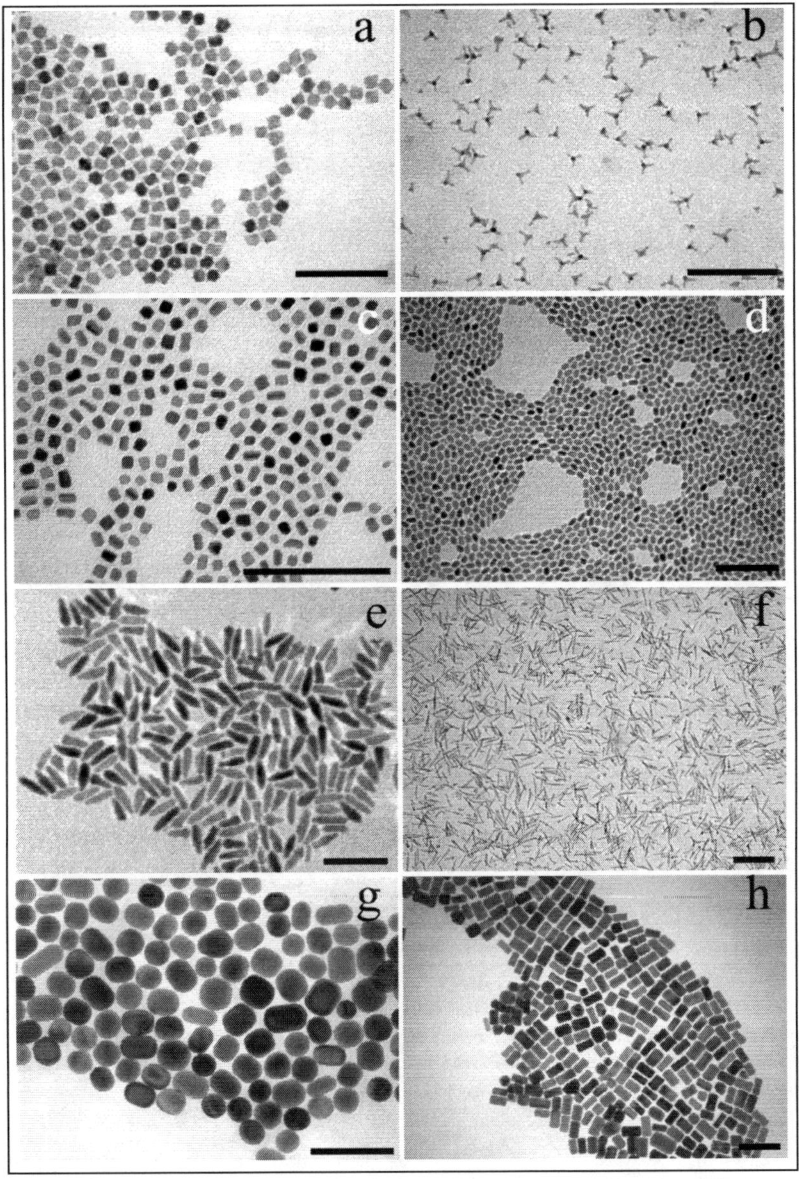

Figure 2. Examples of colloidal NCs with different shapes: TEM images of star-like PbSe NCs (synthesized according to J Phys Chem B 2002; 106:10634) (a), γ-Fe$_2$O$_3$ tetrapods (synthesized according to Nano Lett. 2006, 6, 1986) (b), FePt faceted NCs (synthesized according to J Am Chem Soc 2006; 128:7132) (c), rise-shaped CdSe NCs (synthesized according to ref. 29) (d), pencil-shaped CdS NCs (synthesized according to ref. J Am Chem Soc 2006; 128:7132) (e), high aspect ratio TiO$_2$ nanorods (synthesized according to J Phys Chem B 2005; 109:5389) (f), Ag spheroids (synthesized according to Chem Commun 2001; 7:617) (g), and rectangular-shaped Au nanorods (synthesized according to J Phys Chem B 2005; 109:13857) (h). The scale bar in each panel corresponds to 100 nm.

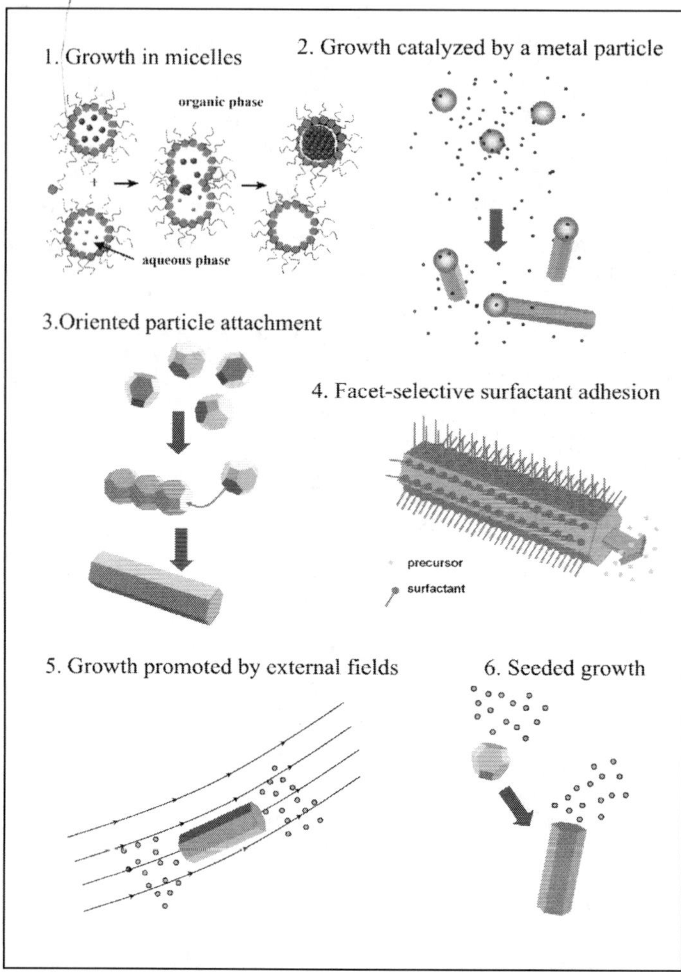

Figure 3. Mechanisms for the growth of colloidal nanocrystals with anisotropic shapes. Reproduced by permission of The Royal Society of Chemistry. Scheme 1; Chem Soc Rev 2006; 35:1195.

Growth in Micelles

A certain degree of shape control is achievable by performing chemical reactions inside micelles, as the latter can indeed evolve from spherical into rod-like or cylindrical aggregates, and planar bilayers, depending on their formation conditions. NCs grown in such confining "nano-reactors" can adopt morphologies resembling those of the micelles themselves (Fig. 3, panel 1). Although rod-, wire- and platelet-shaped NCs have been prepared by this approach,[36,38,70,71] it remains prohibited to materials requiring high temperatures to anneal crystal defects.

Growth in the Presence of a Catalyst Particle

The vapour-liquid-solid (VLS) growth mechanism in which a solid one-dimensional structure departs from a catalyst particle[87-91] has been recently translated into the solution-liquid-solid (SLS) growth of colloidal nanorods and nanowires of several materials, such as of CdSe, InAs,

InP, Si, Ge[47,53,72-74,92] and branched CdSe nanowires[53] (Fig. 3, panel 2).[47,53,72-74,92] In this version, the solution containing the NC molecular precursors is loaded with colloidal metal nanoparticles that act as catalysts, possibly becoming supersaturated with the monomers and offering a preferential site for their deposition. This helps to interrupt the spherical symmetry of growth, thereby promoting unidirectional crystal elongation. Recently, formation of PbSe rods, cubes, stars and multi-pods has been obtained by analogous approach,[78] although in the latter case a heterogeneous nucleation of PbSe on the existing catalyst seeds is more likely to be operative.

Oriented Attachment

In this mechanism,[63,79-82] nearly isotropic NCs are first formed in solution, which then tend to fuse epitaxially along well-defined crystallographic directions as a means to eliminate some high-energy facets; hence, this leads to a decrease in the overall surface energy. Wires,[79] rings,[80] rods,[81] and branched NCs[63] (Fig. 3, panel 3) can be accessed especially when the initial NCs are only weakly passivated by organic ligands, so that dipole-induced inter-particle attractive forces are enhanced and one-directional NC attachment is spontaneously promoted. One peculiar advantage of this method is that it allows materials that crystallize in highly symmetric structures (i.e., PbSe in the rock-salt structure[80] or ZnS in the zinc-blende structure[81]) to grow anisotropically without any catalyst or additive.

Surfactant or Solvent Directed Growth

For many materials, surfactants can be found that are able to adhere with different bonding strengths to the various facets of the NCs, thereby inducing their preferential development along those crystallographic directions which grow the fastest[34,35,41,48,55,56] (Fig. 3, panel 4). Such morphology changes are kinetically controlled processes which occur far from thermodynamic equilibrium when the system is overdriven by a high concentration of monomers.[93] These conditions accentuate the growth rate of the most unstable facets that are usually protected less efficiently by the organic ligands. At low concentration of monomers, the growth is more under thermodynamic control, and the situation can be reversed so that atoms can detach from the most unstable facets and will feed other facets. Over time the overall habit of the crystals evolves toward the shape that minimizes the overall surface energy (such as spheres or cubes).

The surfactant-adhesion growth mechanism produces pronouncedly faceted NCs in materials that crystallize in highly symmetric phases,[94] whereas it leads to strongly anisotropic shapes, such as discs, rods, wires, when the NCs occur in phases with unique axis of symmetry (examples are the hexagonal close packed structure for Co,[56] the wurtzite structure for CdSe,[45] CdS and in some cases for ZnSe,[95] the hematite structure in Fe$_2$O$_3$,[96] the anatase structure for TiO$_2$[97,98]). For instance, in rod-like or platelet-shaped NCs, the unique symmetry axis direction coincides with either the fast or the slow direction of crystal growth, respectively.

Branching can also occur in NCs, leading to multi-armed structures, like tetrapods, in which several rod sections are connected at tetrahedral angles from a branching point. So far, this NC shape has been explained by the formation of planar defects at the junction region between the branching point and each arm.[61,99] There are reported cases, however, in which the nonsymmetric crystal structure is not a key requirement in order to grow anisotropic NCs. PbS in the rock salt phase has been grown in shapes ranging from rods to multipods, stars and cubic shapes, depending on the experimental parameters.[54] Recently, magnetic iron oxide tetrapods in cubic spinel structure of maghemite have been also synthesized.[100] While also in this case the surfactants can play a role in modulating the growth rate of the various facets, kinetic factors are likely to be more relevant.

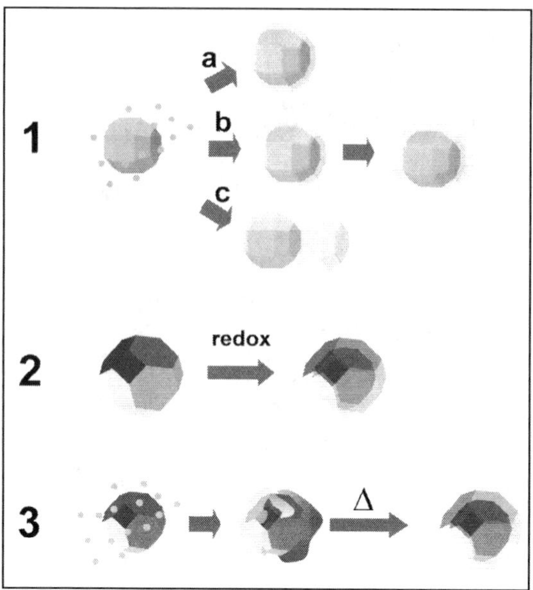

Figure 4. Sketch of possible mechanisms leading to core-shell nanocrystal heterostructures: growth of a single, of a multiple or of an asymmetric shell on nanocrystal cores (paths 1a-c, respectively); shell formation following a redox replacement reaction with the initial core (path 2); formation of a uniform shell upon thermal annealing of an initially amorphous and/or discontinuous coating (path 3). Reproduced by permission of The Royal Society of Chemistry. Scheme 2; Chem Soc Rev 2006; 35:1195.

Anistropic Growth Induced by External Biases

In few cases, it has been possible to induce anisotropic growth of magnetic materials by performing their synthesis in the presence of an external magnetic field. In such circumstances, a preferential elongation of the NCs along the easy magnetization axis direction has been observed.[101,102] (Fig. 3, panel 5). Nanowires have been produced also in the presence of electric fields.[86] These approaches can be advantageous in that one-dimensional NCs can be produced at the locations where they are needed, for instance, between two electrodes, as the anisotropic growth direction is often the one along a field line.

Seeded Growth

Anisotropic growth of noble metal materials (Ag, Au, Pt) has been easily achieved by seed-mediated reaction schemes[12,62,75,76] (Fig. 3, panel 6). In these approaches, preformed (usually spherical) NC "seeds" of the desired metallic material are mixed with metal ion-surfactant complexes and eventually some additives. The seeds behave as efficient red-ox catalysts for metal ion reduction, greatly enhancing heterogeneous nucleation by fast monomer addition to their surface. Under the assistance of facet-selective surfactant adhesion, such process leads to nanorods, nanowires and also branched nanostructures.[62] This method has been also extended to the size and shape control of other materials.[77]

Hybrid Nanocrystals

The most recent developments of colloidal syntheses involve elegant extensions of the above described techniques that allow the fabrication of more elaborate hybrid nanocrystals (HNCs). Such nanostructures are comprised of crystalline domains made of different materials fused together in a single particle without bridging molecules. The increase in the degree of structural complexity is expected to naturally enhance the particle functionality, as the chemical-physical properties characteristic of each material are grouped and/or modulated as a consequence of the mutual interactions among the components.

HNCs can be expected to satisfy both fundamental interests and disparate technological requirements. For instance, HNCs based on combinations of semiconductors, metals and oxides in a centrosymmetric core-shell type configuration often exhibit improved behavior as

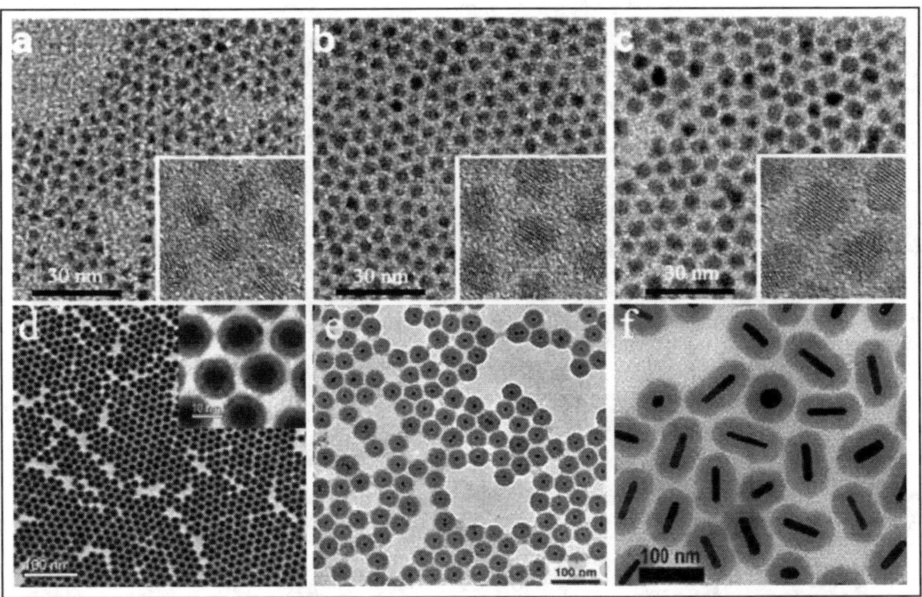

Figure 5. Examples of core/shell nanocrystals. TEM images of: (a) CdSe, (b) CdSe/CdS, and (c) CdSe/CdS/ZnS NCs prepared by consecutively growing CdS and ZnS shells around the same CdSe cores. Reprinted with permission from Figs. 4a-c of J Phys Chem B, 2004; 108:18826. Copyright 2007 American Chemical Society.; (d) Fe/Fe3O4 NCs. Reprinted with permission from Fig. 1A of J Am Chem Soc 2006; 128:10676. Copyright 2007 American Chemical Society.; (e) FePt/SiO$_2$ NCs. Reprinted with permission from Fig. 1B of J Phys Chem B 2006; 110:11160. Copyright 2007 American Chemical Society.; (f) Au nanorod/SiO$_2$ NCs. Reprinted with permission from Fig. 3S of Chem Mater, 2006; 18:2465. Copyright 2007 American Chemical Society.

compared to that inherent to the individual components, such as enhanced or tunable plasmon absorption[103-106] or luminescence,[107-112] modified magnetic behaviour,[113-119] and improved photocatalytic and photoelectrochemical responses.[120-130] In noncentrosymmetric HNCs made of well distinct inorganic sections, such as small groups of spherical and/or rod-like NCs, the properties inherent to each domain, such as magnetism, optical absorption or fluorescence, can be altered to a varying extent because of the contact junction with another material. In semiconductor heterostructures, depending on the relative band gap alignment of the components, the charge carriers can be either localized preferentially in one domain or separated more efficiently, which can have important implications in optoelectronic and photovoltaic applications. Furthermore, HNCs allow for the introduction of anisotropic distributions of surface functional groups, facilitating the site-specific anchoring of biomolecules[131] or the controlled assembly of NCs.[132]

The most widely exploited technique for the synthesis of HNCs relies on a simple principle deriving from the classical nucleation theory, according to which the activation energy for the generation of novel NC nuclei in solution (homogeneous nucleation) is usually higher than the barrier for enlargement of preexisting particles (heterogeneous nucleation). This concept has been translated into a simple and general reaction scheme, referred to as "seed-mediated". In this scheme, one deliberately introduces preformed NCs of one material (the "seeds") into a solution containing the monomers to build a second different material which triggers preferential heterogeneous growth of the latter onto such seeds in contrast to the independent formation of novel embryos. The architectural configuration adopted by resulting hybrid architecture will

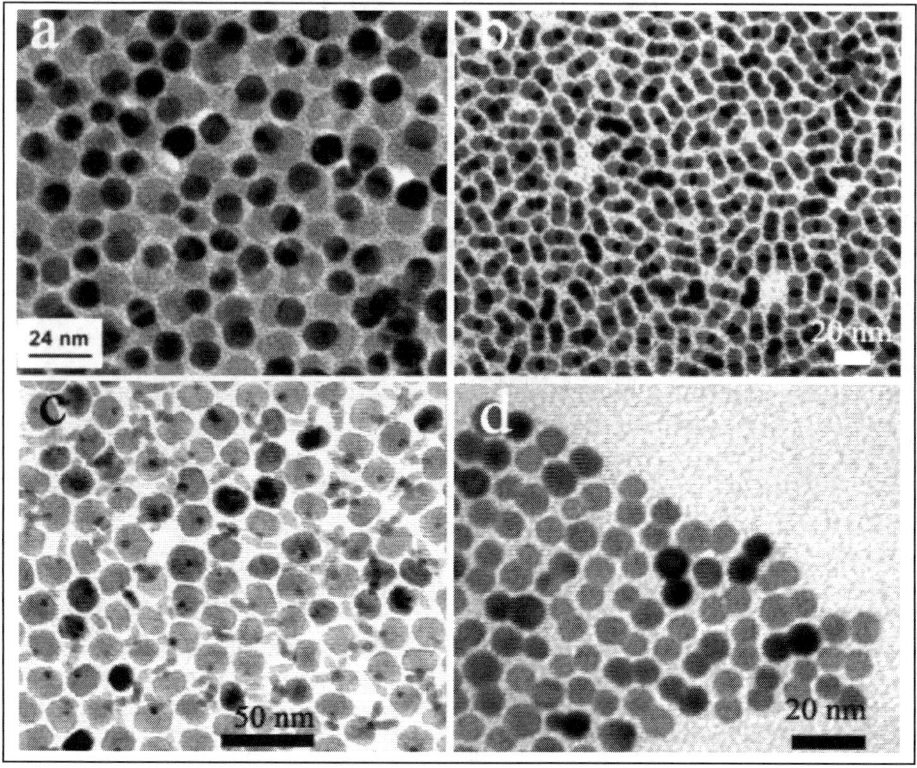

Figure 6. Examples of nanocrystals hetero-oligomers. TEM images of: (a) dumbbell-like Fe3O4-Au dimers. Reprinted with permission from Fig. 1 of Nano Lett, 2005; 5:379. Copyright 2007 American Chemical Society; (b) dumbbell-like PbS-Au-PbS ternary NCs. Reprinted with permission from Fig. 3c of Nano Lett, 2006; 6:875. Copyright 2007 American Chemical Society; (c) PbSe nanorod-Fe$_3$O$_4$-Au ternary NCs (taken from Fig. 2b of Adv Mater 2006; 18:1899); (d) CoPt3-Au dimers (synthesized according to J Am Chem Soc, 2006; 128:6690).

ultimately depend on the interplay among parameters, such as crystal miscibility, interfacial strain between the materials, and facet-selective reactivity of the seeds.

Core-Shell Nanocrystals

A large interface can be shared between two materials if the respective lattice constants do not differ significantly and/or if the interfacial energy is kept low during the synthesis. These conditions lead to the formation of onion-like nanostructures, referred to as core/shell HNCs, in which the starting core is uniformly covered by one or more different inorganic layers. The general strategy to synthesize these objects relies on performing a rather slow addition of the shell molecular precursors to a solution containing the target NC cores at relatively low temperatures, which usually prevents homogeneous nucleation of the second material. Core/shell associations of metal, semiconductor, magnetic and oxide materials have been successfully obtained by various mechanisms, such as: (i) by coreacting all the necessary molecular precursors[133,134] or by alternating deposition of monolayers of each atomic species that will compose the shell[135] (Fig. 4, paths 1a-c); (iii) by replacing the outermost layer of the core with the shell material by means of a sacrificial redox reaction[103,106,136,137] (Fig. 4, path 2); and (iv) by thermally annealing an initially amorphous and/or discontinuous shell[119] (Fig. 4, path 3). In few

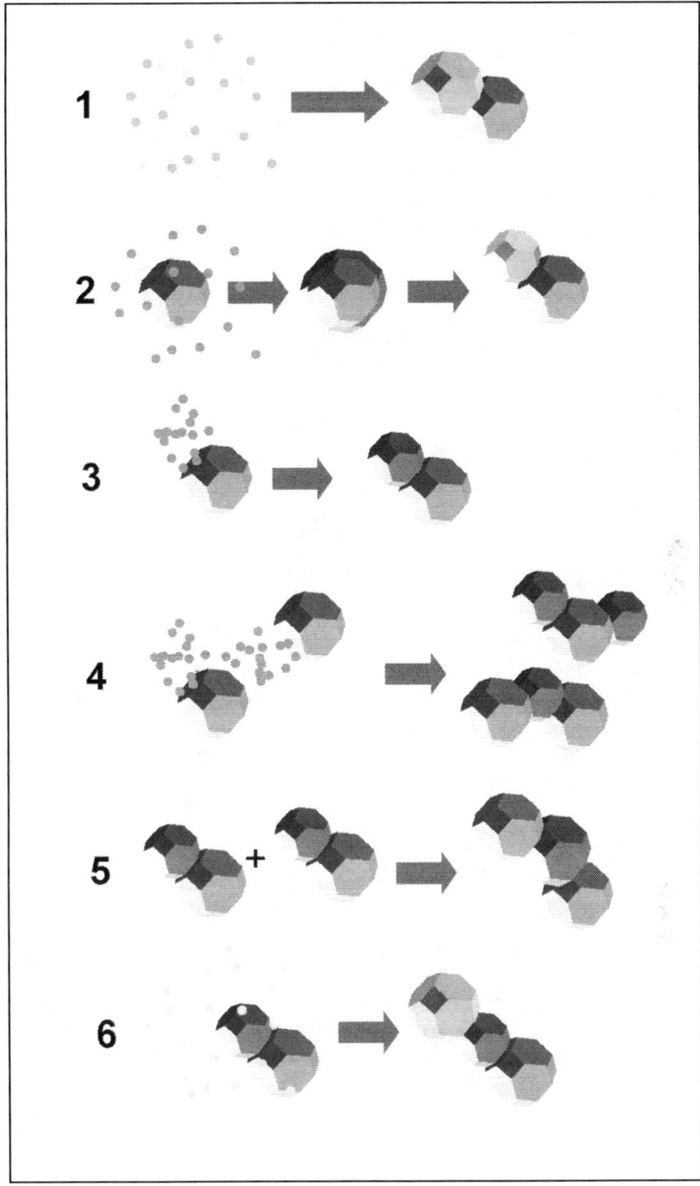

Figure 7. Sketch of possible mechanisms leading to nanocrystal oligomers: formation of a heterodimer by phase segregation of two immiscible materials (path 1), by coalescence of an initially amorphous shell (path 2), or by selective nucleation on a starting seed (path 3); growth of a trimer upon formation of a domain which bridges two preformed NCs (path 4); formation of a trimer by fusion of two reactive domains from distinct dimers (path 5); formation of a trimer by selective nucleation on a preformed heterodimer seed (path 6) Scheme 3 of Chem Soc Rev 2006; 35:1195). Reproduced by permission of The Royal Society of Chemistry.

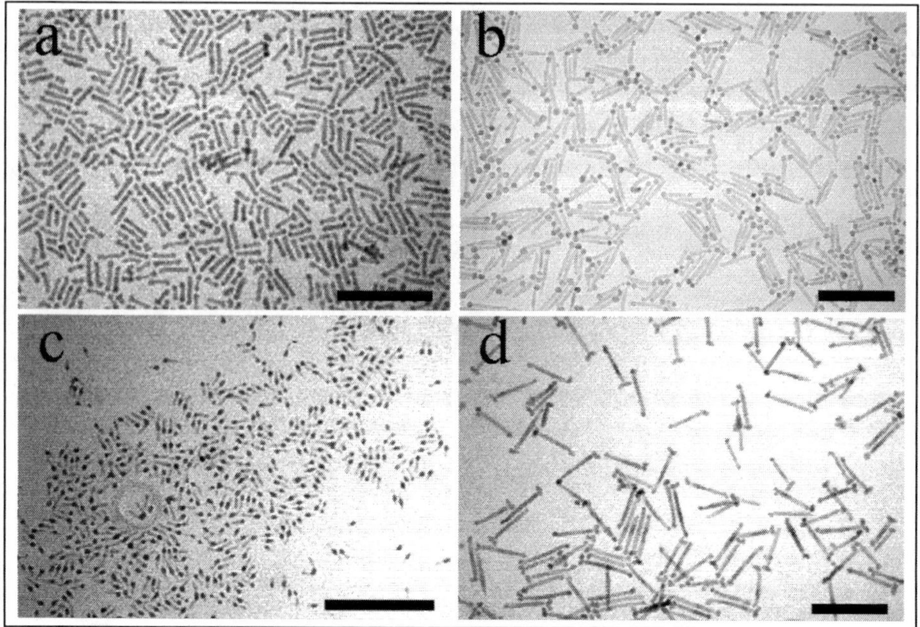

Figure 8. TEM images of nanocrystal heterostructures based on rod-like seeds: (a) dumbbell-like NCs made of a CdSe nanorod with PbSe tips located at both ends (synthesized according to Nano Lett., 2005, 5, 445); (b) matchstick-like NCs made of a CdS nanorod with one PbSe tip (synthesized according to Nano Lett, 2005; 5:445); (c) matchstick-like NCs made of a CdSe nanorod with one Au tip synthesized according to Nat Mater, 2005; 4:855; (d) NCs made of CdS rod with one CdTe tip on one end and a branching point on the opposite end (synthesized according to Nature 2004; 430:190).

cases, a "priming" of the core surface with appropriate molecules is required to activate it for the shell coating procedure.[130] A few examples of typical core/shell NCs are presented in the overview in Figure 5.

Nanocrystal Hetero-Oligomers

Nanocrystal hetero-oligomers are nanostructures comprising two or more nearly spherical inorganic particles connected epitaxially via a small junction area. So far, they have been obtained for a limited number of material combinations only. Some representative examples of this type of HNCs are reported in Figure 6.

When attempting to synthesize an alloy of two compounds characterized by partial miscibility and large interfacial energy, then the two materials can phase-segregate into separate domains, respectively, to form a dimer-like structure (Fig. 7, path 1). This concept has been verified in the growth of Co-Pd and Cu-In sulphide heterodimers by the coreaction of the respective molecular precursors.[138,139] In these cases, the selective nucleation of one material is followed by growth continuation with the second one, which emerges by developing an interface of graded composition.

Several types of nanocrystal hetero-oligomers have been synthesized by various seed-mediated mechanisms. The high interfacial energy between the component materials has explained the formation of FePt-CdS[140] dimers and γ-Fe$_2$O$_3$-MeS (where Me= Zn, Cd, Hg)[59] oligomers by thermal coalescence of an initially deposited amorphous metal sulfide shell into a separate spherical grain attached at one side of the initial seed (Fig. 7, path 2). Au-Fe$_3$O$_4$ dimer- and

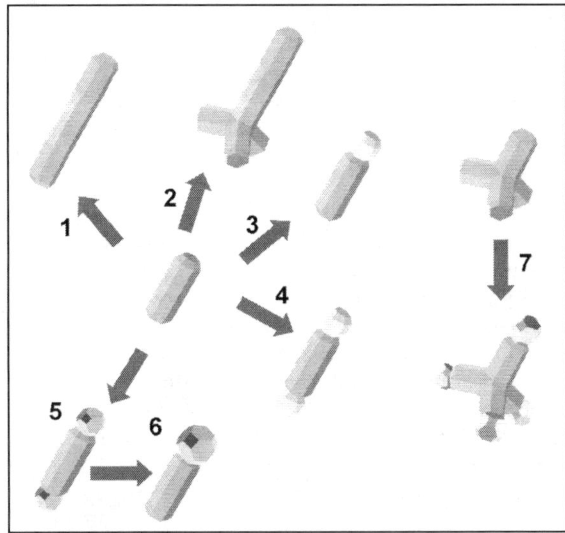

Figure 9. Sketch of possible paths leading to complex nanocrystal heterostructures starting from rod-like seeds: a rod-like seed can be continued either in a linear manner at both ends (path 1), or in a linear manner at just one side and with a branching point at the opposite side (path 2), or with one (path 3) or two spherical terminal tips (path 4); a dumbbell-like heterostructure (path 5) can evolve into a matchstick-like structure via an intra-particle Ostwald ripening scheme (path 6); a tetrapod-shaped nanocrystal can be functionalized with a spherical nanocrystal at each of the four terminations (path 7) (taken from Scheme 4; Chem Soc Rev 2006;35:1195) Reproduced by permission of The Royal Society of Chemistry.

flower-like HNCs have been obtained by heterogeneous growth of preformed Au particle seeds, followed by oxidation under air at room temperature (Fig. 7, path 3).[141] The ability to tune separately the size of the two single domains in a heterodimer has been recently demonstrated for the Au-CoPt$_3$ system.[142] These HNCs were prepared by heterogeneous nucleation of Au on preexisting CoPt$_3$ NCs of different size (Fig. 7, path 3). The latter provided substrates with varying degree of reactivity toward gold, while the relative concentration ratio between the seeds and the Au ions regulated the final size of the Au domains. A series of heterodimers of Fe$_3$O$_4$-Au(Ag), FePt-Ag, and Au-Ag has been achieved in a biphasic aqueous/organic system via ultrasonic emulsification.[143] In this strategy, hydrophobically coated seeds (Fe$_3$O$_4$, FePt, and Au, respectively) provide catalytic sites at the water-oil interface onto which the metal ions (Ag$^+$, AuCl$_4^-$) can be reduced sonochemically and form metallic domains. (Fig. 7, path 3). An aqueous synthesis of Ag-Se hybrid NCs has been also developed,[144] in which sequential reduction of the Ag and Se ions caused the initial nucleation of Ag onto which the reduction of Se ions followed. By increasing the concentration of the ions, it was possible to synthesize hybrid NCs in which either a single or more Se domains were attached to a Ag particle.

A rather comprehensive approach to hybrid NC oligomers comprising several new combinations of materials in a variety of heterostructure geometries (in terms of number of domains, their respective shapes and mutual spatial arrangement) has been reported recently.[145] Two interesting concepts have been introduced. First, NC heterodimers can react with each other, as they were small molecules combining together via their functional moieties to form larger molecules. An example is represented by the heating of peanut-shaped Au-Fe$_3$O$_4$ dimers in the presence of sulfur where Au domains of separate dimers fuse together. This yielded dumbbell-shaped heterostructures made of a single larger Au domain joining two Fe$_3$O$_4$ nanoparticles (Fig. 7, path 5). Second, HNCs themselves can be used as seeds for growing more complex nanostructures. This fact has been proven with the formation of ternary HNCs (Fe$_3$O$_4$-Au-PbSe) in which a rod-like PbSe section grew out of a Fe$_3$O$_4$-Au dimer seed (Fig. 7, path 6).

It has to be remarked that in NC hetero-oligomer growth, the most frequently observed couples of planes which form the interface junctions are usually those that allow the best lattice fit between the respective crystal structures of the two domains. Indeed, such condition allows the interfacial energy to be minimized as predicted by the Coincidence Site Lattice (CSL) theory.[146] In some cases,[141] the solvent can play an important role in the mechanism of hetero-oligomer growth. When a patch of second material domain nucleates on the initial seeds, an induced polarization charge at the seed interface occurs. A relatively polar solvent can facilitate the redistribution of the charge over the seed surface, leading to a higher probability of formation of multiple domains of the second material, resulting in flower-like HNCs or particle hetero-oligomers. Differently, a nonpolar solvent can be expected to sustain mainly single heterogeneous nucleation event.

Hybrid Nanocrystals Based on Rod-Like Sections

Several examples of HNCs based on rod-like domains have become also available. A selected overview of these structures is shown in Figure 8. Most of the structures developed so far have been based on cadmium chalcogenide materials in the wurtzite phase, as these can be easily grown in rod-like or in branched shapes with a high control over their geometrical parameters.[6,61] The rod sections, elongated in the c-axis direction, terminate into polar facets that are the most chemically reactive in such anisotropic NCs. This opens up the possibility of nucleating a second material exclusively at these locations. One additional peculiarity of the wurtzite structure is the absence of a plane of symmetry perpendicular to the *c*-axis, meaning that two basal sides of the rods are chemically dissimilar. Combined with facet-selective surfactant adsorption, this fact translates into the possibility of controlling the growth mode of lateral sections of a second material onto the tips of a starting nanorod.[68] By tuning the monomer reactivity in the solution, it has been possible to decide whether, starting from one rod-like seed, either both ends of the rod are continued in a linear manner (Fig. 9, path 1), or only one termination is continued in linear manner while the opposite one develops into a branching point, from which three distinct arms depart (Fig. 9, path 2). Architectures of this type have been demonstrated in the case of hybrid CdS-CdSe, CdS-CdTe and CdSe-CdTe NCs.[68] The chemical diversity of the basal facets of wurtzite nanorods has been clearly shown with the nucleation of a PbSe domain either on both tips or just on one tip of a CdS rod (Fig. 9, paths 3 and 4).[66]

HNCs resembling either a matchstick or a dumbbell have been synthesized by nucleating a Au domain either on one or on both ends of a CdSe nanorod, respectively (Fig. 9, path 5).[4] Interestingly, it has been observed that the dumbbell-shaped CdSe-Au HNCs are less stable heterostructures, that can evolve into matchstick-like ones (Fig. 9, path 6) via an intra-particle Oswald ripening.[147] The driving force for the latter process arises from the inevitable small difference in the size of the two Au tips, which makes the smaller one more unstable and susceptible to oxidation compared to the larger one when a high concentration of Au ion monomers is present in the solution. During this ripening, the smaller Au tip dissolves progressively, while the released electrons reach the bigger Au tip traveling through the CdSe rod section. Then, reduction of additional gold ions onto the larger tip makes it grow further; this ultimately leads to the matchstick-like CdSe-Au HNCs.

Finally, an extended strategy to access several types of Au-tipped dumbbell-like nanocrystal heterostructures has been developed recently, which involves the selective oxidation of either PbSe or CdTe sacrificial domains, initially grown on CdSe and CdS nanorods, with a Au(III):surfactant complex.[148] This approach is especially advantageous in that Au domains can be grown via an indirect two-step procedure onto the tips of nanorods of materials, like CdS, for which direct metal deposition is impracticable.

Conclusions

The field of nanocrystal preparation by surfactant-based wet-chemistry approaches is currently divided into two main important directions. The first area involves refinement in the ability to generate highly homogeneous nanostructures with precise control over their sizes and/or shapes, as this ultimately dictates their optoelectronic, catalytic, and magnetic behaviour. Anisotropic NCs are especially appealing in that they offer additional degrees of freedom for extending the properties of individual materials and for engineering NC-based devices. The overall progress in this regard has been definitely relevant, even though further investigations are certainly needed to assess more general size and shape guiding synthesis criteria. The other outstanding research direction is related to the programmable fabrication of hybrid NCs in which spherical, rods or branched sections of different materials are connected together in a single particle. The preparation of NC heterostructures is naturally more challenging task, as the ability to tailor specific nanosized materials must be integrated with the understanding of the parameters which dictate heteroepitaxial growth at the nanoscale. This is the reason why just a few types of hybrid architectures have been realized up to date; currently, these structures can not be engineered in such a way to prevent possible detrimental effects on the properties of individual components. The possibility to design and synthesize elaborate NCs with the purposely defined topologically composition appears a fascinating research field which is still in its infancy. It is conceivable that future progress in this synthetic ability will open up access to a completely novel generation of colloidal structures suitable to uncover optoelectronic, magnetic, biomedical, photovoltaic and catalytic applications with a higher level of performance.

References

1. Hyeon T. Chem Commun 2003; (8):927.
2. Schmidt G. Nanoparticles: From theory to applications. Weinheim: Wiley, 2004.
3. Wang X, Zhuang J, Peng Q et al. Nature 2005; 437(7055):121.
4. Ozin GA, Arsenault AC. Nanochemistry: A Chemical Approach to Nanomaterials. London: Royal Society of Chemistry, 2005.
5. Albe V, Jouanin C, Bertho D, Cryst J. Growth 1998; 185:388.
6. Peng XG, Manna L, Yang WD et al. Nature 2000; 404(6773):59.
7. Li LS, Hu JT, Yang WD et al. Nano Lett 2001; 1(7):349.
8. Hens Z, Vanmaekelbergh D, Stoffels E et al. Phys Rev Lett 2002; 88(23), (art. no. 236803).
9. Fonoberov VA, Pokatilov EP, Balandin AA. Phys Rev B 2002; 66(8), (art. no. 085310).
10. Buhro WE, Colvin VL. Nat Mater 2003; 2(3):138.
11. Burda C, Chen XB, Narayanan R et al. Chem Rev 2005; 105(4):1025.
12. Murphy CJ, Sau TK, Gole AM et al. J Phys Chem B 2005; 109(29):13857.
13. Goldstein AN, Echer CM, Alivisatos AP. Science 1992; 256(5062):1425.
14. Alivisatos AP. J Phys Chem 1996; 100(31):13226.
15. Peters KF, Cohen JB, Chung YW. 1998; 57(21):13430.
16. Tolbert SH, Alivisatos AP. Annu Rev Phys Chem 1995; 46:595.
17. Heath JR, Shiang JJ. Chem Soc Rev 1998; 27(1):65.
18. Klimov V. Semiconductor and Metal Nanocrystals. New York: Marcel Dekker, 2004.
19. Rao CNR, Kulkarni GU, Thomas PJ et al. Chem Eur J 2002; 8(1):29.
20. Chen MS, Goodman DW. 2004; 306:252.
21. Narayanan R, El-Sayed M. J Phys Chem B 2005; 109(26):12663.
22. Chen MS, Goodman DW. Catal Today 2006; 111:22.
23. Battle X, Lebarta A. J Phys D Appl Phys 2002; 35:R15.
24. Fiorani D. Surface Effects in Magnetic Nanoparticles. New York: Springer, 2005.
25. Nozik AJ, Physica E. Low-dimens Syst Nanostruct 2002; 14(1-2):115.
26. Sundar VC, Eisler HJ, Bawendi MG. Adv Mater 2002; 14(10):739.
27. Milliron DJ, Gur I, Alivisatos AP. MRS Bull 2005; 30(1):41.
28. Michalet X, Pinaud FF, Bentolila LA et al. Science 2005; 307(5709):538.
29. Grodzinski P, Silver M, Molnar LK. Exp Rev Mol Diag 2006; 6(3):307.
30. Collier CP, Vossmeyer T, Heath JR. Annu Rev Phys Chem 1998; 49:371.
31. Talapin DV, Shevchenko EV, Gaponik N et al. 2005; 17(11):1325.
32. Shevchenko EV, Talapin DV, O'Brien S et al. 2005; 127(24):8741.
33. Desvaux C, Amiens C, Fejes P et al. Nat Mater 2005; 4(10):750.
34. Yin Y, Alivisatos AP. Nature 2005; 437(7059):664.
35. Peng ZA, Peng XG, J Am Chem Soc 2001; 123(7):1389.
36. Pileni MP. Nat Mater 2003; 2(3):145.
37. Lee SM, Cho SN, Cheon J. Adv Mater 2003; 15(5):441.
38. Lisiecki I, J Phys Chem B 2005; 109(25):12231.
39. Donega CD, Liljeroth P, Vanmaekelbergh D. Small 2005; 1(12):1152.
40. Kumar S, Nann T. Small 2006; 2(3):316.
41. Sun YG, Xia YN. Science 2002; 298(5601):2176.

42. Gou LF, Murphy CJ. Nano Lett 2003; 3(2):231.
43. Lifshitz E, Bashouti M, Kloper V et al. Nano Lett 2003; 3(6):857.
44. Yu YY, Chang S, Lee CJ et al. J Phys Chem B 1997; 101(34):6661.
45. Manna L, Scher EC, Alivisatos AP. J Am Chem Soc 2000; 122(51):12700.
46. Cordente N, Respaud M, Senocq F et al. Nano Lett 2001; 1(10):565.
47. Ahrenkiel SP, Micic OI, Miedaner A et al. Nano Lett 2003; 3(6):833.
48. Dumestre F, Chaudret B, Amiens C et al. Angew Chem-Int Edit 2003; 42(42):5213.
49. Busbee BD, Obare SO, Murphy CJ. Adv Mater 2003; 15(5):414.
50. Tang KB, Qian YT, Zeng JH et al. Adv Mater 2003; 15(5):448.
51. Yu H, Buhro WE. Adv Mater 2003; 15(5):416.
52. Yu H, Li JB, Loomis RA et al. J Am Chem Soc 2003; 125(52):16168.
53. Grebinski JW, Hull KL, Zhang J et al. Chem Mat 2004; 16(25):5260.
54. Lee SM, Jun YW, Cho SN et al. J Am Chem Soc 2002; 124(38):11244.
55. Zhao N, Qi LM. Adv Mater 2006; 18(3):359.
56. Puntes VF, Zanchet D, Erdonmez CK et al. J Am Chem Soc 2002; 124(43):12874.
57. Chen SH, Fan ZY, Carroll DL. J Phys Chem B 2002; 106(42):10777.
58. Zhang P, Gao L. J Mater Chem 2003; 13(8):2007.
59. Chen M, Xie Y, Lu J et al. J Mater Chem 2002; 12(3):748.
60. Chen SH, Wang ZL, Ballato J et al. J Am Chem Soc 2003; 125(52):16186.
61. Manna L, Milliron DJ, Meisel A et al. Nat Mater 2003; 2(6):382.
62. Kuo CH, Huang MH. Langmuir 2005; 21(5):2012.
63. Zitoun D, Pinna N, Frolet N et al. J Am Chem Soc 2005; 127(43):15034.
64. Teng X, Yang H. Nano Lett 2005; 5(5):885.
65. Mokari T, Rothenberg E, Popov I et al. Science 2004; 304(5678):1787.
66. Kudera S, Carbone L, Casula MF et al. Nano Lett 2005; 5(3):445.
67. Mokari T, Sztrum CG, Salant A et al. Nat Mater 2005; 4:855.
68. Milliron DJ, Hughes SM, Cui Y et al. Nature 2004; 430(6996):190.
69. Shevchenko EV, Talapin DV, Rogach AL et al. J Am Chem Soc 2002; 124(38):11480.
70. Chen CC, Chao CY, Lang ZH. Chem Mater 2000; 12:1516.
71. Jana NR, Gearheart L, Murphy CJ. J Phys Chem B 2001; 105(19):4065.
72. Trentler TJ, Hickman KM, Goel SC et al. Science 1995; 270(5243):1791.
73. Hanrath T, Korgel BA. J Am Chem Soc 2002; 124(7):1424.
74. Kan SH, Aharoni A, Mokari T et al. Faraday Discuss 2004; 125:23.
75. Murphy CJ, Jana NR. Adv Mater 2002; 14(1):80.
76. Gou LF, Murphy CJ. Chem Mater 2005; 17(14):3668.
77. Hyeon T, Lee SS, Park J et al. J Am Chem Soc 2001; 123(51):12798.
78. Yong KT, Sahoo Y, Choudhury KR et al. Nano Lett 2006; 6(4):709.
79. Tang ZY, Kotov NA, Giersig M. Science 2002; 297(5579):237.
80. Cho KS, Talapin DV, Gaschler W et al. J Am Chem Soc 2005; 127(19):7140.
81. Yu JH, Joo J, Park HM et al. J Am Chem Soc 2005; 127(15):5662.
82. Lee EJH, Ribeiro C, Longo E et al. J Phys Chem B 2005; 109(44):20842.
83. Gautam UK, Ghosh M, Rajamathi M et al. Pure Appl Chem 2002; 74(9):1643.
84. Tanase M, Silevitch DM, Hultgren A et al. J Appl Phys 2002; 91(10):8549.
85. Lee GH, Huh SH, Jeong JW et al. Scripta Mater 2003; 49(12):1151.
86. Cheng CD, Haynie DT. Appl Phys Lett 2005; 87(26).
87. Morales AM, Lieber CM. Science 1998; 279(5348):208.
88. Huang MH, Wu YY, Feick H et al. Adv Mater 2001; 13(2):113.
89. Yang PD, Yan HQ, Mao S et al. Adv Funct Mater 2002; 12(5):323.
90. Gudiksen MS, Lauhon LJ, Wang J et al. Nature 2002; 415(6872):617.
91. Bjork MT, Ohlsson BJ, Sass T et al. Nano Lett 2002; 2(2):87.
92. Holmes JD, Johnston KP, Doty RC et al. Science 2000; 287(5457):1471.
93. Peng ZA, Peng XG. J Am Chem Soc 2002; 124(13):3343.
94. Wang ZL. Adv Mater 1998; 10(1):13.
95. Cozzoli PD, Manna L, Curri ML et al. Chem Mater 2005; 17(6):1296.
96. Sugimoto T, Itoh H, Mochida T. J Colloid Interface Sci 1998; 205(1):42.
97. Cozzoli PD, Kornowski A, Weller H. J Am Chem Soc 2003; 125(47):14539.
98. Jun YW, Casula MF, Sim JH et al. J Am Chem Soc 2003; 125(51):15981.
99. Carbone L, Kudera S, Carlino E et al. J Am Chem Soc 2006; 128(3):748.
100. Cozzoli PD, Snoeck E, Garcia MA et al. Nano Lett 2006; 6(9):1966.
101. Wang J, Zeng C. J Cryst Growth 2004; 270(3-4):729.
102. Wang J, Chen QW, Zeng C et al. Adv Mater 2004; 16(2):137.

103. Yang J, Lee JY, Too HP. J Phys Chem B 2005; 109(41):19208.
104. Srnova-Sloufova I, Vlckova B, Bastl Z et al. Langmuir 2004; 20(8):3407.
105. Rodriguez-Gonzalez B, Burrows A, Watanabe M et al. J Mater Chem 2005; 15(17):1755.
106. Sobal NS, Hilgendorff M, Mohwald H et al. Nano Lett 2002; 2(6):621.
107. Talapin DV, Mekis I, Gotzinger S et al. J Phys Chem B 2004; 108(49):18826.
108. Manna L, Scher EC, Li LS et al. J Am Chem Soc 2002; 124(24):7136.
109. Kim S, Fisher B, Eisler HJ et al. J Am Chem Soc 2003; 125(38):11466.
110. Yu K, Zaman B, Romanova S et al. Small 2005; 1(3):332.
111. Eychmuller A, Mews A, Weller H. Chem Phys Lett 1993; 208(1-2):59.
112. Talapin DV, Koeppe R, Gotzinger S et al. Nano Lett 2003; 3(12):1677.
113. Park JI, Kim MG, Jun YW et al. J Am Chem Soc 2004; 126(29):9072.
114. Ban ZH, Barnakov YA, Golub VO et al. J Mater Chem 2005; 15(43):4660.
115. Mandal M, Kundu S, Ghosh SK et al. J Colloid Interface Sci 2005; 286(1):187.
116. Lyon JL, Fleming DA, Stone MB et al. Nano Lett 2004; 4(4):719.
117. Wang LY, Luo J, Fan Q et al. J Phys Chem B 2005; 109(46):21593.
118. Wang H, BDW, LF et al. Nano Lett 2006; 6(4):827.
119. Kim H, Achermann M, Balet LP et al. J Am Chem Soc 2005; 127(2):544.
120. Kamat PV, Flumiani M, Dawson A. Colloid Surf: A-Physicochem. Eng Asp 2002; 202(2-3):269.
121. Subramanian V, Wolf EE, Kamat PV. J Phys Chem B 2003; 107(30):7479.
122. Dawson A, Kamat PV. J Phys Chem B 2001; 105(5):960.
123. Wood A, Giersig M, Mulvaney P. J Phys Chem B 2001; 105(37):8810.
124. Liz-Marzan LM, Mulvaney P. J Phys Chem B 2003; 107(30):7312.
125. Tom RT, Nair AS, Singh N et al. Langmuir 2003; 19(8):3439.
126. Mayya KS, Gittins DI, Caruso F. Chem Mater 2001; 13(11):3833.
127. Li J, Zeng HC. Angew Chem Int Ed 2005; 44(28):4342.
128. Hirakawa T, Kamat PV. J Am Chem Soc 2005; 127(11):3928.
129. Pastoriza-Santos I, Perez-Juste J, Liz-Marzan LM. Chem Mater 2006; 18(10):2465.
130. Mulvaney P, Liz-Marzan LM, Giersig M et al. J Mater Chem 2000; 10(6):1259.
131. Choi JS, Jun YW, Yeon SI et al. J Am Chem Soc 2006, (http://dx.doi.org/10.1021/ja066547g).
132. Salant A, Amitay-Sadovsky A, Banin U. J Am Chem Soc 2006; 128(31):10006.
133. Talapin DV, Rogach AL, Kornowski A et al. Nano Lett 2001; 1(4):207.
134. Mokari T, Banin U. Chem Mat 2003; 15(20):3955.
135. Li JJ, Wang YA, Guo WZ et al. J Am Chem Soc 2003; 125(41):12567.
136. Park JI, Cheon J. J Am Chem Soc 2001; 123(24):5743.
137. Lee WR, Kim MG, Choi JR et al. J Am Chem Soc 2005; 127(46):16090.
138. Teranishi T, Inoue Y, Nakaya M et al. J Am Chem Soc 2004; 126(32):9914.
139. Choi SH, Kim EG, Hyeon T. J Am Chem Soc 2006; 128(8):2520.
140. Gu HW, Zheng RK, Zhang XX et al. J Am Chem Soc 2004; 126(18):5664.
141. Yu H, Chen M, Rice PM et al. Nano Lett 2005; 5(2):379.
142. Pellegrino T, Fiore A, Carlino E et al. J Am Chem Soc 2006; 128(20):6690.
143. Gu HW, Yang ZM, Gao JH et al. J Am Chem Soc 2005; 127(1):34.
144. Gao XY, Yu LT, MacCuspie RI et al. Adv Mater 2005; 17(4):426.
145. Shi W, Zeng H, Sahoo Y et al. Nano Lett 2006; 6(4):875.
146. Randle V. The role of the coincidence site lattice in grain boundary engineering. Cambridge, England: Woodhead Publishing Limited, 1997.
147. Sugimoto T. Adv Colloid Interface Sci 1987; 28(1):65.
148. Carbone L, Kudera S, Giannini C et al. J Mater Chem 2006; 16(40):3952.

CHAPTER 2

Current Approaches for Engineering Proteins with Diverse Biological Properties

Isaac T.S. Li, Elizabeth Pham and Kevin Truong*

Abstract

In the past two decades, protein engineering has advanced significantly with the emergence of new chemical and genetic approaches. Modification and recombination of existing proteins not only produced novel enzymes used commercially and in research laboratories, but furthermore, they revealed the mechanisms of protein function. In this chapter, we will describe the applications and significance of current protein engineering approaches.

Introduction

The ultimate goal of protein engineering is to create proteins that are perfectly suited to particular biological applications. Proteins, polymers consisting of 20 distinct amino acids, play a pivotal role in many biological activities such as catalysis, signal transduction, structural reinforcement and cell motility, to name a few. The diversity in the choice of amino acids and total length of proteins (ranging from tens to thousands of amino acids) allows for a tremendous combinatorial sequence space. Through evolutionary pressure over time, organisms have found specific sequences of amino acids that fold into unique molecular structures that determine their biochemical functions. The molecular structures of over thirty thousand proteins have been solved and this number continues to grow steadily alongside new initiatives in structural proteomics. Nevertheless, the total number of distinct proteins found in nature represents only a small fraction of the possible polypeptides that could be permutated with the 20 amino acids. By exploring the combinatorial sequence space around these natural proteins, many current approaches in protein engineering can create novel proteins with desirable properties. In this chapter, we will discuss these approaches and their diverse biological applications.

Random Mutagenesis

Random mutagenesis is the process of introducing random point mutations to generate a diverse gene library for screening desired protein properties. It has been successfully used to identify key residues in proteins as well as to improve or alter protein activities.[1-12] Random mutagenesis samples the local sequence space of the starting protein (Fig. 1A). As a result, the properties of the mutants will not diverge far from their parents. If the mutation rate is too low, the diversity in the resulting library may be insufficient, as quite often, biological advantages are conferred from multiple cooccurring mutations. On the other hand, if the mutation rate is too high, potentially advantageous mutations may be silenced by mutations that destroy the protein function altogether. Thus, the rate of mutation is a very important parameter to study and fine tune in mutagenesis experiments. In this section, we summarize three commonly used

*Corresponding Author: Kevin Truong—Institute of Biomaterials and Biomedical Engineering, University of Toronto, 164 College Street, Toronto, Ontario, M5S 3G9, Canada. Email: kevin.truong@utoronto.ca

Bio-Applications of Nanoparticles, edited by Warren C.W. Chan. ©2007 Landes Bioscience and Springer Science+Business Media.

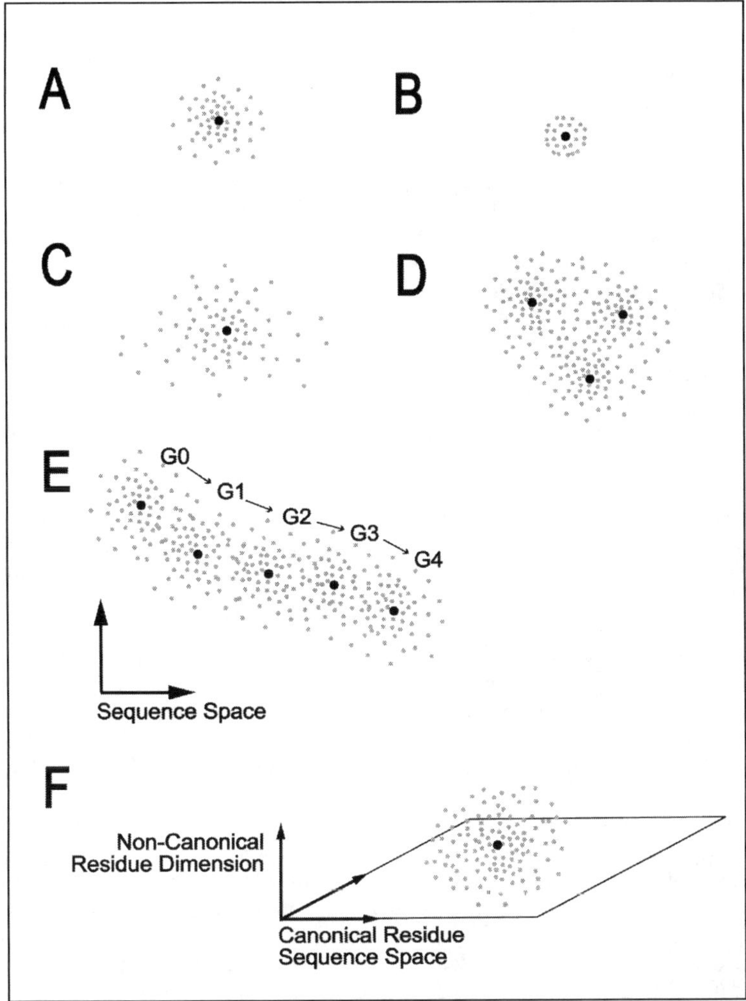

Figure 1. Sequence space representations of different protein engineering approaches. The solid black dots represent the locations of parent genes; the grey dots, mutants. Distance between the dots represents extent of divergence between their corresponding sequences. A) Random mutagenesis samples the sequence space isotropically to generate a library of mutants with a Gaussian distribution centred at the parent gene. B) Site-directed mutagenesis samples small and predefined regions close to the parent gene. C) DNA recombination on mutagenesis libraries results in a larger divergence of mutants around the parent gene than compared to random mutagenesis alone. D) DNA recombination of three distinct genes results in even greater sequence space coverage by including the sequence space between regions. E) Directed evolution can reach far from parent gene by following the path of generations of improved sequences. Starting from the parent gene at G0 (Generation 0), mutagenesis and recombination is used to generate a library of mutants. The mutant with the most improved properties is selected as the parent gene for the next generation—G1. This process is repeated for subsequent generations until the desired improvement of enzymatic properties is achieved. F) Introduction of noncanonical amino acids adds new dimensions to the existing sequence space. The horizontal plane shows the canonical sequence space, where the vertical axis illustrates the new dimension created by noncanonical amino acids.

techniques in random mutagenesis: error-prone polymerase chain reaction (PCR), bacteria mutator strains and chemical mutagens.

Error-prone PCR is an in vitro technique to increase the DNA copy error rate by altering the PCR conditions. For example, high Mg^{2+} concentration stabilizes base-pair mismatch, the addition of Mn^{2+} reduces base-pair specificity, imbalanced nucleotides (dNTP) enhance the mismatching probability during PCR and a higher concentration of *Taq* DNA polymerase encourages elongation of mistakenly primed termini. Statistics show that under normal conditions, PCR reactions using *Taq* DNA polymerase has an error rate of ~0.01%, while under the error-prone PCR conditions, the error rate increases to ~1%. Of the total mutations, 90% are nucleotide substitutions and of these mutations, ~30% are silent and ~70% result in amino acid substitutions. The remaining 10% of mutations are insertions or deletions, which usually result in frame shifts and subsequently, incorrect gene expressions. Multiple cycles of error-prone PCR can further control the amount of mutations introduced to the template. This technique has recently helped identify key residues in the M3 muscarinic acetylcholine receptor, epidermal growth factor receptor (EGFR) and *Saccharomyces cerevisiae* Hal3 salt tolerance regulator.[1,2,4] From the engineering/design aspect, PCR random mutagenesis has been used to improve enzymatic properties of subtilisins E, *Thermus aquaticus* amylomaltase, cyclodextrin glucanotransferase and *Rhizopus niveus* lipase.[5-8]

Random mutations can also be introduced in vivo by mutator bacteria strains such as mutD5-FIT and XL1-Red (Stratagene).[13,14] These *Escherichia coli* strains have mutations in genes (such as the *mutD*, *mutS*, *mutL*, *mutH* and *mutT*) found in the DNA repair pathway which increase the rate of mutation to over a thousand fold higher than the wild type strain.[15,16] It is estimated that the error rate after 30 cell generations is ~0.05%. The advantage of this technique is in its simplicity—transform and harvest. Although the mutation rate is not as high as error-prone PCR, it is proven sufficient to generate diversity in many studies.[3,9,17-19] Random mutagenesis with mutator strains was used to identify the essential residues responsible for the function of Erm(B) rRNA methyltransferase.[3] A 5-fold improvement of enzymatic activity using this technique was demonstrated for polyhydroxyalkanoate (PHA) synthase.[9] In another example, mutator strains also helped alter protein properties as seen in the conversion of oxidosqualene-cycloartenol synthase specificity to that of oxidosqualene-lanosterol cyclase.[12]

A third way to introduce random mutations in DNA is by chemical mutagens. One class of chemical mutagens are DNA base analogs such as 5-bromo-deoxyuridine (5BU), which exist in two tautomeric forms, mimicking cytosine (C) and thymine (T). Therefore, when mutagens in this class are introduced to the cells during DNA replication, the mutagens are likely to be incorporated into the DNA, causing mutations to occur. Another class of mutagens are alkylators such as ethyl methane sulfonate (EMS), which reacts with certain bases on the DNA. Other mutagens such as sodium bisulphite converts cytosine to uracil, achieving substitution directly on the target DNA.[20]

Site-Directed Mutagenesis

Site-directed mutagenesis allows the introduction of a mutation to a specific location on a gene. It is commonly used for studying the effects of individual point mutations and designing novel proteins with key mutations (Fig. 1B). There are in vivo and in vitro site-directed mutagenesis techniques and among them, two common techniques will be summarized here: oligonucleotide mismatch mutagenesis and PCR based mutagenesis.[21-23] The in vivo oligonucleotide mismatch mutagenesis was originally proposed by Michael Smith in 1979, who later won the Nobel Prize in 1993. This technique incorporates a point mutation to an oligonucleotide whose sequence is otherwise complementary to the region of mutation on the gene of interest. This oligonucleotide is then annealed to the gene of interest and the full-length plasmid DNA is synthesized with mismatching double strands at the point of mutation. Then, cells are transformed with this plasmid DNA, which replicate the mismatching gene into two

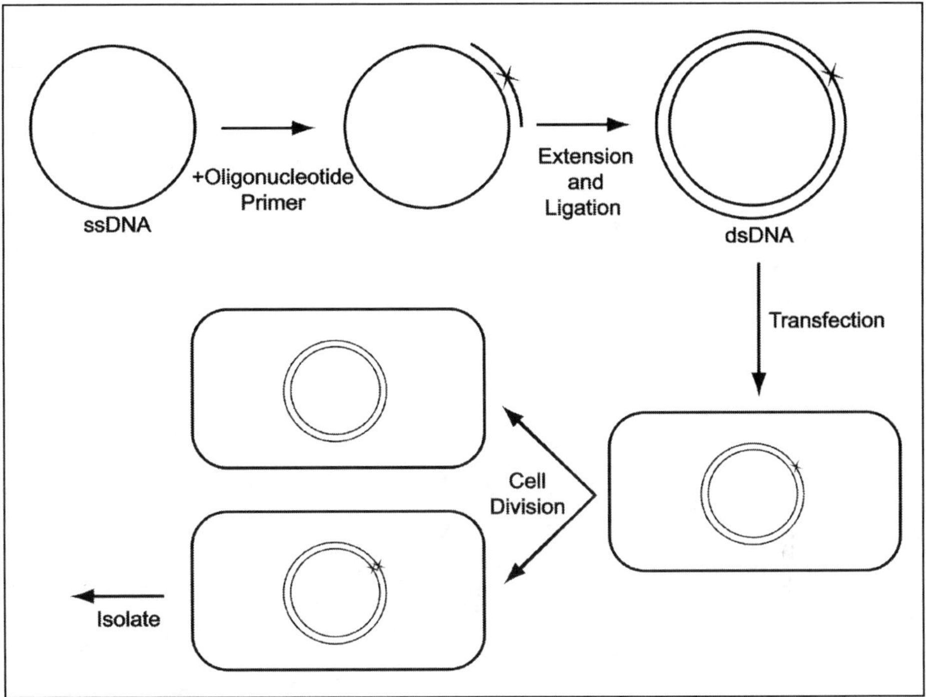

Figure 2. Oligonucleotide mismatch mutagenesis. The cross "×" donates a point mutation on a primer or strand of DNA.

species of proteins—one containing the mutation and the other retaining the natural form (Fig. 2). Site-directed mutagenesis can also be performed in vitro by PCR. Primers with desired point mutations first amplify regions of the parental DNA on both sides of the mutation site, which creates two portions of the parental DNA with overlapping regions (Fig. 3). Then a PCR is performed using the mixture of the two overlapping portions to create the full length DNA, which now contains the point mutation.

Site-directed mutagenesis studies provide valuable information on the structure-function relationship of a protein as shown in numerous recent examples.[24-29] In addition to learning about proteins, site-directed mutagenesis is applied to the rational design of proteins based on the knowledge of structure-function relationships. For example, improvements on catalytic activities were shown in the *Aspergillus niger* NRRL 3135 phytase and *Bacillus* protease by targeted mutations of residues in the catalytic sites revealed by their molecular structures.[30,31] Alteration of enzymatic properties has been demonstrated in the 150-fold change of relative substrate specificity of the oncoprotein v-Fps by mutating the recognition element at the arginine residue at position 1130 (Arg1130).[32] Similarly, the excitation and emission spectra of green fluorescent protein (GFP) was red-shifted by replacing the Thr203 with Tyr or His to promote aromatic interactions with the fluorophore.[33]

Non-Canonical Amino Acid Substitution

The introduction of noncanonical amino acids provides a new perspective to engineering proteins by expanding the combinatorial sequence space (Fig. 1F). Most natural organisms can only synthesize and incorporate the 20 canonical amino acids as the building blocks for their proteins. By substituting noncanonical amino acids for natural residues, we overcome this limitation. In general, there are two types of noncanonical amino acid substitutions:

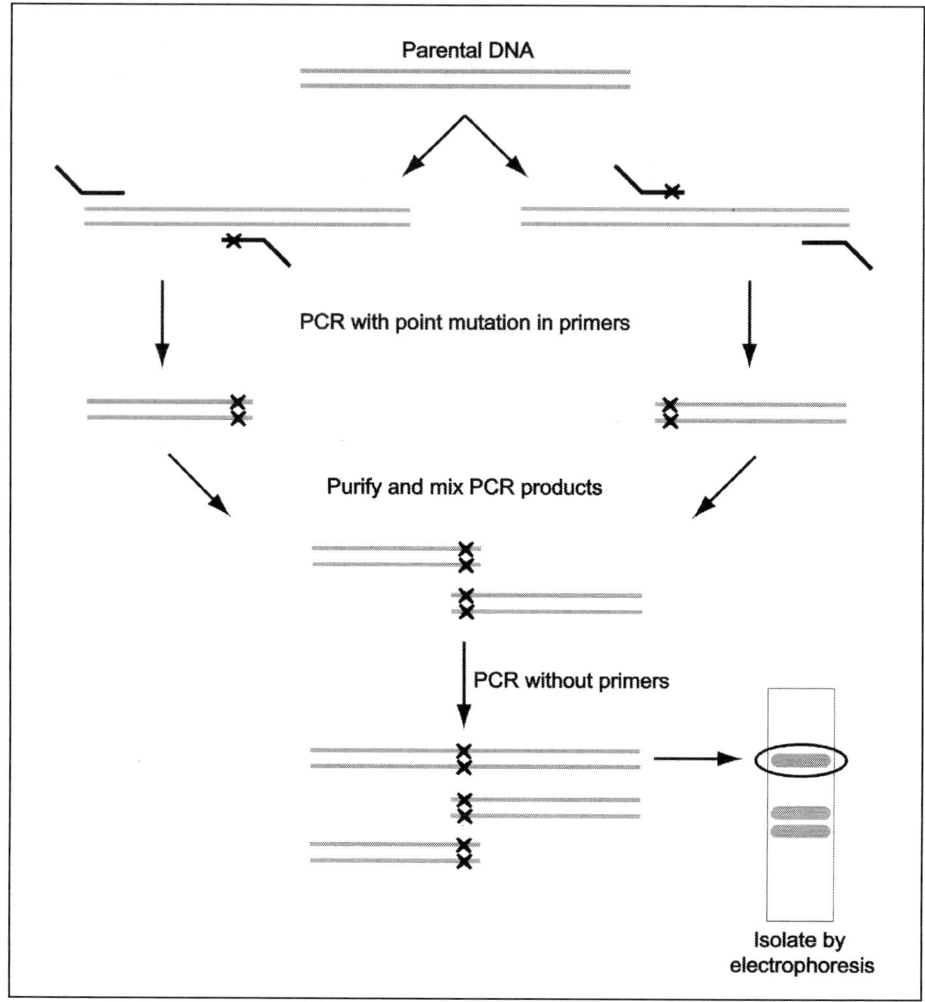

Figure 3. PCR-based site-directed mutagenesis. The cross "×" donates a point mutation a primer or strand of DNA.

residue-specific and site-specific. In residue-specific substitution, a specific type of amino acid is replaced by a noncanonical one. For instance, the replacement of tryptophan in the barstar protein by a noncanonical pH sensitive aminotryptophan created a pH sensor.[34] In site-specific substitution, the amino acid at a particular location is replaced by a noncanonical amino acid. Similar to site directed mutagenesis, this facilitates protein engineering based on structure-function relationships. The challenge of noncanonical amino acid substitution lies in the synthesis of the modified proteins. Measures must be taken when synthesizing proteins with noncanonical residues in vivo to ensure the proteins of the expression host do not acquire the noncanonical amino acids, which could be fatal to the host cell. Furthermore, in site-specific substitution, one must also ensure that only specific amino acids on the protein are targeted.

Common approaches in residue-specific substitution rely on gene expression in an auxotroph grown in a media containing the noncanonical amino acid.[35] Auxotrophic bacteria lack the ability to synthesize a certain amino acid, hence relying on the supply of that amino acid in the growth media. To maximize expression of target protein with noncanonical amino acids, auxotroph bacteria are first grown in media with the natural amino acids. During this stage, the expression of the target protein is strongly repressed to minimize its expression with natural amino acids. After the cell density threshold is reached, the expression of the target protein is induced in a new growth media containing the noncanonical amino acids. Using this technique, noncanonical amino acids were incorporated into enhanced cyan fluorescent protein (ECFP) to produce "gold" fluorescent protein (GdFP) that had an emission spectrum red-shifted by 69 nm.[36] The noncanonical protein expression level can be further improved by over-expressing the corresponding aminoacyl-tRNA synthase (aaRS) to the replaced natural amino acid. The 20 aaRS, one for each type of amino acid, create tRNA amino acid complexes between particular amino acids and tRNAs according to the genetic code. Since the rate at which aaRS incorporates noncanonical amino acids is over a thousand-fold less than that for natural amino acids, over-expression of the particular aaRS would produce more tRNA amino acid complexes and hence allow the cell to incorporate noncanonical amino acids in target proteins more efficiently.

There are other techniques for incorporating noncanonical amino acids to target proteins including modifications to aaRS that favor recognition of noncanonical amino acids and inaccurate hydrolytic editing.[37] Alternatively, heterologous aaRS and tRNA from other organisms that have different codon usage can be introduced into the host organism.[38] Hence, the codon degeneracy is broken as long as the noncanonical amino acids and tRNA do not interfere with the tRNA synthesis pathway of the host. This technique allows for a codon-specific substitution. Global noncanonical amino acid substitution can also be achieved by in vitro post-translational chemical modifications.[39]

To achieve site-specific substitution of a noncanonical amino acid, one approach is by breaking the codon degeneracy using 4-base or 5-base codons to code for the noncanonical amino acid through in vitro protein synthesis.[40-44] The specificity and efficiency is increased by using these 4- and 5-base codon systems. One example is the incorporation of a fluorescent noncanonical amino acid to a specific site of streptavidin by 4-base codons.[40,41] The fluorescence of the incorporated amino acid is highly sensitive to the binding of biotin, which can be used for low concentration detection. In another example, the Tyr66 of GFP was mutated to 18 distinct noncanonical aromatic amino acids by the 4-base CCCG codon tRNA, resulting in two blue-shifted mutants.[43]

DNA Recombination

Novel protein characteristics can be created by swapping the regions of one gene with another in a process called DNA recombination. This process samples a larger sequence space compared to mutagenesis alone (Fig. 1D). In many cases, mutagenesis and recombination are used together to create gene libraries with even greater diversity. It can enhance or alter protein activities by preserving advantageous mutations while removing disadvantageous ones.[45,46] Here, we discuss two DNA recombination techniques: DNA shuffling and the staggered extension process.

In 1994, William Stemmer introduced the DNA shuffling technique to create a chimera of β-lactamase with a 32,000-fold increase in tolerance to its inhibitor.[47,48] In DNA shuffling, a library of homologous genes is first randomly digested by DNase I (Fig. 4). Then, DNA fragments with sizes from 10 to 50 base pairs are isolated. PCR amplification of the gene library with these isolated fragments as primers will produce many PCR products from different homologous templates. For this step to be effective there must be significant homology (> 70%) in the gene library. Next, an additional PCR step is performed to extract the desired genes using primers that define the start and end of a full-length gene. Finally, these PCR products

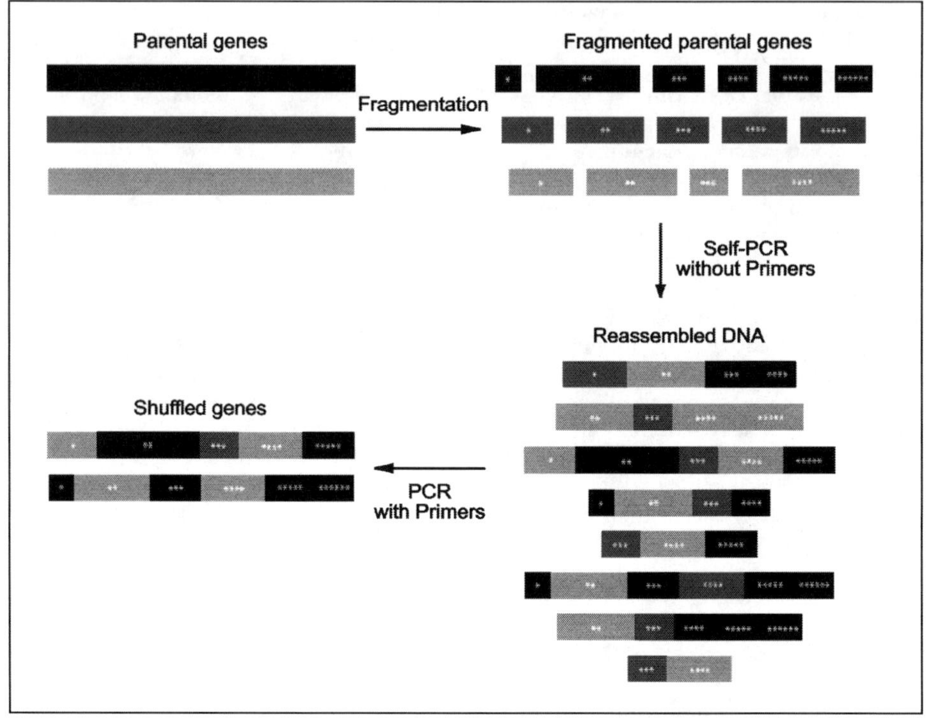

Figure 4. DNA shuffling. There are three homologous parental genes identified by their distinct shades. The white dots labelled on each gene fragments represent their relative order to the parental gene, which is preserved after reassembly from the self-PCR process. Only the full length reassembled genes can be amplified using primers that specify the two ends of the gene.

are inserted into vectors, expressed and screened for desired properties. Using 3 cycles of DNA shuffling on a GFP mutant library, a combination of three mutations were discovered that improved folding properties and increased fluorescence intensity by 45-fold.[49] A family of serum paraoxonases were recombined and screened to create a variant with 40-fold higher enzymatic activity and over 2000-fold specificity.[50] Applying DNA shuffling to eight cefE homologous genes in combination with directed evolution created chimeras with 118-fold increased activity.[51] In another example, Lipase B protein homologs from three organisms were subjected to DNA shuffling, resulting in a chimera with improved stability and a 20-fold higher hydrolysis activity towards diethyl 3-(3',4'-dichlorophenyl)glutarate.[52]

In 1998, Zhao and colleagues proposed another in vitro recombination technique called the staggered extension process (StEP).[53] The recombination was achieved in one PCR reaction with optimized conditions. Similar to DNA shuffling, a library of homologous genes is used as the template. The annealing and extension steps in each PCR cycle are replaced by an extremely short annealing step (~1s) (Fig. 5). This short annealing step amplifies only a fraction of the template. Since a mixture of templates is used, the partly extended primer will randomly anneal to a different template in the next thermal cycle. Consequently, the primers are extended to full length by fragments from each of the homologous genes. Next, these fragments are inserted into the expression vector and screened. StEP was used to recombine the aroA gene from *Salmonella typhimurium* and *Escherichia coli*, creating 4 chimeras with a 2 to 10 fold increase in activity.[54] In another study, by combining random mutagenesis and StEP, the 1,6 regioselectivity of AgaB gene from *Bacillus stearothermophilus* was removed and 1,3 regioselectivity was retained.[55]

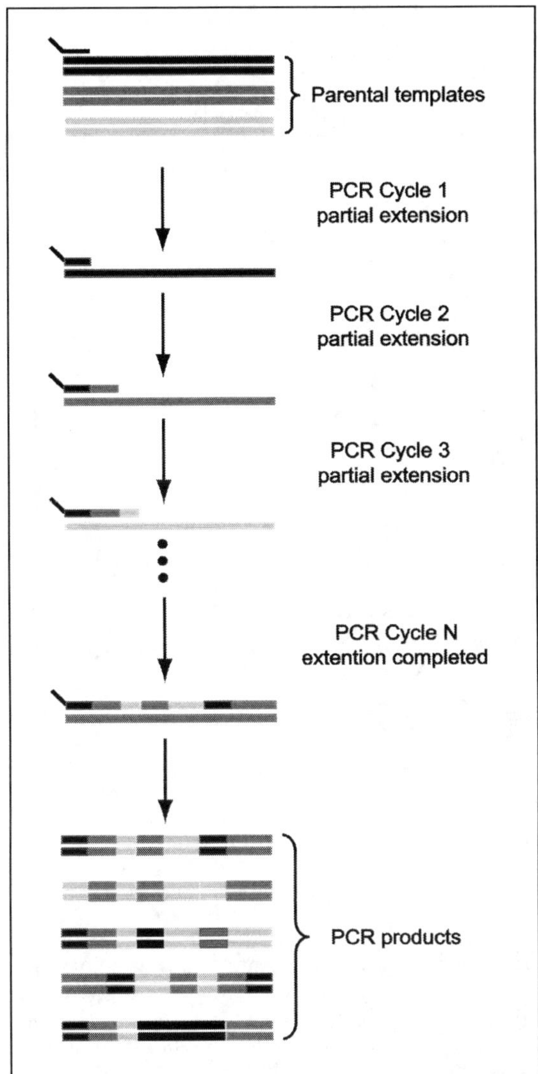

Figure 5. Staggered Extension Process (StEP). There are three double-stranded homologous parental genes that are distinguished by their shade and used as PCR templates. In the first PCR cycle, the primer anneals to the single-stranded parental template after melting. Following annealing, the extension time of the PCR reaction is short enough to allow only partial extension of the primer on the parental gene. The same process is repeated in subsequent cycles until the extended primer reaches full length as the parental genes.

Directed Evolution

DNA mutagenesis and recombination alone are often insufficient to yield desired enzymatic properties because the sequence space is sampled around the parent genes. However, the coding sequence of the desired enzymes may be significantly divergent from the parent. Directed evolution creates a path in the sequence space to guide mutagenesis and recombination (Fig. 1E). There are four essential steps involved in directed evolution (Fig. 6). First, a large

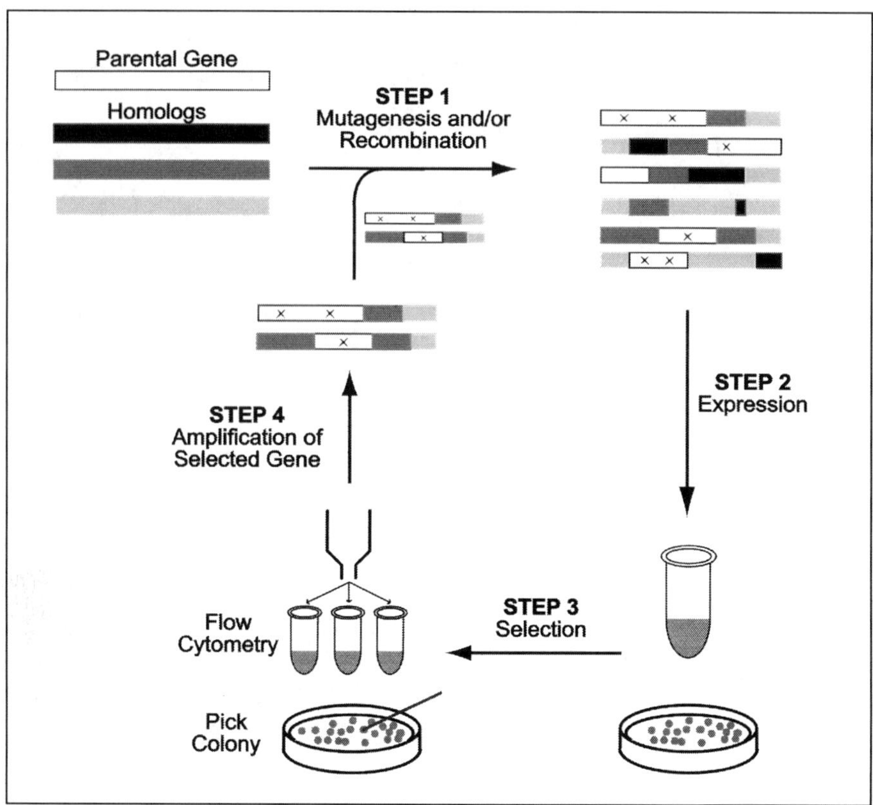

Figure 6. Directed evolution. The cross "×" donates a point mutation; a grey dot, colony on a culture plate. The first generation of genes contains the parental gene and/or its homologs as identified by distinct shades. Then, mutants are created, expressed, selected and amplified for subsequent generations.

and diverse gene library is generated by mutagenesis or recombination from parent genes. Second, the library of genes is expressed by cells or phages. Third, cells exhibiting more desired properties are selected through either a manual process of choosing colonies from agar plates or an automated process such as flow cytometry. Note here that it is important to devise an evolutionary pressure in the selection process that is strong enough to reduce the selection cycles, yet sensitive enough to discriminate between proteins with small differences in properties. Lastly, the genes from the selected cells are amplified and extracted before they are used as parents in the next cycle. Several cycles of this process may be necessary to generate proteins with significantly enhanced properties.

Directed evolution has been successfully applied to enhance or create enzymatic activity, cellular pathways, molecular switches and fluorescence properties. Recently, directed evolution has been used to enhance enzymatic activity and thermal stability.[56-63] In other examples, directed evolution was applied to alter properties of proteins involved in cellular pathways. For instance, modified proteins were introduced into cells to create cellular pathways that form disulfide bonds; in another case, proteins in the carotenoid biosynthetic pathways were altered to create new carotenoid products.[64,65] Molecular switches were also created by evolving the hybrid of maltose binding protein and β-lactamase.[66-68] After applying mutagenesis and 3 generations of directed evolution to the *EcoRV* restriction endonuclease, a variant was created with altered DNA substrate specificity.[11] Recently, a superfolder GFP was created with a 50-fold brighter fluorescence and a higher tolerance to circular permutation and chemical denaturants.[45] This superfolder GFP was created by screening for brighter GFP mutants that were fused with a poorly folded peptide, which effectively targeted the folding properties of GFP.

Fusion Proteins

Fusion proteins are created by joining two or more proteins at their N- and C- termini (Fig. 7A), thereby forming a single protein with the combined functions of the components. By exploiting the intramolecular interactions among components, fusion proteins can create biosensors and molecular switches. As genetic engineering of fusion proteins can be tedious, a fluorescent cassette technique was introduced to make fusion proteins more efficiently.[69]

Fluorescent protein biosensors represent a major class of fusion protein applications. For example, the N- or C-terminal fusion of a fluorescent protein (FP) to a target protein (Fig. 7B) allows tracking of the target protein's expression, localization and movement inside cells. By fusing a sensor protein to the middle of a FP gene, researchers have created split-FP biosensors whose fluorescence is modulated by stimulus-induced conformational change of the sensor protein (Fig. 7C). For example, when maltose is bound, the adjacent N- and C-terminal of maltose binding protein (MBP) allows two halves of FP to fold together, hence fluorescence increases; when maltose is released, the large conformational change of MBP separates the two halves of FP and decreases fluorescence.[70] Another class of biosensors is based on the principle of fluorescence resonance energy transfer (FRET) where the distance and orientation between a donor and an acceptor FP governs their fluorescence emission spectral properties. Typically, the donor and acceptor FP are fused to the N- and C-termini of a sensor protein that responds to a biological event with a large protein conformational change. This conformational change alters the relative position between the donor and acceptor FP and hence a change in FRET efficiency is correlated to the biological event (Fig. 7D,E). FRET biosensors based on the intrinsic conformational change of the sensor proteins have been demonstrated using IP_3 receptors, cGMP-dependent protein kinase I, MBP and bacterial periplasmic binding proteins (Fig.7D).[71-74] Alternatively, conformational change caused by the conditional binding of a protein to its peptidal substrate can also induce FRET efficiency changes (Fig. 7E). A classic example is the Ca^{2+} biosensor. Calmodulin and its Ca^{2+}-dependent binding peptide M13 are fused and sandwiched between cyan fluorescent protein (CFP) and yellow fluorescent protein (YFP). When Ca^{2+} increases, the binding of calmodulin to M13 shortens the distance between CFP and YFP and hence FRET efficiency increases.[75-77] Using similar concepts, FRET biosensors have also been created to detect protein kinase activities.[78-81]

Protein fusion was also used to create molecular switches. Since the function of a protein is closely related to its conformation, the simplest approach to engineer a protein switch is by an allosteric effect (Fig. 7F). This can be achieved by fusing an effector protein with a sensor protein. For example, circularly permutated β-lactamase (BLA) was inserted into specific locations in MBP such that the enzymatic site of BLA is distorted by the conformational change of MBP upon the release of maltose, resulting in the loss of BLA activity.[66-68]

One drawback of fusion proteins is that the functions of proteins are not always preserved. Fusing proteins together may cause one or more of the fusion partners to fold improperly. It has been experimentally observed that well folded proteins make the fusion protein fold better, whereas poorly folded proteins could destabilize the entire fusion. Steric restrictions between fusion partners may also cause the loss of function. In addition, fusion proteins are more prone to aggregate due to their large size and flexibility, destroying their activities as well. Besides the limited guidelines provided by semi-rationally studying the protein structure of each fusion partner, fusion protein engineering is still a trial-and-error process.

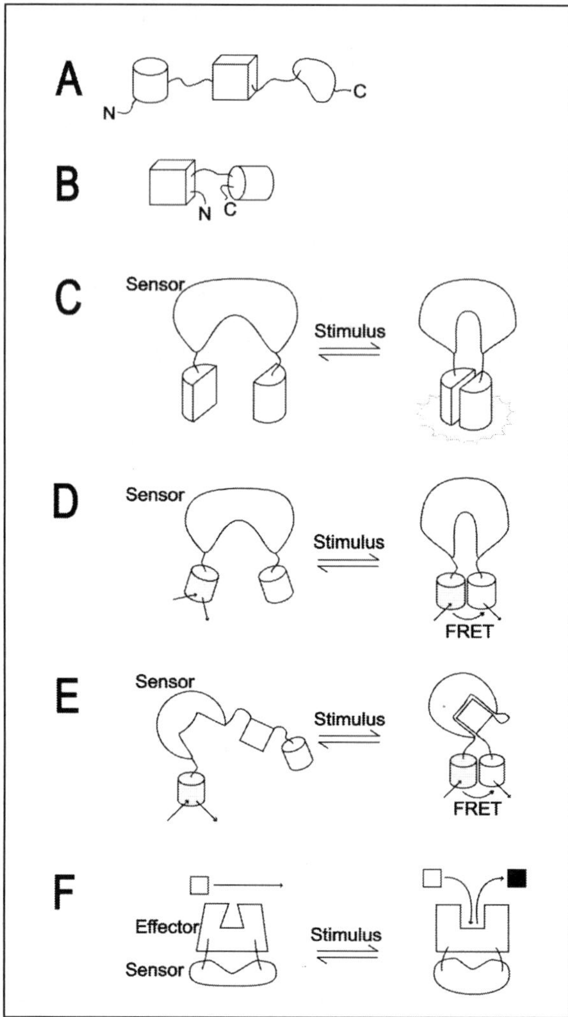

Figure 7. Fusion protein and application mechanisms. The grey cylinder represents CFP, while the white, YFP. Arrows entering the cylinder represent excitation light, while arrows leaving the cylinder represent emission light. A) The concept of a fusion protein. B) FP tagged fusion protein. C) Split-FP biosensor modulates fluorescence by the conformational change of the sensor protein. D) FRET of the biosensor is increased by the conformational change of the sensor protein that brings the FRET pairs together. E) FRET of the biosensor is created when the sensor binds to the substrate and vice versa. F) An example of the protein switching mechanism. The white and black boxes are the substrate and converted substrate of the effector enzyme. Stimulation changes the conformation of the sensor protein, which alters the enzymatic site of the effector enzyme that allows enzymatic activities to be carried out and hence turns it "on".

Circular Permutation

Circular permutation of a protein creates new N- and C- termini at nonnatural locations on the protein by fusing together the natural N- and C- termini (Fig. 8A). This can be achieved easily by fusing two copies of the target gene in tandem and then amplifying at specific circular permutation locations by PCR (Fig. 8B).[82] Circular permutation has been used to study the

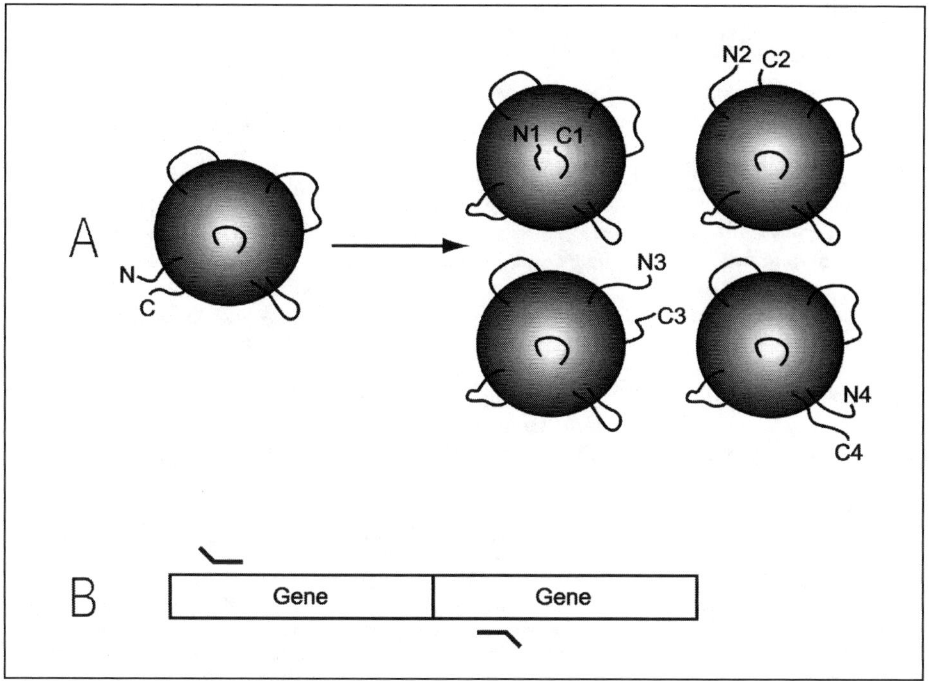

Figure 8. Circular permutation. A) Four possible circular permutations from the original protein. N1-4 and C1-4 are the newly created termini. B) A technique to generate circular permutations of a gene. Each pair of primers with the same shade will be used in a PCR reaction to create circular permutation at that location.

biochemical properties of proteins such as folding and functional elements. In one example, dihydrofolate reductase was circularly permutated systematically to break its backbone at each amino acid and the resulting variants were tested for functionalities. The correlation between the circular permutation location and the activity of the variant helped to determine the regions crucial for folding and function.[83,84] In addition to identifying critical regions of a protein, circular permutation was used to engineer proteins with enhanced properties. The catalytic activity of *Candida antarctica* Lipase B was improved in its circularly permutated variants due to better active site accessibility and backbone flexibilities.[85] In another example, circular permutation of yellow fluorescent protein improved the dynamic range of FRET Ca^{2+} biosensor because it changes the orientation factor between the donor and acceptor fluorescent proteins.[86]

Conclusions and Perspectives

Nature tinkers with new protein designs by random processes that alter genomic DNA such as mutations, insertions, deletions, inversions, gene fusions and translocations. Then through natural selection arising from environmental conditions, the best protein designs are discovered. These natural evolutionary mechanisms have inspired the techniques discussed in this chapter. Random mutagenesis and DNA recombination can create random gene libraries that can then be screened through successive generations by directed evolutionary techniques for incrementally better designs. Protein fusion, another natural evolutionary mechanism, can combine functions from two different proteins. Furthermore, by hijacking the natural processes for protein translation, we can expand protein design by introducing noncanonical amino acids.

Finally, with a strong understanding of protein structure-function relationships, site directed mutagenesis provides the most direct route to protein design.

These techniques, however, are restricted by the starting templates—the natural proteins. This limits the functions of engineered proteins to those similar to natural proteins. In order to engineer proteins with truly novel structures and customized functions, we need to be able to predict the structure and function of a protein from its primary structure. This requires overcoming two major hurdles in understanding: first, how proteins fold into unique three-dimensional structures and second, how structure determines biochemical function. As a deeper insight is gained into sequence-structure-function relationships, rational protein design will become more feasible. Looking ahead, protein engineering will play increasingly important roles in making new molecular tools, in designing better industrial enzymes, in producing biomaterials with novel properties and in rewiring cellular pathways related to human diseases.

Acknowledgements

This work was supported by grants from the Canadian Foundation of Innovation (CFI) and the National Science and Engineering Research Council (NSERC).

References

1. Li B, Nowak NM, Kim SK et al. Random mutagenesis of the M3 muscarinic acetylcholine receptor expressed in yeast: Identification of second-site mutations that restore function to a coupling-deficient mutant M3 receptor. J Biol Chem 2005; 280(7):5664-5675.
2. Chao G, Cochran JR, Wittrup KD. Fine epitope mapping of anti-epidermal growth factor receptor antibodies through random mutagenesis and yeast surface display. J Mol Biol 2004; 342(2):539-550.
3. Farrow KA, Lyras D, Polekhina G et al. Identification of essential residues in the Erm(B) rRNA methyltransferase of Clostridium perfringens. Antimicrob Agents Chemother 2002; 46(5):1253-1261.
4. Munoz I, Ruiz A, Marquina M et al. Functional characterization of the yeast Ppz1 phosphatase inhibitory subunit Hal3: A mutagenesis study. J Biol Chem 2004; 279(41):42619-42627.
5. Takahashi M, Hasuura Y, Nakamori S et al. Improved autoprocessing efficiency of mutant subtilisins E with altered specificity by engineering of the pro-region. J Biochem (Tokyo) 2001; 130(1):99-106.
6. Fujii K, Minagawa H, Terada Y et al. Use of random and saturation mutageneses to improve the properties of Thermus aquaticus amylomaltase for efficient production of cycloamyloses. Appl Environ Microbiol 2005; 71(10):5823-5827.
7. Shim JH, Kim YW, Kim TJ et al. Improvement of cyclodextrin glucanotransferase as an antistaling enzyme by error-prone PCR. Protein Eng Des Sel 2004; 17(3):205-211.
8. Kohno M, Enatsu M, Funatsu J et al. Improvement of the optimum temperature of lipase activity for Rhizopus niveus by random mutagenesis and its structural interpretation. J Biotechnol 2001; 87(3):203-210.
9. Amara AA, Steinbuchel A, Rehm BH. In vivo evolution of the Aeromonas punctata polyhydroxyalkanoate (PHA) synthase: Isolation and characterization of modified PHA synthases with enhanced activity. Appl Microbiol Biotechnol 2002; 59(4-5):477-482.
10. Delagrave S, Hawtin RE, Silva CM et al. Red-shifted excitation mutants of the green fluorescent protein. Biotechnology (NY) 1995; 13(2):151-154.
11. Lanio T, Jeltsch A, Pingoud A. Towards the design of rare cutting restriction endonucleases: Using directed evolution to generate variants of EcoRV differing in their substrate specificity by two orders of magnitude. J Mol Biol 1998; 283(1):59-69.
12. Wu TK, Griffin JH. Conversion of a plant oxidosqualene-cycloartenol synthase to an oxidosqualene-lanosterol cyclase by random mutagenesis. Biochemistry 2002; 41(26):8238-8244.
13. Greener A, Callahan M, Jerpseth B. An efficient random mutagenesis technique using an E. coli mutator strain. Mol Biotechnol 1997; 7(2):189-195.
14. Greener A, Callahan M, Jerpseth B. An efficient random mutagenesis technique using an E. coli mutator strain. Methods Mol Biol 1996; 57:375-385.
15. Kunkel TA, Erie DA. DNA mismatch repair. Annu Rev Biochem 2005; 74:681-710.
16. Brakmann S, Lindemann BF. Generation of mutant libraries using random mutagenesis. In: Brakmann S, Schwienhorst A, eds. Evolutionary methods in biotechnology: Clever tricks for directed evolution. Weinheim: Wiley-VCH, 2004:5-12.

17. Bornscheuer UT, Altenbuchner J, Meyer HH. Directed evolution of an esterase: Screening of enzyme libraries based on pH-indicators and a growth assay. Bioorg Med Chem 1999; 7(10):2169-2173.
18. Bornscheuer UT, Altenbuchner J, Meyer HH. Directed evolution of an esterase for the stereoselective resolution of a key intermediate in the synthesis of epothilones. Biotechnol Bioeng 1998; 58(5):554-559.
19. Henke E, Bornscheuer UT. Directed evolution of an esterase from Pseudomonas fluorescens. Random mutagenesis by error-prone PCR or a mutator strain and identification of mutants showing enhanced enantioselectivity by a resorufin-based fluorescence assay. Biol Chem 1999; 380(7-8):1029-1033.
20. Shortle D, Nathans D. Local mutagenesis: A method for generating viral mutants with base substitutions in preselected regions of the viral genome. Proc Natl Acad Sci USA 1978; 75(5):2170-2174.
21. Shimada A. PCR-based site-directed mutagenesis. Methods Mol Biol 1996; 57:157-165.
22. Storici F, Lewis LK, Resnick MA. In vivo site-directed mutagenesis using oligonucleotides. Nat Biotechnol 2001; 19(8):773-776.
23. Hutchison IIIrd CA, Phillips S, Edgell MH et al. Mutagenesis at a specific position in a DNA sequence. J Biol Chem 1978; 253(18):6551-6560.
24. Saint-Joanis B, Souchon H, Wilming M et al. Use of site-directed mutagenesis to probe the structure, function and isoniazid activation of the catalase/peroxidase, KatG, from Mycobacterium tuberculosis. Biochem J 1999; 338(Pt 3):753-760.
25. Ohtsubo K, Imajo S, Ishiguro M et al. Studies on the structure-function relationship of the HNK-1 associated glucuronyltransferase, GlcAT-P, by computer modeling and site-directed mutagenesis. J Biochem (Tokyo) 2000; 128(2):283-291.
26. Farh L, Hwang SY, Steinrauf L et al. Structure-function studies of Escherichia coli biotin synthase via a chemical modification and site-directed mutagenesis approach. J Biochem (Tokyo) 2001; 130(5):627-635.
27. Alam M, Vance DE, Lehner R. Structure-function analysis of human triacylglycerol hydrolase by site-directed mutagenesis: Identification of the catalytic triad and a glycosylation site. Biochemistry 2002; 41(21):6679-6687.
28. Sviridov D, Hoang A, Huang W et al. Structure-function studies of apoA-I variants: Site-directed mutagenesis and natural mutations. J Lipid Res 2002; 43(8):1283-1292.
29. Gupta RP, He YA, Patrick KS et al. CYP3A4 is a vitamin D-24- and 25-hydroxylase: Analysis of structure function by site-directed mutagenesis. J Clin Endocrinol Metab 2005; 90(2):1210-1219.
30. Mullaney EJ, Daly CB, Kim T et al. Site-directed mutagenesis of Aspergillus niger NRRL 3135 phytase at residue 300 to enhance catalysis at pH 4.0. Biochem Biophys Res Commun 2002; 297(4):1016-1020.
31. Tobe S, Shimogaki H, Ohdera M et al. Expression of Bacillus protease (Protease BYA) from Bacillus sp. Y in Bacillus subtilis and enhancement of its specific activity by site-directed mutagenesis-improvement in productivity of detergent enzyme. Biol Pharm Bull 2006; 29(1):26-33.
32. Konkol L, Hirai TJ, Adams JA. Substrate specificity of the oncoprotein v-Fps: Site-specific mutagenesis of the putative P+1 pocket. Biochemistry 2000; 39(1):255-262.
33. Ormo M, Cubitt AB, Kallio K et al. Crystal structure of the Aequorea victoria green fluorescent protein. Science 1996; 273(5280):1392-1395.
34. Budisa N, Rubini M, Bae JH et al. Global replacement of tryptophan with aminotryptophans generates noninvasive protein-based optical pH sensors. Angew Chem Int Ed Engl 2002; 41(21):4066-4069.
35. Cohen GN, Cowie DB. Total replacement of methionine by selenomethionine in the proteins of *Escherichia coli*. C R Hebd Seances Acad Sci 1957; 244(5):680-683.
36. Hyun Bae J, Rubini M, Jung G et al. Expansion of the genetic code enables design of a novel "gold" class of green fluorescent proteins. J Mol Biol 2003; 328(5):1071-1081.
37. Kobayashi T, Sakamoto K, Takimura T et al. Structural basis of nonnatural amino acid recognition by an engineered aminoacyl-tRNA synthetase for genetic code expansion. Proc Natl Acad Sci USA 2005; 102(5):1366-1371.
38. Kwon I, Kirshenbaum K, Tirrell DA. Breaking the degeneracy of the genetic code. J Am Chem Soc 2003; 125(25):7512-7513.
39. Plettner E, Khumtaveeporn K, Shang X et al. A combinatorial approach to chemical modification of subtilisin Bacillus lentus. Bioorg Med Chem Lett 1998; 8(17):2291-2296.
40. Taki M, Hohsaka T, Murakami H et al. A novel fluorescent nonnatural amino acid that can be incorporated into a specific position of streptavidin. Nucleic Acids Res Suppl 2002; (2):203-204.
41. Murakami H, Hohsaka T, Ashizuka Y et al. Site-directed incorporation of fluorescent nonnatural amino acids into streptavidin for highly sensitive detection of biotin. Biomacromolecules. Spring 2000; 1(1):118-125.

42. Sisido M, Hohsaka T. Introduction of specialty functions by the position-specific incorporation of nonnatural amino acids into proteins through four-base codon/anticodon pairs. Appl Microbiol Biotechnol 2001; 57(3):274-281.

43. Kajihara D, Hohsaka T, Sisido M. Synthesis and sequence optimization of GFP mutants containing aromatic nonnatural amino acids at the Tyr66 position. Protein Eng Des Sel 2005; 18(6):273-278.

44. Hohsaka T, Sisido M. Incorporation of nonnatural amino acids into proteins by using five-base codon-anticodon pairs. Nucleic Acids Symp Ser 2000; (44):99-100.

45. Pedelacq JD, Cabantous S, Tran T et al. Engineering and characterization of a superfolder green fluorescent protein. Nat Biotechnol 2006; 24(1):79-88.

46. Castle LA, Siehl DL, Gorton R et al. Discovery and directed evolution of a glyphosate tolerance gene. Science 2004; 304(5674):1151-1154.

47. Stemmer WP. Rapid evolution of a protein in vitro by DNA shuffling. Nature 1994; 370(6488):389-391.

48. Stemmer WP. DNA shuffling by random fragmentation and reassembly: In vitro recombination for molecular evolution. Proc Natl Acad Sci USA 1994; 91(22):10747-10751.

49. Crameri A, Whitehorn EA, Tate E et al. Improved green fluorescent protein by molecular evolution using DNA shuffling. Nat Biotechnol 1996; 14(3):315-319.

50. Aharoni A, Gaidukov L, Yagur S et al. Directed evolution of mammalian paraoxonases PON1 and PON3 for bacterial expression and catalytic specialization. Proc Natl Acad Sci USA 2004; 101(2):482-487.

51. Hsu JS, Yang YB, Deng CH et al. Family shuffling of expandase genes to enhance substrate specificity for penicillin G. Appl Environ Microbiol 2004; 70(10):6257-6263.

52. Suen WC, Zhang N, Xiao L et al. Improved activity and thermostability of Candida antarctica lipase B by DNA family shuffling. Protein Eng Des Sel 2004; 17(2):133-140.

53. Zhao H, Giver L, Shao Z et al. Molecular evolution by staggered extension process (StEP) in vitro recombination. Nat Biotechnol 1998; 16(3):258-261.

54. He M, Yang ZY, Nie YF et al. A new type of class I bacterial 5-enopyruvylshikimate-3-phosphate synthase mutants with enhanced tolerance to glyphosate. Biochim Biophys Acta 2001; 1568(1):1-6.

55. Dion M, Nisole A, Spangenberg P et al. Modulation of the regioselectivity of a Bacillus alpha-galactosidase by directed evolution. Glycoconj J 2001; 18(3):215-223.

56. Vamvaca K, Butz M, Walter KU et al. Simultaneous optimization of enzyme activity and quaternary structure by directed evolution. Protein Sci 2005; 14(8):2103-2114.

57. Gould SM, Tawfik DS. Directed evolution of the promiscuous esterase activity of carbonic anhydrase II. Biochemistry 2005; 44(14):5444-5452.

58. Wong DW, Batt SB, Lee CC et al. High-activity barley alpha-amylase by directed evolution. Protein J 2004; 23(7):453-460.

59. Kim YW, Lee SS, Warren RA et al. Directed evolution of a glycosynthase from Agrobacterium sp. increases its catalytic activity dramatically and expands its substrate repertoire. J Biol Chem 2004; 279(41):42787-42793.

60. Johannes TW, Woodyer RD, Zhao H. Directed evolution of a thermostable phosphite dehydrogenase for NAD(P)H regeneration. Appl Environ Microbiol 2005; 71(10):5728-5734.

61. Hoseki J, Yano T, Koyama Y et al. Directed evolution of thermostable kanamycin-resistance gene: A convenient selection marker for Thermus thermophilus. J Biochem (Tokyo) 1999; 126(5):951-956.

62. Hao J, Berry A. A thermostable variant of fructose bisphosphate aldolase constructed by directed evolution also shows increased stability in organic solvents. Protein Eng Des Sel 2004; 17(9):689-697.

63. Zhang N, Suen WC, Windsor W et al. Improving tolerance of Candida antarctica lipase B towards irreversible thermal inactivation through directed evolution. Protein Eng 2003; 16(8):599-605.

64. Masip L, Pan JL, Haldar S et al. An engineered pathway for the formation of protein disulfide bonds. Science 2004; 303(5661):1185-1189.

65. Umeno D, Tobias AV, Arnold FH. Diversifying carotenoid biosynthetic pathways by directed evolution. Microbiol Mol Biol Rev 2005; 69(1):51-78.

66. Guntas G, Mitchell SF, Ostermeier M. A molecular switch created by in vitro recombination of nonhomologous genes. Chem Biol 2004; 11(11):1483-1487.

67. Guntas G, Ostermeier M. Creation of an allosteric enzyme by domain insertion. J Mol Biol 2004; 336(1):263-273.

68. Guntas G, Mansell TJ, Kim JR et al. Directed evolution of protein switches and their application to the creation of ligand-binding proteins. Proc Natl Acad Sci USA 2005; 102(32):11224-11229.

69. Truong K, Khorchid A, Ikura M. A fluorescent cassette-based strategy for engineering multiple domain fusion proteins. BMC Biotechnol 2003; 3:8.

70. Jeong J, Kim SK, Ahn J et al. Monitoring of conformational change in maltose binding protein using split green fluorescent protein. Biochem Biophys Res Commun 2006; 339(2):647-651.
71. Fehr M, Frommer WB, Lalonde S. Visualization of maltose uptake in living yeast cells by fluorescent nanosensors. Proc Natl Acad Sci USA 2002; 99(15):9846-9851.
72. Nikolaev VO, Gambaryan S, Lohse MJ. Fluorescent sensors for rapid monitoring of intracellular cGMP. Nat Methods 2006; 3(1):23-25.
73. Remus TP, Zima AV, Bossuyt J et al. Biosensors to measure inositol 1,4,5-trisphosphate concentration in living cells with spatiotemporal resolution. J Biol Chem 2006; 281(1):608-616.
74. Deuschle K, Okumoto S, Fehr M et al. Construction and optimization of a family of genetically encoded metabolite sensors by semirational protein engineering. Protein Sci 2005; 14(9):2304-2314.
75. Miyawaki A, Llopis J, Heim R et al. Fluorescent indicators for Ca^{2+} based on green fluorescent proteins and calmodulin. Nature 1997; 388(6645):882-887.
76. Truong K, Sawano A, Mizuno H et al. FRET-based in vivo Ca^{2+} imaging by a new calmodulin-GFP fusion molecule. Nat Struct Biol 2001; 8(12):1069-1073.
77. Miyawaki A, Griesbeck O, Heim R et al. Dynamic and quantitative Ca^{2+} measurements using improved cameleons. Proc Natl Acad Sci USA 1999; 96(5):2135-2140.
78. Braun DC, Garfield SH, Blumberg PM. Analysis by fluorescence resonance energy transfer of the interaction between ligands and protein kinase C delta in the intact cell. J Biol Chem 2005; 280(9):8164-8171.
79. Schleifenbaum A, Stier G, Gasch A et al. Genetically encoded FRET probe for PKC activity based on pleckstrin. J Am Chem Soc 2004; 126(38):11786-11787.
80. Sato M, Umezawa Y. Imaging protein phosphorylation by fluorescence in single living cells. Methods 2004; 32(4):451-455.
81. Wang Y, Botvinick EL, Zhao Y et al. Visualizing the mechanical activation of Src. Nature 2005; 434(7036):1040-1045.
82. Chiang JJ, Li I, Truong K. Creation of circularly permutated yellow fluorescent proteins using fluorescence screening and a tandem fusion template. Biotechnol Lett 2006; 28(7):471-475.
83. Nakamura T, Iwakura M. Circular permutation analysis as a method for distinction of functional elements in the M20 loop of Escherichia coli dihydrofolate reductase. J Biol Chem 1999; 274(27):19041-19047.
84. Iwakura M, Nakamura T, Yamane C et al. Systematic circular permutation of an entire protein reveals essential folding elements. Nat Struct Biol 2000; 7(7):580-585.
85. Qian Z, Lutz S. Improving the catalytic activity of Candida antarctica lipase B by circular permutation. J Am Chem Soc 2005; 127(39):13466-13467.
86. Nagai T, Yamada S, Tominaga T et al. Expanded dynamic range of fluorescent indicators for Ca^{2+} by circularly permuted yellow fluorescent proteins. Proc Natl Acad Sci USA 2004; 101(29):10554-10559.

CHAPTER 3

Past, Present, and Future of Gold Nanoparticles

Travis Jennings and Geoffrey Strouse*

Abstract

Colloidal gold nanoparticles have been around for centuries. Historically, the use of gold nanoparticles has been predominantly found in the work of artists and craftsman because of their vivid visible colors. However, through research, the size, shape, surface chemistry, and optical properties of gold nanoparticles are all parameters which are under control and has opened the doors to some very unique and exciting capabilities. The purpose of this chapter is to review some of the important discoveries and give background in regard to gold nanoparticles. First, the most common wet chemical methods toward their synthesis are reviewed, specifically discussing routes toward spherical colloidal synthesis and controllable rod formation. Next, because many applications of gold nanoparticles are a result of their magnificent interactions with light, some of the basic optical-electronic properties and the physics behind them are elucidated. Finally, by taking advantage of the optical-electronic properties, numerous proven applications for gold nanoparticles are discussed, as well as their predicted applications in the future.

Introduction to Gold Nanoparticles

The future success of nanotechnology is akin to capturing a wild horse: powerful and full of potential but must be tamed before it will be useful. The taming of any beast requires a deep understanding of the basic fundamental traits that govern behavior, which, for a nanomaterial, is a combination of primarily composition, size, and shape. In order to advance nanotechnology for applications in bioengineered devices and high speed electronics, development of methods to understand and control the behaviours of nanomaterials is needed.

A nanomaterial may be defined as any material (insulator, conductor, semiconductor), which has been controllably synthesized on the size range of roughly 1 to 100 nm. At this size and dimensional range, essentially any material will exhibit different properties than it would as an atomic cluster or as the larger bulk material. In fact, a change in the nanomaterial's size confers dynamic properties upon the system, and is commonly used to manipulate the light absorption and photoluminescence properties, as well as magnetization and electrical conductivity. Metallic gold nanoparticles have long been a popular choice of nanomaterial for many scientists to work with due to their facile methods of synthesis, the high degree of control over shape and size, their long-term stability in a wide variety of solvents and pH conditions, and their conducive nature toward surface-molecule (ligand) modification. The ability to controllably choose the ligand attached to a nanoparticle (NP) surface, such as proteins or nucleic acids, introduces

*Corresponding Author: Geoffrey Strouse—Department of Chemistry, Florida State University, Tallahassee, FL 32306-4390, U.S.A. Email: strouse@chem.fsu.edu

Bio-Applications of Nanoparticles, edited by Warren C.W. Chan. ©2007 Landes Bioscience and Springer Science+Business Media.

a means by which the nanoparticle may be directed to a targeted site in a body or in a cell for either detection, diagnostics, or therapeutic purposes. Further, the effective use of a nanomaterial probe requires an understanding of both the relationship that exists between the material type chosen and the physical dimensions of the nanomaterial as well as the effect of ligands on the optical-electronic properties of the NP.

The purpose of this chapter is to provide an overview of the wide variety of different gold nanomaterials and their optical, electronic, and magnetic properties. At the end of this discussion, a review of their use in medicine and diagnostic applications will be discussed.

Synthetic Routes of Materials Synthesis

In comparison to most inorganic nanomaterials, gold nanoparticles can be synthesized in the largest variety of shapes and sizes. The most common shape, and arguably one of the easiest to synthesize, is the simple sphere. In the synthesis, gold chloride is added to aqueous solvents in the presence of a reducing agent. Gold nanoparticles are quickly formed, as observed simply by the change in the color of the solution (light yellow goes to red). Growth is rapid and no crystallographic face is favored over others when a neutral stabilizing agent such as citrate is used to bind the surface. In most solution-phase synthetic methods, stabilizing agents such as phosphine, thiol, carboxylate, or amine containing molecules are often used. These molecules are important primarily in mediating the interactions between the NP surface and the surrounding bath. Without stabilizing agents, particles in solution are more susceptible to either oxidation or the attractive interparticle Van der Waal's forces which cause them to aggregate and precipitate from solution. The most successful approaches to controlling nanoparticle size and dispersity are therefore often dependent upon the choice of ligand, surfactant, or stabilizing agent.

Spherical Metal Nanoparticle Synthesis

Figure 1 outlines a scheme for the size-controlled growth of gold nanoparticles where the ultimate size is chosen by the ratio of tannic acid to citrate used to reduce Au(III) into metallic gold nanoparticles. Tannic acid is a more aggressive reducing agent than citrate and will nucleate gold centers rapidly, followed by the more mild citrate reduction onto the existing nuclei. Geuze et al. used this difference in reduction activity to selectively grow gold nanoparticles at sizes from ~3 to 17 nm with very narrow size distributions.[1] With a smaller amount of tannic acid, upon mixing with a solution of Au(III), a small number of nuclei are formed, leaving the citrate to finish reducing the rest of the Au(III) onto these cores. The extra reducing step results in larger nanoparticles. However, the injection of higher concentrations of tannic acid results could lead to smaller nanoparticles because this higher concentration leads to the rapid formation of a greater number of nuclei. This method could also be used to prepare larger sized gold nanoparticles (up to ~50 nm diameter),however this could lead to greater heterogeneity in size distribution.

The formation of smaller particles, below 3 nm diameter, requires even more aggressive reducing agents. Murray et al. and Hutchison et al. have both shown separately that the use of sodium borohydride at room temperature is aggressive enough to nucleate and form small nanoparticles from 1.5 nm[2] up to 5.2 nm.[3] For studies such as these, stabilizing agents such as phosphines or alkanethiols are important to control the reaction and maintain narrow size distributions.

Gold nanoparticles are stable in aqueous media and require no special handling or storage for up to many months. In order to clean the nanoparticles from the reaction solution, it is common to precipitate the nanoparticles through the addition of salt. Cleaning refers to the purification of gold nanoparticles from the reducing agents, excess ligand, or other impurities in the solvent. The increased salt concentration helps screen charge between neighboring particles in solution so that Van der Waal's attractive forces take over and the aggregated particles will precipitate out. The aggregation of colloidal gold nanoparticles is easily characterized by

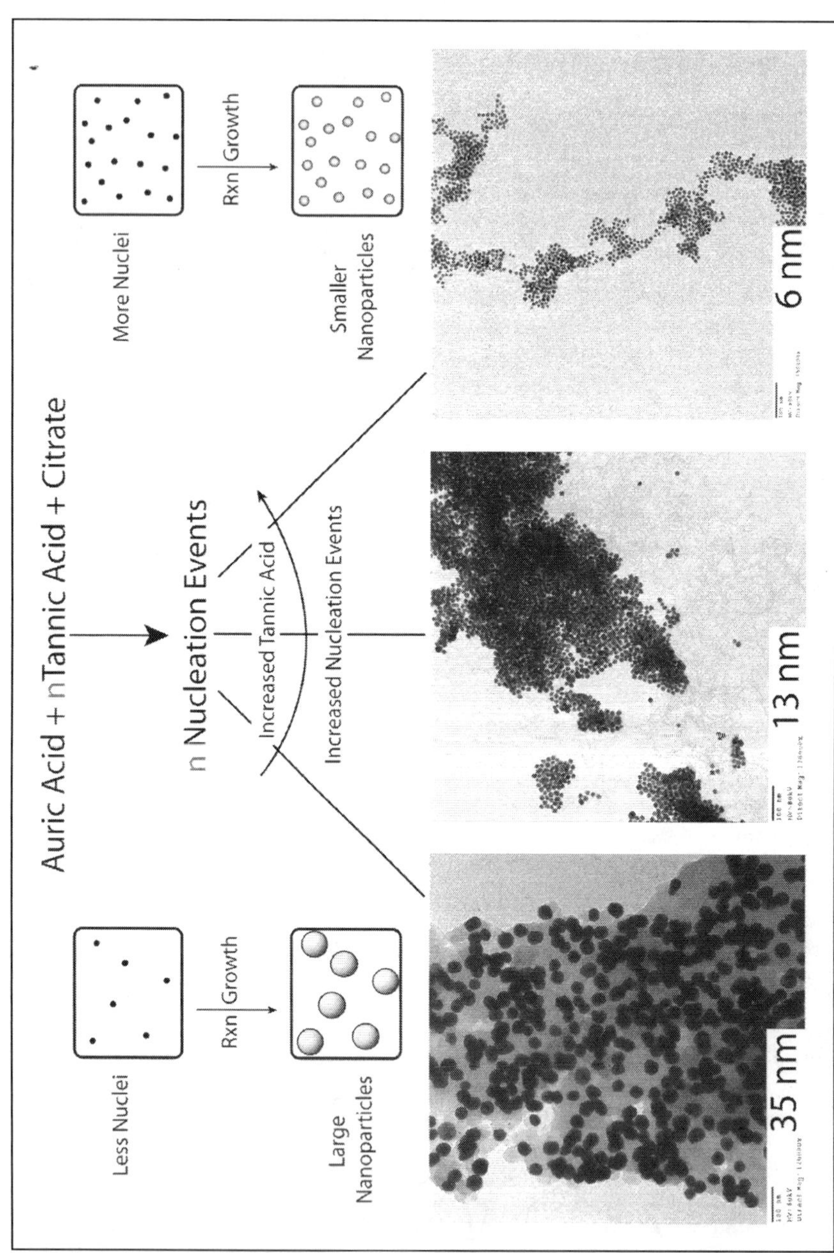

Figure 1. Growth of gold nanoparticles using the rapid reduction and slow growth method. The number of initially formed nuclei determine the eventual size of the nanoparticles once growth is complete.

the visible shift in color from ruby red to a dark blue. A researcher may run into trouble, however, if citrate is the only stabilizing ligand available in solution because the particles may never resuspend after the first precipitation. Therefore, it is generally a good idea to perform a ligand exchange using a ligand with greater binding affinity to the metal surface, such as a phosphine or a thiol, prior to cleaning procedures.

Rod-Shaped Metal Nanoparticle Synthesis

Metallic nanorods differ from spheres because they have a particular aspect ratio in which the material is elongated along a single dimension, keeping the other two dimensions approximately equal. In order to synthesize nanorods, it is necessary to limit growth direction to a single axis. Physically, this is accomplished by reducing metal salts inside either rod-shaped micelles such as those formed in solution by cetyl trimethyl ammonium bromide (CTAB), or by a hard template approach such as anodized porous alumina.[4] The use of a spherical gold nanoparticle as a seed may aid the rod growth inside a micelle to form gold and silver nanorods of high uniformity and controllable aspect ratio in solution.[5] The advantage of hard templated routes for nanorod synthesis is the ease of adjusting the length and hence the aspect ratio of the rod. Further, this method can be used to prepare rods up to microns in length. Micellar templates, on the other hand, require less work or special equipment and may be more practical for large scale synthesis but have greater difficulty in controlling the morphology and aspect ratio of the rod.

Optical and Electronic Properties

An excellent example of how colloidal metal particles interact with light is the stained glass windows from any European Gothic cathedral up to the 15th century. The artisans of the day achieved the brilliant blues, reds, yellows, and greens in the glass windows by mixing metal chlorides into molten glass before pouring it. The metal chlorides nucleated and formed nanoparticles in the molten glass before cooling, making art one of the first uses for nanotechnology.

Although bulk gold and silver are widely known for their lustrous surfaces and colors, there is a drastic color difference when the metal reduces in dimensions. As mentioned above, gold and silver nanoparticles are responsible for transmitting the brilliant colored light through stained glass windows, yet these colors do not resemble the characteristic yellow or bluish reflective surfaces of the bulk metals. Although the artisans did not know it at the time, the mixing of the metal chlorides with molten glass led to the formation of metallic nanoparticles of different shape and size, hence the physical dimensions of the metal nanoparticles had interesting interactions with light and produced visibly beautiful colors. How do these nanoparticles interact with light? The answer to this question lies in the understanding of surface plasmon resonance (SPR) phenomena, which are deeply affected by the nanometal shape and environment.

Figure 2 shows the differing electronic interactions that NP structures may exhibit with passing photons of light. The oscillating electric field of a light ray propogating near a colloidal gold nanoparticle will interact with the free electrons. This causes a concerted oscillation of electron charge that is in resonance with the light frequency. These resonant oscillations are known as surface plasmons, which may also be thought of as giant electronic dipoles whose strength and energy depend upon the both shape and size of the NP. For example, colloidal rods display two plasmon modes that differ in energy due to the length and width of the nanoparticle. A sphere is symmetric about any perspective, and may only be characterized by a transverse plasmon mode, while a rod will display longitudinal modes as well as transverse due to the elongation about a single axis. The longitudinal plasmon mode is shifted to lower and lower energy as the aspect ratio of the rod increases, while the transverse mode is consistently centered around 520 nm absorption wavelength. Indeed, this effect may be seen colorimetrically because a well-dispersed spherical colloid solution of gold nanoparticles will appear a magnificent ruby red, while a dispersion of gold nanorods will take on a brownish color to the

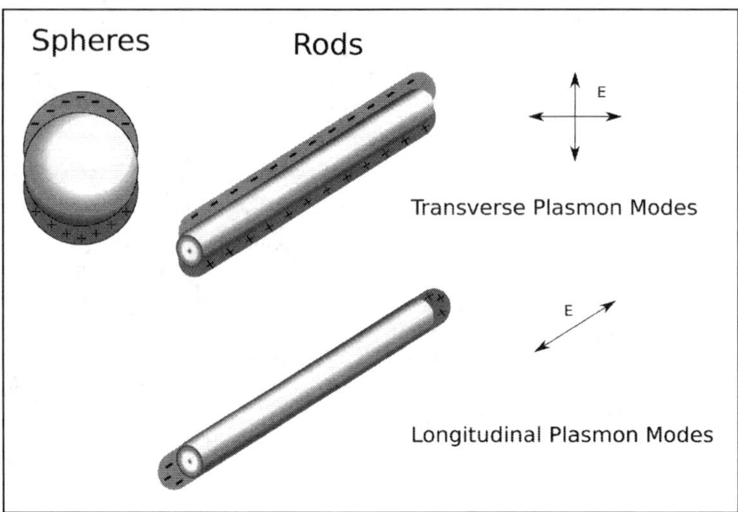

Figure 2. Free electron clouds will interact with the electric field of a glancing light ray. The type and magnitude of this interaction will depend upon the size and shape properties of the nanometal.

eye as the longitudinal plasmon absorption removes red wavelengths from the transmitted light. Aside from the characteristic longitudinal plasmon shifts with differently shaped structures, NP size will yield a pronounced affect upon the transverse plasmon mode. To better understand the affect of nanoparticle size on the optical-electronic properties of gold nanospheres, we will approach the scenario from a "bottom up" view.

The properties of bulk solids can often be described through an inspection of the evolution of electronic states from atoms to clusters to bulk solids. Metals are no different. Fermi-Dirac statistics state that no two electrons may occupy the same quantum state at the same time. The ramifications of Fermi-Dirac statistics into band structures of matter can be illustrated with a simple state-splitting model (see Fig. 3). As more and more atoms are added to any system, that system will go from atomic-like electronic energy levels to more complex levels, yet still discrete energy splitting in the cluster form, until the difference between neighboring energy levels becomes indistinguishable. In this case, the energy levels are considered to be a continuous "band" of energy and no longer discrete. Although all matter types exhibit this behavior, band structure is most commonly split into 3 relativistic categories: Insulators, Semi-Conductors, and Metals. The difference between each of these categories relies upon the relative energy difference between the bottom of the conduction band and the top of the valence band. For metals, the Fermi energy, E_f (defined as the highest energy level achieved for all fermions at absolute zero), overlaps with the conduction band; this allows movement of energetic electrons within the solid. This property of electronic states for metals gives rise to their conduction properties, which directly influences the optical properties.

In the past couple of decades, research has shown an intensified interest in the electronic and optical properties of materials as their dimensions are reduced to the nanometer scale. The electronic properties are changed drastically as the density of states decreases and the spatial dimensions of electronic motion become confined. These changes in properties are also translated into dynamic optical properties because the interactions of a material with light are primarily electronic in nature. Surface plasmon modes are one of the most interesting dynamic properties of metals as dimensions become confined.

The physical origins of SPR bands come from the coherent interaction of conduction electrons with an incident light wave. When the frequency of the external electric field (incoming

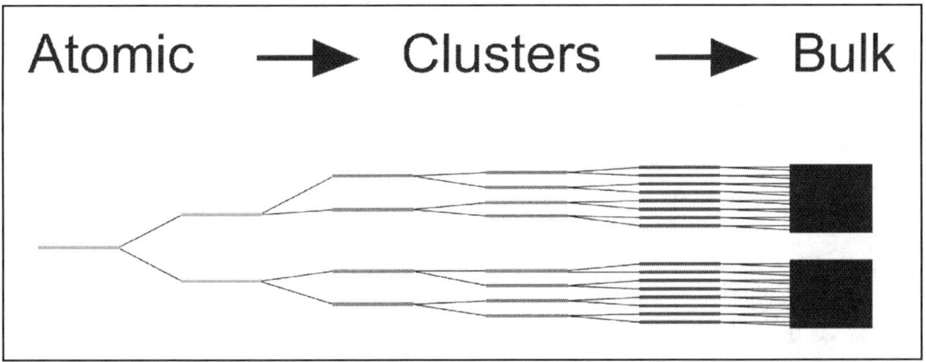

Figure 3. Scheme of energy splitting from the atomic level up to bulk solid. Electrons follow Fermi-Dirac statistics and therefore split levels as more atomic states are added. Eventually there are enough states that the energy splitting between states becomes indistinguishable and is considered to be a continuous band of energy. This demonstrates band formation for a semiconductor where there is an energy gap between the valence and conduction bands. For a bulk metal system, these bands form one continuum.

light frequency) becomes resonant with the coherent electronic motion, the glancing light ray is strongly absorbed by the metal electrons. In general, a passing electromagnetic field may interact with the conduction electrons in any metal as described by absorption features, however the interaction is greatly enhanced within the bandwidth of surface plasmon modes. In the planar metal case, the metal film must be illuminated through a material of higher dielectric for the observation of surface plasmons.[6] However, for metal NPs this condition is relaxed and there are no special requirements for the generation of SPR.[7,8] This is the reason that colloidal noble metal NPs display vivid colors in solution under ambient conditions.

When Mie solved for the optical extinction of colloidal metal particles, he performed his calculations under the Drude model. This assumes the positive atom cores are immobile while the conduction electrons are treated as a gas, characterized by similar properties such as the relationship between pressure and the mean-free path of a particle. Pressure for a gas is determined by the density of particles and the average kinetic motion for the system. The mean-free path of a particle in such a system is therefore directly affected by these parameters, where an increase in density or temperature inversely affects the mean-free path of a particle. The Fermi function gives the probability that a particular energy state, E, will be occupied at a given temperature. The function has the form:

$$f(E) = \frac{1}{e^{(E-E_f)/kT} + 1} \tag{1}$$

where E_f is the Fermi energy of the metal, k is Boltzmann's constant, and T is the temperature.

Metals have within some volume (considered to be finite but inconsequential for the bulk system), a certain density of free-electrons above the Fermi level for a given temperature. Equation 1 relates the number of free electrons to the temperature of the system where an increase in the temperature increases the number of free electrons in the same volume. The increased density of electrons at higher temperature will decrease the electron mean-free path, and therefore the conductivity, because the electrons will tend to collide more frequently with each other. Although this relation between temperature and decreased conductivity has been understood for a long time, physical scientists have begun to ask questions regarding the optical and electronic properties when the dimensions of a metal fall below the mean-free electron path.[8-11] Figure 4 demonstrates the difference in absorption features that occur as the diameter of the

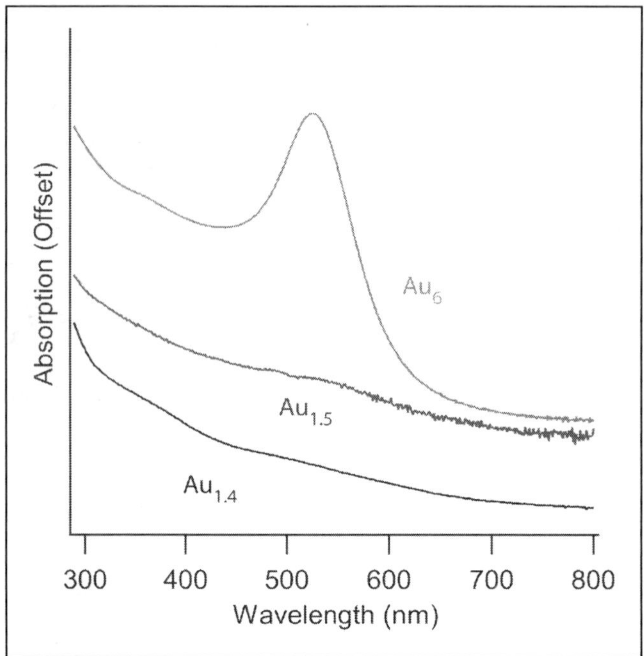

Figure 4. Colloidal gold absorption data for d = 1.4 nm (black), 1.5 nm (blue), and 6 nm (red) NPs in 20 mM phosphate buffer saline. Below ~2 nm diameter, gold NP's exhibit mostly featureless absorption data with an onset beginning at 1.7 eV (730 nm) due to interband absorption and rising to the UV. When d >2.0 nm the NP displays SPR behavior, giving rise to the large absorption band centered at 525 nm.

NP moves from very small (d = 1.4 nm) to the larger sizes where SPR behavior begins to dominate the region from 450 - 650 nm, (2.8 - 1.9 eV).

Generally, for metallic NPs larger than 20 nm it is assumed that the particle is much smaller than the wavelength of interacting light. A mathematical equation describing the absorption of a colloidal system's interaction with light was solved by Mie in 1908,[12] which provided a relationship between absorption and scattering and the observation of distinct solution colors of metallic NPs. When NPs are much smaller than wavelength of the interacting light (<20 nm), only the dipole induced by the electromagnetic wave contributes significantly to the absorption cross section. Mie's theory can then be reduced to the form:

$$\sigma = \frac{9 \cdot V \cdot \varepsilon_m^{3/2}}{c} \cdot \frac{\omega \cdot \varepsilon_2(\omega)}{\left[\varepsilon_1(\omega) + 2\varepsilon_m\right]^2 + \varepsilon(\omega)^2} \tag{2}$$

where V is the particle volume, $\omega°$ is the angular frequency of the exciting light ($2\pi\nu$), c is the speed of light, ε_m is the dielectric of the surrounding medium, and $\varepsilon(\omega)=\varepsilon_1(\omega)+i\varepsilon_2(\omega)$ is the dielectric of the metal. The mean-free path of an electron is the average distance that each electron can move inside the metal structure before colliding with either another electron, a phonon (vibration), or the surface potential. For bulk gold, the mean free path has been measured and is reported as 430 Å,[13] which begs the question: What happens to the optical properties and electronic states of a gold NP as its dimensions drop below 43 nm? Kriebig suggested that once the dimensions are reduced below the mean-free path, the surface scattering of electrons may begin to dominate the optical response.[14] Kriebig et al. and Alvarez et al. applied theoretical dielectric functions to the experimental results of small gold NPs to as an attempt to model the light absorption of gold NPs as size is decreased.[9,14]

For metals, the dielectric constant ε_m is a combination of real and complex values, which take into account the lagging response of conduction electrons in an accelerating field due to a smeared positive background (ionic core atoms). The Drude model is often used to describe dielectrics of metals (see equation 3):

$$\varepsilon_{1D}(\omega) = 1 - \frac{\omega_p^2}{\omega^2 + \omega_0^2} + \frac{i\omega_p^2\omega_0^2}{\omega(\omega^2 + \omega_0^2)} \tag{3}$$

where $\omega_p^2 = [ne^2/(\varepsilon_0\, m_{eff})]$ is the bulk plasmon frequency (8.89 eV for Au) in terms of the free-electron density, n, the fundamental electron charge, e, the permittivity of a vacuum ε_0, and the electron effective mass m_{eff}. The frequency of inelastic collisions, ω_0 from electron-electron, electron-phonon, electron-defect, etc. is usually very small, however can become a dominant term for very small NPs. The interband transitions account for the response of the 5d electrons and are nearly constant, but response of the Drude free-electrons can be greatly affected by the decreased size and electron-surface scattering. Therefore, the use of a surface damping γ_0 term for the small sizes has been suggested in place of ω_0:[9,15]

$$\gamma(r) = \gamma_0 + \frac{Av_F}{R} \tag{4}$$

where A is called the proportionality factor which includes details of the scattering process. Its value ranges from 1 if isotropic, to 3/4 if diffusive, and zero if elastic. v_F is the Fermi velocity of an electron in the metal, (gold = 1.4×10^8 cm/s). Using Equation 4 as the dominant scattering term in Equation 3 yields a full expression for the size dependent (R) dielectric constant.

$$\varepsilon(\omega_i R) = 1 - \frac{\omega_p^2}{\omega^2 + (\gamma_0 + Av_F/R)^2} + \frac{\omega_p^2(\gamma_0 + Av_F/R)}{\omega^2 + (\gamma_0 + Av_F/R)^2} \tag{5}$$

The inclusion of this surface damping term gave good agreement between experimental and theoretical results for Kriebig[14] for particle sizes down to 3.1 nm but the smaller NP sizes did not fit into this equation. Alvarez et al. developed the equation for NPs below 3.1 nm.

Alvarez et al. compared the optical absorption spectra for a series of gold NPs from d = 1.7 nm to 2.5 nm by separating out the interband and Drude components from the total dielectric function:

$$\varepsilon_1(\omega) = \varepsilon_{1IB}(\omega) + \varepsilon_{1D}(\omega) \tag{6a}$$

$$\varepsilon_2(\omega) = \varepsilon_{2IB}(\omega) + \varepsilon_{2D}(\omega) \tag{6b}$$

where the Drude-like terms may be calculated by:

$$\varepsilon_{1D}(\omega) = 1 - \frac{\omega_p^2}{\omega^2 + \omega_0^2} \tag{7a}$$

$$\varepsilon_{2D}(\omega) = \frac{\omega_p^2 \cdot \omega_0}{\omega(\omega^2 + \omega_0^2)} \tag{7b}$$

By substituting Equation 4 in as the damping term, ω_0, these equations have size-dependence built into them, exactly the same as Kriebig has done.

Figure 5 demonstrates the behavior of each dielectric as a function of optical energy. The dielectrics were calculated using the values for optical constants of gold published by Johnson and Christy[13] together with the relationships: $\varepsilon_1 = n^2-k^2$ and $\varepsilon_2 = nk$. The separation of the interband and Drude terms from each other was accomplished by fitting the Drude segments to spline graphs, analogous to the efforts of Alvarez. Doing this allows the recalculation of the

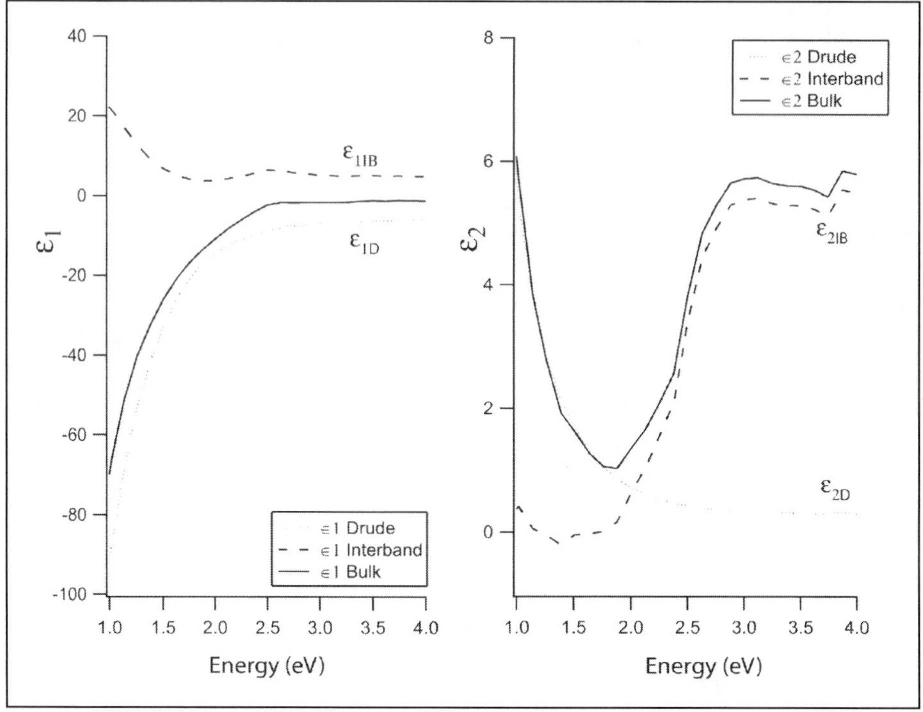

Figure 5. (Left) Real part of the dielectric constant for gold as a function of optical energy. (Right) Imaginary part of the dielectric. The bulk values represented by the solid line (—) have been extracted and plotted from the published values of Johnson and Christy. The Drude contributions (…) have been spline fit and subtracted to isolate the interband contributions (- - -).

Drude contributions, which has a strong relationship to size, using Equations 7a and 7b. Because the interband contributions (neighboring energy levels inside a band) are constant, even at the smallest measured sizes of 1.4 nm, only the Drude terms (free electrons contributions) need to be recalculated.

The extinction spectrum can be calculated, therefore, using these new size-dependent dielectric functions together with Equation 2. Figure 6 shows three calculated extinction spectra: one for bulk gold, one for a 6 nm NP, and another for a 1.5 nm NP. The disappearance of the plasmon absorption is obvious at smaller sizes, which is consistent with a lower density of free electrons. Although the density of electrons is dropping at these smaller sizes, Alvarez et al. realized some very interesting results when trying to fit the theoretical results to the experimentally measured values.

The original attempts to fit the experimentally measured optical absorption spectra for small gold NP's resulted in poor agreement with the theory. Allowing the volume of the particle to float mathematically as a variable in their extinction calculations (Equation 2) resulted in greatly improved fits to experimental data. Essentially, this has the effect of changing the electron density and if ligands chemisorbed onto the metal surface, the ligands can donate electrons into the metal. This means gold NPs from 1.7 - 2.5 nm in diameter may have free electron densities higher than would normally be predicted and the theoretical extinction within the dipole approximation is best obtained when electron density is allowed to float mathematically.

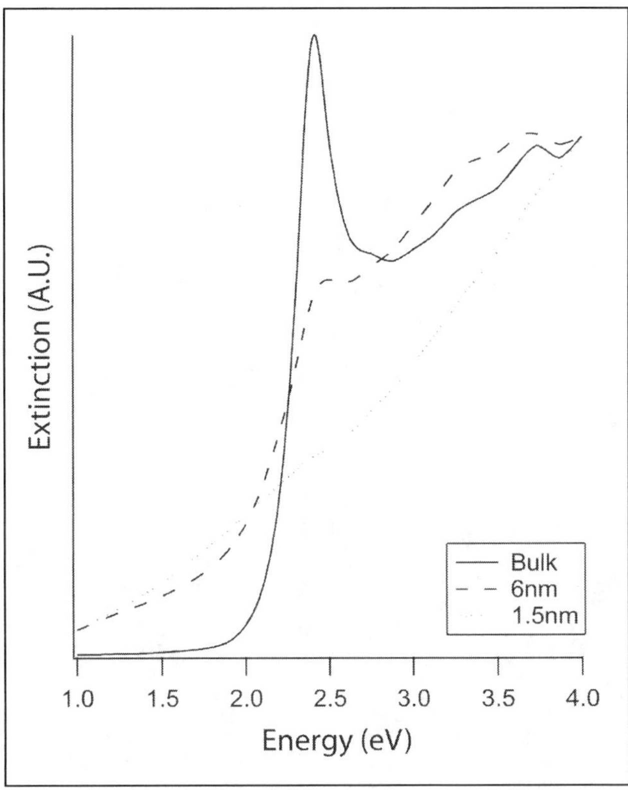

Figure 6. Theoretically generated gold extinction spectra using the dielectric terms calculated exactly analogous to Alvarez et al. The solid line (—) is the spectrum for bulk gold calculated using exactly the values published by Johnson and Christy. Also calculated are exctinction spectra for a 6 nm gold NP (- - -) and for a 1.5 nm NP (···). All spectra were normalized at 4.0 eV.

Metal Nanoparticles in Energy Quenching

In the mid 1960s, Karl Drexhage performed some very interesting experiments in which dipole emitters were suspended at well-controlled distances above planar metal surfaces.[16-18] He noticed some very surprising results from these studies in which the fluorescence quantum yield was heavily quenched at close distances (<1-30 nm) and then began to oscillate at longer distances more on the wavelength scale of the emitted radiation, sometimes even increasing the quantum yield above its native value! These experiments prompted a host of theoreticians to attempt explanations for the observed behavior of fluorescent molecules near conducting metal surfaces. Notable among the resulting theories is the work of Chance, Prock, and Silbey as well as Persson and Lang for explaining the quenching effect of metals on fluorescent dyes.[19,20] The models set up by these theoreticians explain that an oscillating excited state dipole of a dye molecule is also an oscillating electric field which can interact with the free electrons of a nearby metal. As Figure 7 illustrates, time-dependent changes in an electric field may induce a current in the metal, or even form electron-hole pairs in appropriate conditions. Persson and Lange proposed a relation in separation distance between the gold surface and the emitting molecule which is related to the inverse distance to the fourth power. Also, just as important in the Persson and Lang theory is a dependence on electron scattering within the metal surface to quench the photoluminescence of a nearby molecule.

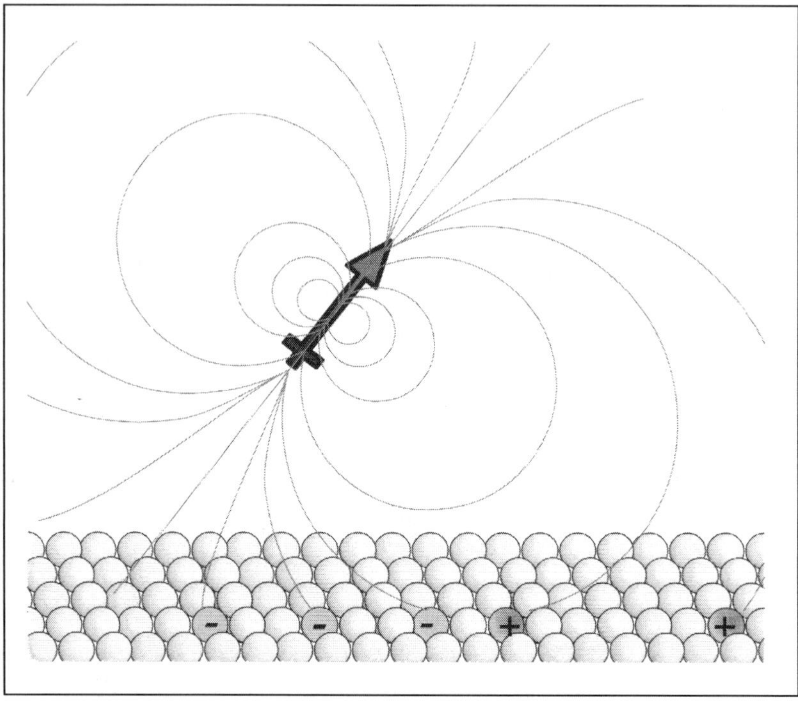

Figure 7. An excited state dipole suspended over a metal surface will assert an oscillating electric field over the surrounding environment. This can lead to the creation of electron-hole pairs in metallic interband transitions.

The theories of Persson and Lang are profound if proven to be true, because this means that metal nanoparticles may be used as molecular beacons and rulers in many applications (e.g., determining molecular distances and kinetics of reactions). This is akin the popular of Förster Resonance Energy Transfer (FRET) technique, which is one of the key technologies in molecular biology and bioengineering. Strouse et al. compared the theory to experimental data using double-stranded DNA for well-defined separation distances between 1.4-1.5 nm gold particles and florescein dye.[21,22] The experiments used very small gold nanoparticles in which the transverse surface plasmon mode is not pronounced and yet achieved fluorescence quenching that followed a $1/d^4$ separation dependence as predicted by Persson and Lang. The implications of this study are that gold nanoparticles as small as 1.4 nm diameter maintain band structure similar to bulk gold, but display increased electron scattering frequencies against the surface potential, as explained by Alvarez et al. which enables them to quench fluorescence emission according to the predictions of Persson and Lang. The term, Nanometal Surface Energy Transfer (NSET) was coined to describe the process of dye quenching by metal surfaces and set it apart from FRET, which is more dependent upon separation distances and energy overlap between donor acceptor pairs than is NSET.[21] The quenching of molecular dyes by gold nanoparticles is now well documented and described both theoretically and experimentally, has progressed beyond the physical characterization steps, and is being utilized to monitor molecular distances and reaction rates.[23]

All of the unique optical and electronic properties of metallic nanoparticles provide the basis for the use of metallic NPs in biomedical applications. The unique relationship between the SPR, NSET and NP size turns out to be useful in many biomedical engineering applications such as the diagnosing or treating of disease in the near future.

Practical Uses of Gold Nanomaterials

Many of the properties of gold nanometal colloids make them particularly amenable to biological applications, namely: they are highly customizeable in size, shape, and surface chemistry, they are stable in a wide variety of environments, they are inert, non-toxic, and they have controllable optical-electronic properties. In this section, we will discuss the use of these materials for biological sensing and medical therapeutic applications.

Gold NPs demonstrate massive dipole interactions with light rays, and, like we mentioned earlier, these SPR bands are very sensitive to their environment. Plasmon resonance is the basis behind a popular analytical technique for biomolecular detection, where the index of refraction changes upon biomolecular binding (e.g., protein) to thin gold films. Mirkin and coworkers used such a technique to monitor nucleic acid interactions through shifts in colloidal gold SPR bands.[24]

As previously described, the aggregation of gold nanoparticles may be characterized by a drastic shift in color from red to blue. The color shift is a result of the interacting electric fields of neighboring particles and has a tendency to lower the resonant frequency of plasmon oscillations (lower energy SPR absorption band.) The closer particles get to one another, the greater the magnitude of this shift. Shifting SPR energies therefore give the basis for colorimetric tests using capture-target biomolecular binding schemes. Figure 8 (left) demonstrates this colorimetric scheme for the detection of specific DNA strands. When gold nanoparticles containing a specific recognition strand of DNA are introduced to a solution in which the complementary strand is also bound to gold nanoparticles, the proper hybridization will draw the nanoparticles together and shift the SPR frequency to lower energy. A change in the color of the solution leads to the detection of the DNA strands.

The use of gold nanoparticles in biomolecular detection schemes have advanced rapidly by Mirkin's group to include not only shifting in the SPR bands for colorimetric detection but to incorporate silver ion reduction into the detection platform for enhanced detection sensitivity via darker color intensity or surface-enhanced Raman spectroscopy (SERS).[25] Using a surface-based capture assay, a gold NP containing a specific sequence of single-stranded DNA will bind to a patterned surface of different complementary sequences. After washing the slide to remove non-specifically bound NPs, only the patterned regions of interest contain gold NPs via DNA hybridization. The NPs can then be detected either colorimetrically by reducing silver ions onto the gold NPs for an enhanced color change, or using SERS. The use of many spots combined with the unique DNA sequence on each spot can be used for high-throughput analysis and detection of genes or for profiling hereditary disease or genomic-based pathogen detection.

In addition to Mirkin's group, Libchaber et al. demonstrated the use of gold nanoparticles to detect single basepair mismatches of complementary DNA by using the electronic properties of gold particles not in an absorption based colorimetric assay, but as the quenching species of a fluorescent molecule.[26] Figure 8 (right) demonstrates the experiment performed by Libchaber et al. in which a hairpin loop of single-stranded oligonucleotide sequence was synthesized to contain a small gold NP at one end and a fluorescent dye at the other. When the oligonucleotide is wrapped up in the hairpin loop, the fluorescent dye and the NPs are very close together and the dye is incapable of emitting photons of light. However, if a matching oligonucleotide sequence, or sequence to be detected, with a perfect complementary match is introduced into solution, the hairpin structure will unravel to form a linear double-stranded structure that is more stable than the hairpin conformation. As a result, the dye and NP are spatially separated and this leads to the dye emitting photons of light. The output output correlates to the concentration of the sequence to be detected, which can be from cells or tissues associated with a disease. Although the ability for a gold NP to quench the fluorescence of a dye molecule is clear, the mechanism responsible for this observation requires further elucidation.

Aside from colorimetric assays and NSET fluorescence quenching, the SPR of spheres, rods, and even gold shells is being hotly pursued for the potential use in photothermal therapy. The SPR band, as described earlier, is a gigantic dipole, with extinction coefficients ranging

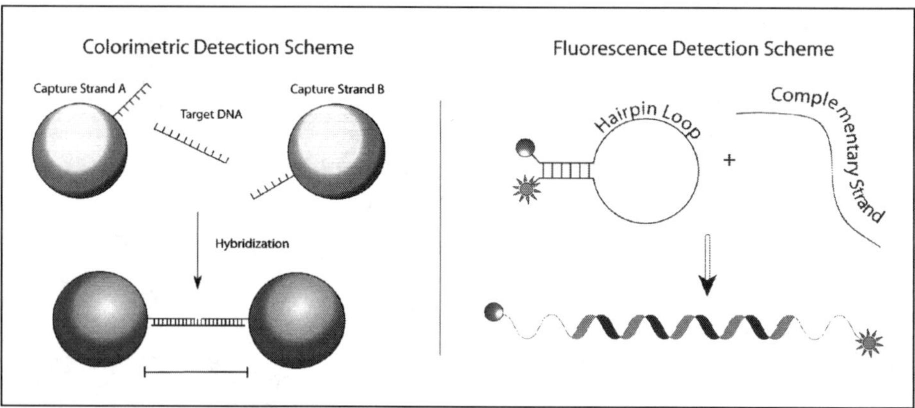

Figure 8. Colorimetric detection scheme as developed by Mirkin et al. The hybridization of complementary surface-bound DNA strands brings gold nanoparticles together, which affects SPR behavior and will change the visible color of the solution.

into the billions ($M^{-1}cm^{-1}$) if the particles are large enough. It was realized that this massive absorption feature could extend metallic nanoparticles toward therapeutic purposes. Gold NPs are (mostly) non-photoluminescent structures where the absorbed energy by the NP are mostly dissipated to the surrounding environment as heat. If the NP exists in an environment such as a tumor cell, then the heat dissipated by the NP through light absorption will locally destroy its environment, leaving the healthy neighboring cells alone. To accomplish this goal, of course, the gold NPs must first target specific cancerous cells and not internalize into the healthy ones. The use of specific targeting antibodies, such as Herceptin's affinity for breast cancer, may be employed toward this end. Studies to understand the cell's uptake dependence on shape and size are being performed,[27] as well as the efficiency of photothermal therapy to destroy specific cells while leaving healthy cells in tact are being performed in-vitro.[28-30] Rod-shaped NPs may prove to be particularly useful for photothermal therapy because the longitudinal plasmon mode in the near IR wavelength range will be more accessible for external illumination through skin tissue or blood than is the transverse plasmon mode.

We have provided some descriptions of the applications of metallic NPs. Since the application of this technology is so broad (and difficult to describe all applications), we recommend the interested reader to some excellent recent reviews on this topic.[31-33]

Conclusions

Although not considered as sexy as their brightly-emitting semiconductor quantum dot neighbors, gold NPs are perhaps finding the widest variety of applications in biology and medicine. This is mostly due to their ease of synthesis and control over shape, size, surface chemistry, and particularly their stability in almost any biologically relevant environment. While many researchers struggle to gain control and understanding of the semiconductor nanomaterials, gold NPs have been reliably present, offering their unique optical properties for assistance, which has led to their unique involvement in unconventional treatments and assays. As stated in the introduction, the usefulness of any physical material lies within the ability to both understand and control it, which makes the gold NP "beast" possibly the most tamed of all nanomaterials.

References

1. Slot JW, Geuze HJ. A new method of preparing gold probes for multiple-labeling cyto-chemistry. Eur J Cell Biol 1985; 38(1):87-93.
2. Weare WW et al. Improved synthesis of small (d(CORE) approximate to 1.5 nm) phosphine-stabilized gold nanoparticles. J Am Chem Soc 2000; 122(51):12890-12891.

3. Hostetler MJ et al. Alkanethiolate gold cluster molecules with core diameters from 1.5 to 5.2 nm: Core and monolayer properties as a function of core size. Langmuir 1998; 14(1):17-30.

4. Whitney TM et al. Fabrication and magnetic-properties of arrays of metallic nanowires. Science 1993; 261(5126):1316-1319.

5. Gole A, Murphy CJ. Seed-mediated synthesis of gold nanorods: Role of the size and nature of the seed. Chem Mater 2004; 16(19):3633-3640.

6. Moskovits M. Surface-enhanced spectroscopy. Reviews of Modern Physics 1985; 57(3):783-826.

7. Lakowicz JR. Radiative decay engineering 5: Metal-enhanced fluorescence and plasmon emission. Anal Biochem 2005; 337(2):171-194.

8. Link S, El-Sayed MA. Optical properties and ultrafast dynamics of metallic nanocrystals. Annu Rev Phys Chem 2003; 54:331-366.

9. Alvarez MM et al. Optical absorption spectra of nanocrystal gold molecules. J Phys Chem B 1997; 101(19):3706-3712.

10. Link S, El-Sayed MA. Spectral properties and relaxation dynamics of surface plasmon electronic oscillations in gold and silver nanodots and nanorods. J Phys Chem B 1999; 103(40):8410-8426.

11. Logunov SL et al. Electron dynamics of passivated gold nanocrystals probed by subpicosecond transient absorption spectroscopy. J Phys Chem B 1997; 101(19):3713-3719.

12. Mie G. Beitrage zur optik truber medien, speziell kolloidaler metallosungen. Annalen Der Physik 1908; 25:376-445.

13. Johnson PB, Christy RW. Optical constants of the noble metals. Phys Rev B 1972; 6(12):4370.

14. Kreibig U, Genzel L. Optical-absorption of small metallic particles. Surface Science 1985; 156(Jun):678-700.

15. Hövel H et al. Width of cluster plasmon resonances: Bulk dielectric functions and chemical interface damping. Phys Rev B 1993; 48(24):18178.

16. Drexhage KH. Analysis of light fields with monomolecular dye layers. J Opt Soc Am 1970; 60(11):1541.

17. Drexhage KH et al. Beeinflussung der fluoreszenz eines europiumchelates durch einen spiegel. Berichte Der Bunsen-Gesellschaft Fur Physikalische Chemie 1966; 70(9-10):1179.

18. Drexhage KH, Kuhn H, Schafer FP. Variation of fluorescence decay time of a molecule in front of a mirror. Berichte Der Bunsen-Gesellschaft Fur Physikalische Chemie 1968; 72(2):329.

19. Chance RR, Prock A, Silbey RJ. Molecular fluorescence and energy transfer near interfaces. Advances in Chemical Physics 1978; 37:1.

20. Persson BNJ, Lang ND. Electron-hole-pair quenching of excited-states near a metal. Phys Rev B 1982; 26(10):5409-5415.

21. Jennings TL, Singh MP, Strouse GF. Fluorescent lifetime quenching near d=1.5 nm gold nanoparticles: Probing NSET validity. J Am Chem Soc 2006; 128(16):5462-5467.

22. Yun CS et al. Nanometal surface energy transfer in optical rulers, breaking the FRET barrier. J Am Chem Soc 2005; 127(9):3115-3119.

23. Jennings TL et al. NSET molecular beacon analysis of hammerhead RNA substrate binding and catalysis. Nano Letters 2006; 6(7):1318-1324.

24. Taton TA, Mirkin CA, Letsinger RL. Scanometric DNA array detection with nanoparticle probes. Science 2000; 289(5485):1757-1760.

25. Cao YWC, Jin RC, Mirkin CA. Nanoparticles with Raman spectroscopic fingerprints for DNA and RNA detection. Science 2002; 297(5586):1536-1540.

26. Dubertret B, Calame M, Libchaber AJ. Single-mismatch detection using gold-quenched fluorescent oligonucleotides. Nature Biotechnol 2001; 19(4):365-370.

27. Chithrani BD, Ghazani AA, Chan WCW. Determining the size and shape dependence of gold nanoparticle uptake into mammalian cells. Nano Letters 2006; 6(4):662-668.

28. Hirsch LR et al. Nanoshell-mediated near-infrared thermal therapy of tumors under magnetic resonance guidance. Proc Natl Acad Sci USA 2003; 100(23):13549-13554.

29. Huang XH et al. Cancer cell imaging and photothermal therapy in the near-infrared region by using gold nanorods. J Am Chem Soc 2006; 128(6):2115-2120.

30. Loo C et al. Immunotargeted nanoshells for integrated cancer imaging and therapy. Nano Letters 2005; 5(4):709-711.

31. Parak WJ et al. On the development of colloidal nanoparticles towards multifunctional structures and their possible use for biological applications. Small 2005; 1(1):48-63.

32. Daniel MC, Astruc D. Gold nanoparticles: assembly, supramolecular chemistry, quantum-size-related properties, and applications toward biology, catalysis and nanotechnology. Chem Rev 2003; 104(1):293-346.

33. Ferrari M. Cancer nanotechnology: opportunities and challenges. Nat Rev Cancer 2005; 5(3):161-171.

CHAPTER 4

Multi-Functional Gold Nanoparticles for Drug Delivery

Gang Han, Partha Ghosh and Vincent M. Rotello*

Abstract

Multi-functional gold nanoparticles have been demonstrated to be highly stable and versatile scaffolds for drug delivery due to their unique size, coupled with their chemical and physical properties. The ability to tune the surface of the particle provides access to cell-specific targeting and controlled drug release. This chapter describes current developments in the area of drug delivery using gold nanoparticles as delivery vehicles for multiple therapeutic purposes.

Introduction

Nanomaterials are of great interest in biology and medicine, owing to their numerous applications including DNA/ protein detection,[1] biomolecular regulators,[2] cell imaging[3] and cancer cell diagnostics.[4] Currently, the use of nanomaterials as drug delivery systems has become an emerging area in the field of nanomedicine. A wide variety of nanomaterials, such as nanotubes,[5] nanorods,[6] and nanoparticles,[7] have been explored as carriers for delivering "small-molecule" drugs, proteins, and genetic materials, exploiting their unique dimensions and specific physical and chemical properties. These novel drug delivery systems offer the opportunity to improve poor solubility, limited stability, biodistributions, and pharmacokinetics of drugs as well as offering the potential ability to target specific tissues and cell types.

Multi-functional gold nanoparticles are attractive organic-inorganic hybrid materials composed of an inorganic metallic gold core surrounded by an organic and/or biomolecular monolayer. They provide many desirable attributes for the creation of drug delivery systems. First, the core materials of gold are chemically inert and nontoxic.[8] Second, essentially monodisperse nanoparticles can be fabricated with tunable core shape and size, providing a wide range of attractive properties, including controllable plasmon resonance for photo-thermal therapeutic treatments using a light at the visible or near infrared (NIR) regions.[9] Third, the unique nanoscale dimension of gold nanoparticle provides a large surface area for efficient conjugation and protection of drugs and targeting ligands.[10] The attachment of payload can be readily achieved by either noncovalent interaction (e.g., DNA, RNA or proteins via electrostatic interaction) or covalent chemical conjugation of "small molecule" drugs. Finally, the well-defined surface chemistry allows modulation of monolayer properties of nanoparticles in highly divergent fashion with a wide range of ligand functionality. This provides an effective drug delivery system to ensure cellular uptake, controlled payload release, and specific cells targeting.

*Corresponding Author: Vincent M. Rotello—Department of Chemistry, University of Massachusetts, 710 North Pleasant Street, Amherst, MA, 01003, U.S.A. Email: Rotello@Chem.umass.edu

Bio-Applications of Nanoparticles, edited by Warren C.W. Chan. ©2007 Landes Bioscience and Springer Science+Business Media.

Recently, a number of reviews concerning nanoparticle-based biological applications have been published, focusing on biosensing,[11] diagnostic applications,[12] and nanoparticle-biomolecule assemblies.[13] In this chapter, we will detail the advances made in multi-functional gold nanopaticles as drug delivery systems. Focus is given to the use of these systems for controllable and targeted release of the drug.

Gold Nanoparticles as Nucleic Acid Delivery Vehicles

Gene therapy is a highly promising approach for the treatment of genetic disorders.[14] One of the current limitations with this process is in the design of effective gene-delivery vectors for transporting plasmid DNA, small interfering RNA (siRNA), or antisense oligonucleotides controllably and specifically into living cells. Although viral vectors are very effective,[15] they have raised many safety concerns such as unpredictable cytotoxicity and immune responses.[16] In this regard, nonviral synthetic materials have been developed as gene delivery carriers.[17] To be effective, those materials must meet several requirements, such as the ability to condense DNA into compact complexes which can be readily taken up by cells, the efficient protection of DNA from degradation by nucleases, and the release of DNA in functional form.

Nanoparticles provide attractive scaffolds for the creation of DNA delivery vectors due to their tunable attributes, such as size, shape, and surface functionality. Rotello et al have demonstrated that gold nanoparticles covered with quaternary ammonium groups interact with plasmid DNA through electrostatic interactions,[18] resulting in effective protection of DNA from enzymatic digestion.[19] These highly stable DNA-nanoparticle complexes provide an effective vector for gene delivery. The effects of varying nanoparticle monolayer (e.g., positive charge coverage and various lengths of unfunctionalized alkane thiols) as well as DNA-to-nanoparticle ratios were systemically investigated to establish the optimal parameters on the efficiency of transfection of mammalian 293T cells (Fig. 1).[20] The results showed that an excess amount of nanoparticles was required to enable the DNA-nanoparticle complex overall positive charge for effective cellular uptake. It is also found that amphiphilic particles were superior to purely cationically-functionalized systems on the transfection efficacy, which was ~8-fold more efficient than polyethyleneimine (PEI), a widely used gene delivery agent, and 60-fold more efficient than reported silica nanoparticles.[21]

In a subsequent study, Klibanov et al have modified the surface of gold nanoparticles using branched 2 kDa polyethylenimine (PEI) to investigate their transfection efficiencies into monkey kidney (Cos-7) cells.[22] It was found that the transfection efficiency varied with the PEI:gold molar ratio in the conjugates, with the best one being a dozen times more potent than the unmodified polycation. Their study also suggested that increasing the hydrophobicity of the transfection agent could enhance cellular internalization.

Very recently, Nagasaki and coworkers described a novel cytoplasmic delivery system of siRNA using gold nanoparticles.[23] The thiolated siRNA was coated on the surface of gold nanoparticles with poly (ethylene glycol)-blockpoly(2-(N,N-dimethylamino)ethyl methacrylate) copolymer (PEG-PAMA). These PEGylated gold nanoparticles containing thiols on the siRNA showed a significant RNAi activity in HuH-7 cells.

Positively charged synthetic materials are generally used as nonviral transfection vectors to ensure cellular uptake. Mirkin et al have recently shown that gold nanoparticles (Au NP) attached with single stranded oligodeoxynucleotides that feature negative charges can be used for gene therapy (Fig. 2).[24] Fluorescence images obtained by incubating the cells with DNA (labeled with a fluorophore) conjugated nanoparticles revealed their cellular uptake. The DNA on particle surface binds with its complementary strand with higher affinity compared to the unmodified DNA. Moreover, the DNA strand conjugated with the particle is less susceptible to nuclease activity. These properties render the NP-DNA conjugates excellent intracellular gene regulation agents. By conjugation of antisense DNA with gold nanoparticle, the protein expression level was controlled. EGFP-expressing C166 cells were incubated with antisense DNA functionalized nanoparticles (ASNP). Due to EGFP gene knockdown, reduction in

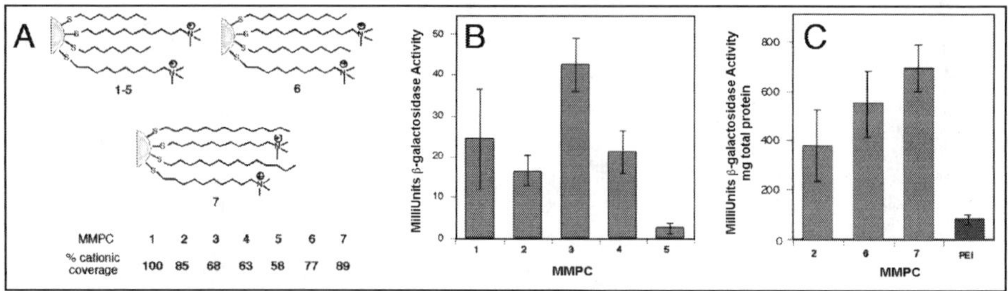

Figure 1. A) MMPCs used for transfection. B) Transfection of β-galactosidase mediated by formation of various MMPC-DNA complexes at 2200:1 nanoparticle/DNA ratio. C) Transfection with MMPCs 5, 6, 7 (2200:1 nanoparticle/DNA ratio) and PEI (60 kDa). All transfections were performed in the presence of chloroquine (100 μM) and serum (10%).

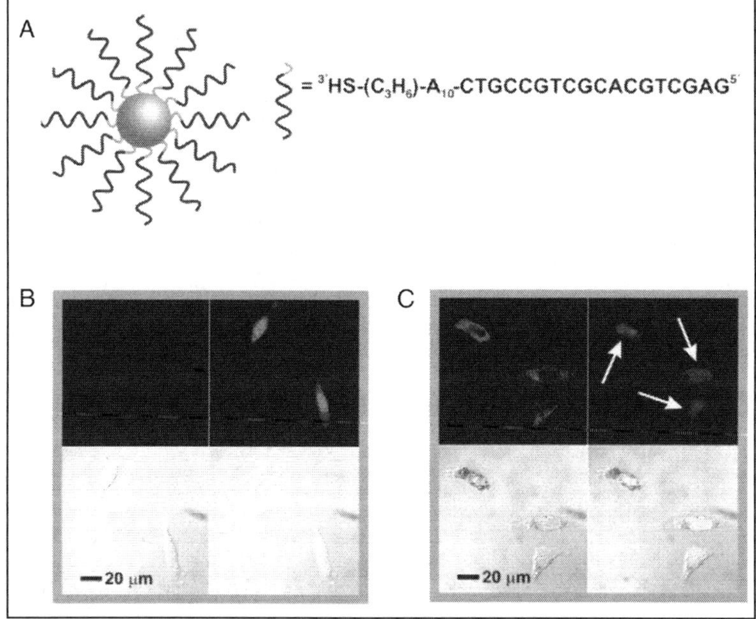

Figure 2. A) Antisense DNA modified gold nanoparticle (ASNP). B) Before and C) after incubation with antisense DNA functionalized nanoparticles for EGFP-expressing C166 cells. (Rosi NL, et al. Science 2006; 312:1027-1030.[24] Reprinted with permission from AAAS.)

fluorescence intensity was observed in cells incubated with ASNP compared to untreated cells from two photon excitation confocal laser microscopy experiments. This approach provides a highly efficient gene regulator in terms of high loading of the antisense DNA with little or no toxicity at the concentrations studied.

Protein and Peptide Delivery Using Gold Nanoparticles

Gold nanoparticle can likewise deliver proteins and peptides of interest. Insulin dependent diabetes mellitus (IDDM) is characterized by the marked inability of the pancreas to secrete insulin because of autoimmune destruction of the beta cells of the islets of langerhans. Current treatment for IDDM includes insulin administration via subcutaneous route. Research is underway for systems to deliver insulin transmucosally and avoid the traumatic subcutaneous and intramuscular routes. In recent works, Sastry et al have demonstrated that gold nanoparticles have the capability of binding with insulin and hence can be used as insulin carriers.[25] 3.5 nm core bare nanoparticle and aspartic acid capped nanoparticle were loaded quantitatively with insulin presumably via covalent attachment and hydrogen bonding, respectively. Subsequently, they have shown administration of insulin-loaded nanoparticle in diabetic Wistar rats via oral and intranasal routes resulted in a substantial diminution of blood glucose levels.

Controlled Drug Release by Gold Nanoparticle

Glutathione (GSH) mediated release represents a nonenzymatic approach to the release of therapeutic agents in a controlled fashion after delivery to the cells. This strategy is based on the huge difference in intracellular GSH concentration (1-10 mM[26,27]) compared to intercellular GSH levels (e.g., 2 μM in red blood plasma).[28] Recently, delivery vehicles based on disulfide linkages have been reported the advantage of this method to release DNA and drug molecules inside cells.[29,30] Although this approach can be effective, it is a challenging task to tune the reactivity of the disulfide linkage. Additionally, thiol-disulfide exchange reactions can occur with cysteines on the surface of proteins in bloodstream, thus altering the drug carrier's pharmacokinetic profiles. However, monolayer protected nanoparticle drug carriers can be expected to resist the exchange with proteins because of the steric shielding of the organic thiol monolayers on the surface. More importantly, the tunable chain lengths and headgroup functionalities on the monolayers can be used to manipulate the release of payload in response to GSH.

In recent studies, an intracellular concentration of glutathione (GSH) has been employed as a trigger to restore the transcription of cationic nanoparticle-bound DNA.[31] Trimethylammonium-functionalized mixed monolayer protected clusters (MMPCs) 1 and 2 (Fig. 3A), with different sidechain lengths bind with DNA molecules strongly through electrostatic complementarity. Therefore, the transcription of DNA by T7 RNA polymerase is completely inhibited at an appropriate stoichiometry. The DNA-particle complexes were stable at extracellular GSH concentrations but labile at intracellular GSH levels, showing dose-dependent recovery of DNA transcription (Fig. 3B). These phenomena are attributed to the place-exchange of the cationic residues of the nanoparticles with anionic GSH which diminishes the overall positive charge of the particles and subsequently attenuate its electrostatic affinity towards the DNA backbone. Because MMPC 2 with a longer tether was more stable against ligand-exchange than MMPC 1, MMPC2 showed significant recovery of DNA transcription only at higher GSH concentrations. Therefore, the release of DNA and recoveries of DNA transcription can be tuned through the choice of monolayer coverage and intracellular GSH levels.

Recently, Rotello et al have demonstrated GSH-mediated release of a fluorophore model of "small molecule" drug from a monolayer protected gold nanoparticle. The gold nanoparticles (AuNP) used in this work feature a 2-nm core and a mixed monolayer composed of a tetra(ethylene glycol)lyated cationic ligand TTMA and a thiolated Bodipy dye, HSBDP.[32] The TTMA ligand is used to generate a cationic surface to enhance cellular uptake. The dye molecule doped into the particle monolayer provides an analog for hydrophobic drugs and allows facile detection of payload release. Moreover, gold nanoparticles provide excellent fluorescent quenchers, allowing the timely observation of release of HSBDP from nanoparticles.

GSH-mediated release was first observed in Human liver cells (Hep G2). After the cells incubated with AuNP, strong fluorescence from Bodipy was observed (Fig. 4A). The results clearly show that multifunctional cationic nanoparticles efficiently penetrated cell membranes and that the payload dye molecules were successfully released in living cells. The release of payload from AuNP surfaces arose from GSH, which was confirmed by using glutathione monoester (GSH-OEt) as an external stimulus to trigger HSBDP discharge (Fig. 4B). GSH-OEt can offer a method to transiently manipulate intracellular GSH concentrations. In

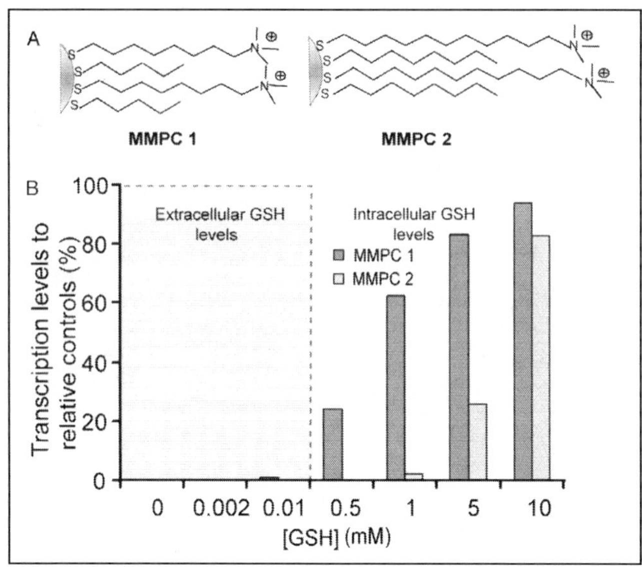

Figure 3. A) Structure of the MMPC1 and MMPC2. B) Normalized MMPCs bound DNA transcriptions levels in presence of different concentrations of glutathione.

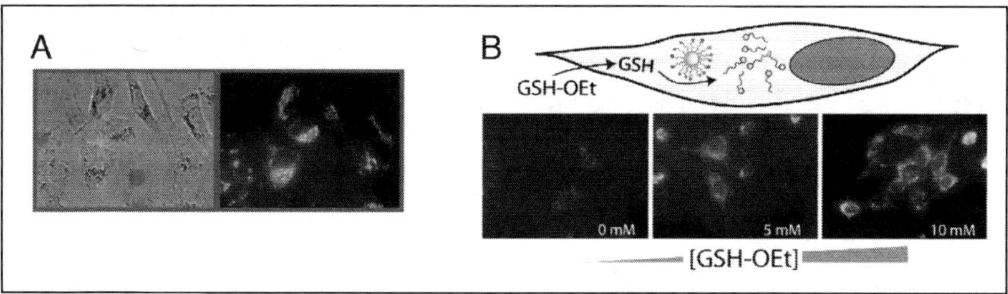

Figure 4. A) Representative bright field and fluorescence images of Human Hep G2 cells incubated with AuNP for 96 hours. B) Schematic representation and fluorescence images using GSH-OEt as an external stimulus to release HSBDP from AuNP.

this experiment, mouse embryonic fibroblast cells containing ~50% lower GSH levels than Hep G2, were first treated with varied concentrations of GSH-OEt, and followed by incubating with AuNP. Increased fluorescence intensity was clearly observed with increasing GSH-OEt concentration (Fig. 4B). This dose-dependent increase in fluorescence effectively demonstrates that GSH is responsible for releasing dye molecules from the AuNP carrier.

Several groups have taken advantage of surface chemistry on the MMPCs to realize the small molecular releasing. For example, Schoenfisch and coworkers showed the efficient release of NO was established by means of diazeniumdiolate-modified MMPCs.[33] However, photochemical release of materials represents a useful orthogonal dissociation mechanism. This approach provides external control over payload release in a unique site- and time-specific fashion.[34] To provide transport and photorelease of DNA, Rotello developed a photolabile nanoparticle that can be converted from cationic to anionic upon irradiation of light, thus releasing DNA (Fig. 5).[35] This monolayer features a photosensitive *o*-nitrobenzyl ester linker

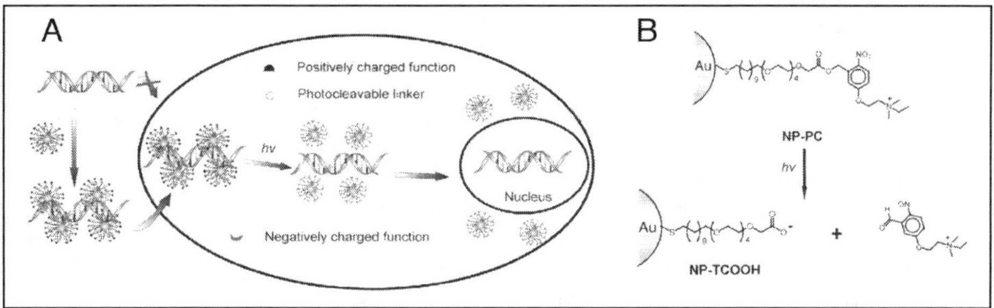

Figure 5. A) Schematic illustration for the release of DNA from NP-PC-DNA complex upon UV irradiation within the cell. B) Schematic presentation of light-induced surface transformation of NP-PC.

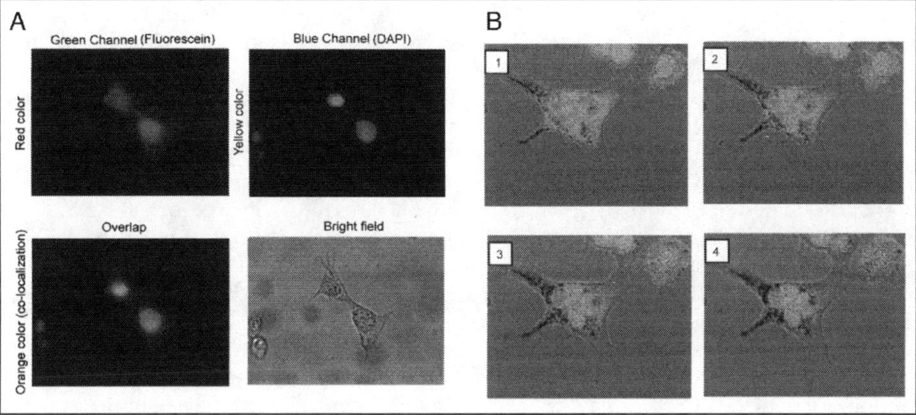

Figure 6. A) Fluorescence and bright field microscopy images after photo-triggered DNA release from NP-PC-DNA complex. To clarify the overlap of F-DNA and nuclei stain DAPI; green (Fluorescein) and blue (DAPI) channel are depicted with red and yellow color, respectively. B) Confocal microscopy image illustrating that the photo-released DNA accumulates inside the nucleus. Panel 1, 2, 3 and 4 show four consecutive slices of middle sections of z-series confocal images (Interval = 1.0 μm).

with a quaternary amine headgroups for DNA binding. Near-UV irradiation cleaves the nitrobenzyl linkage, releasing the positively charged alkyl amine leaving behind a negatively charged carboxylate group. Effective release was established in vitro using a T7 RNA polymerase assay. DNA transcription was restored upon light irradiation. Cell culture studies demonstrated effective uptake and release of FITC-labeled DNA after incubation for six hours followed by UV irradiation. More importantly, a high degree of nuclear localization was observed for the released DNA (Fig. 6), one of basic criteria for DNA delivery.

Physical properties of nanoparticles can also be used as a tool for drug delivery in living cells. Skirtach et al have demonstrated the release of the encapsulated drug inside a nanoengineered polyelectrolyte-multilayer capsules in a controlled fashion at a remote place.[36] Fluorescently labeled polymer was taken as a model for drug molecule and doping of the microcapsules walls with metallic nanoparticles helps to discharge the encapsulated material. Nanoparticles absorb the energy from a laser beam of biologically friendly near-infrared (NIR) region. The absorbed energy causes local heating, and results in disruption of the shells of multilayer-capsules and hence release of the encapsulated material.

Targeted Drug Delivery

Site-specific drug delivery can be achieved through a transmembrane receptor mediated endocytosis pathway. This can be accomplished by conjugating a drug delivery vehicle with a ligand that specifically recognizes the receptor. Transferrin (TF), a protein, can be employed as a ligand since many tumor cells overexpress transferrin receptors on their surface. He et al have reported the enhanced uptake of tansferrin-conjugated gold nanoparticles by tumor cells.[37] AFM images were taken to visualize the endocytosis process of TF-gold nanoparticles in human nasopharyngeal carcinoma cells. A "bumpy surface" was observed on the cells after incubation with Au-TF nanoparticles for 5 h. The cellular uptake of Au-TF conjugates was confirmed by transferrin competition experiments using confocal scanning laser microscopy. Images were taken after incubating the cells with nanoparticles alone, fluorophore-labeled Au-TF and Au-albumin nanoparticles (Fig. 7). Only cells treated with Au-TF nanoparticles showed considerable fluorescence.

Conjugation of folic acid on the surface of drug delivery systems (DDS) provides another strategy for targeted drug delivery since folic acid receptors are upregulated in various tumor cells.[38] Recently, Andres and coworkers have successfully demonstrated the specific uptake of folate-conjugated gold nanoparticles by folate receptor-positive tumor cells.[39] A polyethylene glycol (PEG) chain was anchored by thiooctic acid and folic acid on opposite ends and the conjugate was grafted onto 10 nm diameter gold nanoparticle to provide a biocompatible DDS. These nanoparticles exhibited excellent stability over a wide range of pH (2-12) and electrolyte concentration (0-0.5 M NaCl). The possibility of cellular uptake was investigated by TEM imaging of KB cells, which actively express folate receptors on their membrane, and WI-38 cells, as negative control, after incubation with various AuNP constructs for 2 h. Significant uptake of folic acid conjugated AuNP was observed into KB cells, whereas negligible cellular uptake was detected in three controls: (1) incubation of KB cells with PEG-linked AuNP lacking folate, (2) exposure of KB cells with folate conjugated AuNP in presence of excess free folic acid, and (3) exposure to cells that do not overexpress folate receptor.

To date, a few research groups have studied the interaction of gold nanoparticle in vivo. Paciotti et al investigated the effect of injecting TNF bound to 26 nm gold nanoparticles into tumor-bearing mice.[40] They found that the particles preferentially accumulated in the tumors

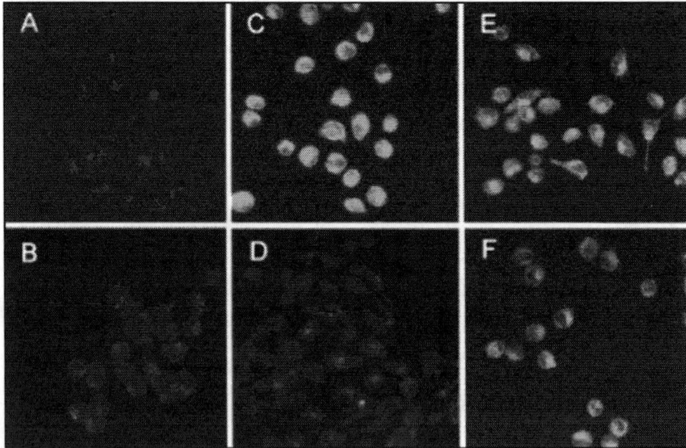

Figure 7. Confocal cell images of cells treated with various gold nanoparticles. A) Cells without any NP; B) cells with unmodified Au NPs; C) cells treated with transferrin conjugated Au-NP; D) cells incubated with Au-albumin NPs; E,F) cells incubated with different ratios of Au-TF versus holo-TF (1:2 and 1:5, respectively). Reprinted with permission from American Chemical Society, © 2005.[37]

and that the particle-bound TNF was somewhat more effective at diminishing tumor mass than free TNF. The preferential accumulation of particles in tumors is a promising result indicating potential for the therapeutic-bound particles. More recent studies by this group[41] have demonstrated enhanced efficacy with particles featuring TNF and a thiolated paclitaxel prodrug.

Conclusions

In this chapter, we have reviewed recent advances on the applications of drug delivery by means of gold nanoparticles. Gold nanoparticles have unique chemical and physical properties (e.g., tunable core size, mono-dispersity, low toxicity, large surface to volume ratio, facile fabrication and multi-functionalization) that facilitate their use in drug delivery applications. At the same time, the release of payloads on gold nanoparticles can be controlled by intracellular glutathione or external light. Despite the many advances that have been made, there are still many significant questions awaiting to be answered through systemic study of the delivery process such as cellular uptake, payload release rate and in vivo immune response and toxicity.

Acknowledgement

This research was supported by the NIH (R21 EB004503-01) and the NSF Center for Hierarchical Manufacturing at the University of Massachusetts (NSEC, DMI-0531171).

References

1. Liu JW, Lu Y. Fast colorimetric sensing of adenosine and cocaine based on a general sensor design involving aptamers and nanoparticles. Angew Chem Int Edit 2006; 45:90-94.
2. You CC, Arvizo RR, Rotello VM. Regulation of alpha-chymotrypsin activity on the surface of substrate-functionalized gold nanoparticles. Chem Commun 2006; 2905-2907.
3. Groneberg DA, Giersig M, Welte T et al. Nanoparticle-based diagnosis and therapy. Current Drug Targets 2006; 7:643-648.
4. Huang XH, El-Sayed IH, Qian W et al. Cancer cell imaging and photothermal therapy in the near-infrared region by using gold nanorods. J Am Chem Soc 2006; 128:2115-2120.
5. Wu W, Wieckowski S, Pastorin G et al. Targeted delivery of amphotericin B to cells by using functionalized carbon nanotubes. Angew Chem Int Edit 2005; 44:6358-6362.
6. Salem AK, Searson PC, Leong KW. Multifunctional nanorods for gene delivery. Nat Mater 2003; 2:668-671.
7. Luo D, Saltzman WM. Enhancement of transfection by physical concentration of DNA at the cell surface. Nat Biotechnol 2000; 18:893-895.
8. Connor EE, Mwamuka J, Gole A et al. Gold nanoparticles are taken up by human cells but do not cause acute cytotoxicity. Small 2005; 1:325-327.
9. Pissuwan D, Valenzuela SM, Cortie MB. Therapeutic possibilities of plasmonically heated gold nanoparticles. Trends In Biotechnology 2006; 24:62-67.
10. Verma A, Rotello VM. Surface recognition of biomacromolecules using nanoparticle receptors. Chem Commun 2005; 303-312.
11. Chang MMC, Cuda G, Bunimovich YL et al. Nanotechnologies for biomolecular detection and medical diagnostics. Curr Opin Chem Biol 2006; 10:11-19.
12. Rosi NL, Mirkin CA. Nanostructures in biodiagnostics. Chem Rev 2005; 105:1547-1562.
13. You CC, Verma A, Rotello VM. Engineering the nanoparticle-biomacromolecule interface. Soft Matter 2006; 2:190-204.
14. Miller AD. Human gene-therapy comes of age. Nature 1992; 357:455-460.
15. Yeh P, Perricaudet M. Advances in adenoviral vectors: From genetic engineering to their biology. Faseb J 1997; 11:615-623.
16. Check E. Gene therapy: A tragic setback. Nature 2002; 420:116-118.
17. Thomas M, Klibanov AM. Nonviral gene therapy: Polycation-mediated DNA delivery. Appl Microbiol Biotechnol 2003; 62:27-34.
18. McIntosh CM, Esposito EA, Boal AK et al. Inhibition of DNA transcription using cationic mixed monolayer protected gold clusters. J Am Chem Soc 2001; 123:7626-7629.
19. Han G, Martin CT, Rotello VM. Stability of gold nanoparticle-bound DNA toward biological, physical, and chemical agents. Chem Biol Drug Des 2006; 67:78-82.
20. Sandhu KK, McIntosh CM, Simard JM et al. Gold nanoparticle-mediated transfection of mammalian cells. Bioconjugate Chem 2002; 13:3-6.

21. Kneuer C, Sameti M, Bakowsky U et al. A nonviral DNA delivery system based on surface modified silica-nanoparticles can efficiently transfect cells in vitro. Bioconjugate Chem 2000; 11:926-932.
22. Thomas M, Klibanov AM. Conjugation to gold nanoparticles enhances polyethylenimine's transfer of plasmid DNA into mammalian cells. Proc Natl Acad Sci USA 2003; 100:9138-9143.
23. Oishi M, Nakaogami J, Ishii T et al. Smart PEGylated gold nanoparticles for the cytoplasmic delivery of siRNA to induce enhanced gene silencing. Chemistry Letters 2006; 35:1046-1047.
24. Rosi NL, Giljohann DA, Thaxton CS et al. Oligonucleotide-modified gold nanoparticles for intracellular gene regulation. Science 2006; 312:1027-1030.
25. Joshi HM, Bhumkar DR, Joshi K et al. Gold nanopartncles as carriers for efficient transmucosal insulin delivery. Langmuir 2006; 22:300-305.
26. Hassan SSM, Rechnitz GA. Determination of glutathione and glutathione-reductase with a silver sulfide membrane-electrode. Anal Chem 1982; 54:1972-1976.
27. Anderson ME. Glutathione: An overview of biosynthesis and modulation. Chem-Biol Interact 1998; 112:1-14.
28. Jones DP, Carlson JL, Samiec PS et al. Glutathione measurement in human plasma Evaluation of sample collection, storage and derivatization conditions for analysis of dansyl derivatives by HPLC. Clin Chim Acta 1998; 275:175-184.
29. Mahajan SS, Paranji R, Mehta R et al. A glutathione-based hydrogel and its site-selective interactions with water. Bioconjugate Chem 2005; 16:1019-1026.
30. Saito G, Swanson JA, Lee KD. Drug delivery strategy utilizing conjugation via reversible disulfide linkages: Role and site of cellular reducing activities. Adv Drug Deliv Rev 2003; 55:199-215.
31. Han G, Chari NS, Verma A et al. Controlled recovery of the transcription of nanoparticle-bound DNA by intracellular concentrations of glutathione. Bioconjugate Chem 2005; 16:1356-1359.
32. Hong R, Han G, Fernandez JM et al. Glutathione-mediated delivery and release using monolayer protected nanoparticle carriers. J Am Chem Soc 2006; 128:1078-1079.
33. Rothrock AR, Donkers RL, Schoenfisch MH. Synthesis of nitric oxide-releasing gold nanoparticles. J Am Chem Soc 2005; 127:9362-9363.
34. Berg K, Selbo PK, Prasmickaite L et al. A Photochemical drug and gene delivery. Curr Opin Mol Ther 2004; 6:279-287.
35. Han G, You CC, Kim BJ et al. Light-regulated release of DNA and its delivery to nuclei by means of photolabile gold nanoparticles. Angew Chem Int Edit 2006; 45:3165-3169.
36. Skirtach AG, Javier AM, Kreft O et al. Laser-induced release of encapsulated materials inside living cells. Angew Chem Int Edit 2006; 45:4612-4617.
37. Yang PH, Sun XS, Chiu JF et al. Transferrin-mediated gold nanoparticle cellular uptake. Bioconjugate Chem 2005; 16:494-496.
38. Lee RJ, Low PS. Folate-mediated tumor-cell targeting of liposome-entrapped doxorubicin in vitro. Biochim Biophys Acta-Biomembr 1995; 1233:134-144.
39. Dixit V, Van den Bossche J, Sherman DM et al. Synthesis and grafting of thioctic acid-PEG-folate conjugates onto Au nanoparticles for selective targeting of folate receptor-positive tumor cells. Bioconjugate Chem 2006; 17:603-609.
40. Paciotti GF, Myer L, Weinreich D et al. Colloidal gold: A novel nanoparticle vector for tumor directed drug delivery. Drug Deliv 2004; 11:169-183.
41. Paciotti GF, Kingston DGI, Tamarkin L. Colloidal gold nanoparticles: A novel nanoparticle platform for developing multifunctional tumor-targeted drug delivery vectors. Drug Dev Res 2006; 67:47-54.

CHAPTER 5

Quantum Dots for Cancer Molecular Imaging

Xiaohu Gao* and Shivang R. Dave

Abstract

Quantum dots (QDs), tiny light-emitting particles on the nanometer scale, are emerging as a new class of fluorescent probes for biomolecular and cellular imaging. In comparison with organic dyes and fluorescent proteins, quantum dots have unique optical and electronic properties such as size-tunable light emission, improved signal brightness, resistance against photobleaching, and simultaneous excitation of multiple fluorescence colors.[1] These properties are most promising for improving the sensitivity of molecular imaging and quantitative cellular analysis by 1-2 orders of magnitude. Recent advances have led to multifunctional nanoparticle probes that are highly bright and stable under complex in-vivo conditions. A new structural design involves encapsulating luminescent QDs with amphiphilic block copolymers, and linking the polymer coating to tumor-targeting ligands and drug-delivery functionalities. Polymer-encapsulated QDs are essentially nontoxic to cells and small animals, but their long-term in-vivo toxicity and degradation need more careful studies. Nonetheless, bioconjugated QDs have raised new possibilities for ultrasensitive and multiplexed imaging of molecular targets in living cells and animal models.

Introduction

Since 1999, cancer has been the leading cause of death for Americans under the age of 85, and the eradication of this disease has been the long-sought-after goal of scientists and physicians.[2] Clinical outcome of cancer diagnosis is strongly related to the stage at which the malignancy is detected, and therefore early screening has become desirable, especially for breast and cervical cancer in women and colorectal and prostate cancer in men. However most solid tumors are currently only detectable once they reach approximately 1 centimeter in diameter, at which point the mass constitutes millions of cells that may already have metastasized. The most commonly used cancer diagnostic techniques in clinical practice are medical imaging, tissue biopsy, and bioanalytic assay of bodily fluids, all of which are, at present, insufficiently sensitive and/or specific to detect most types of early stage cancers, let alone precancerous lesions.

Once cancer has been detected, the next challenge is to classify that specific tumor into one of various subtypes, each of which can have drastically different prognoses and preferred methods of treatment. Diagnosis of cancer subtypes is vitally important, yet many types of cancer do not yet have reliable tests to differentiate between highly invasive types and less fatal types, and

*Corresponding Author: Xiaohu Gao—Department of Bioengineering, University of Washington, William H Foege Building N530M, Campus Box 355061, Seattle, WA 98195-5061, U.S.A. Email: xgao@u.washington.edu

Bio-Applications of Nanoparticles, edited by Warren C.W. Chan. ©2007 Landes Bioscience and Springer Science+Business Media.

the final judgment is commonly left to the expert opinion of a pathologist who studies the tumor biopsy. With the advent of high-throughput data analysis of genomic and proteomic classifications of cancer tissues, it is becoming apparent that many subtypes are only distinguished by differences as small as the concentration of a specific protein on a cell's surface. Identification of a cancer by its molecular expression profile, rather than by one specific biomarker, might be necessary to thoroughly classify cancer subtypes and understand their pathophysiology. One cancer subtype may also be heterogeneous over patient populations, making personalized medicine highly desirable in order to treat a patient uniquely for his or her distinct cancer phenotype. Personalized medicine, however, cannot succeed without developing tools to sensitively detect cancer and reveal clinical biomarkers that can distinguish specific cancer types.

Nanotechnology has been heralded as a new field that has the potential to revolutionize medicine, as well as many other seemingly unrelated subjects, like electronics, textiles, and energy production. The heart of this field lies in the ability to shrink the size of tools and devices to the nanometer size range, and to assemble atoms and molecules into larger structures with useful properties, while maintaining their dimensions on the nanometer-length scale. The nanometer-scale is also the scale of biological function, i.e., the same size range as enzymes, DNA, and other biological macromolecules and cellular components. Many nanotechnologies are predicted to soon become translational tools for medicine, and move quickly from discovery-based devices to clinically useful therapies and medical tests. Among these, quantum dots (QDs) are unique in their far-reaching possibilities in many avenues of medicine. A QD is a fluorescent nanoparticle that has the potential to be used as a sensitive probe for screening cancer markers in fluids, as a specific label for classifying tissue biopsies, and as a high-resolution contrast agent for medical imaging, capable of detecting even the smallest tumors. These particles have the unique ability to be sensitively detected on a wide range of length scales, from macroscale visualization, down to atomic resolution using electron microscopy.[3] Most importantly for cancer detection, QDs hold massive multiplexing capabilities for the detection of many cancer markers simultaneously, which holds tremendous promise for unraveling the complex gene expression profiles of cancers and for accurate clinical diagnosis. This chapter will summarize how QDs have recently been used in encouraging experiments for future clinical diagnostic tools for the early detection and classification of cancer.

Quantum Dot Photophysics and Chemistry

QDs are nearly spherical, fluorescent nanocrystals composed of semiconductor materials which bridge the gap between individual atoms and bulk semiconductor solids.[4,5] Owing to this intermediate size, which is typically between 2-8 nm in diameter or hundreds to thousands of atoms, QDs possess unique properties unavailable in either individual atoms or bulk materials. In their biologically useful form, QDs are colloids with similar dimensions to large proteins, dispersed in an aqueous solvent and coated with organic molecules to stabilize their dispersion.

Photophysical Properties

Because QDs are composed of inorganic semiconductors, they contain electrical charge carriers, which are negatively charged electrons and positively charged holes (an electron and hole pair is called an exciton). Bulk semiconductors are characterized by a composition-dependent bandgap energy, which is the minimum energy required to excite an electron to an energy level above its ground state. Excitation can be initiated by the absorption of a photon of energy greater than the bandgap energy, resulting in the generation of charge carriers. The newly created exciton can relax back to its ground state through recombination of the constituent electron and hole, which may be accompanied by the conversion of the bandgap energy into an emitted photon, which is the mechanism of fluorescence. Because of the small size of QDs, these generated charge carriers are confined to a space that is smaller than their natural size in bulk semiconductors. This "quantum confinement" of the exciton is the principle that causes the optoelectronic properties of the QD to be dictated by the size of the QD.[6-8] Decreasing the size of a QD results in a higher degree of confinement, which produces an exciton of higher energy, thereby increasing the bandgap energy. The most important consequence of this property is that the bandgap and emission wavelength of a QD may be tuned by adjusting its size, with smaller particles emitting at shorter wavelengths (Fig. 1). By adjusting their size and composition, QDs can now be prepared to emit fluorescent light from the ultraviolet, throughout the visible, and into the infrared spectra (400-4000 nm).[9-14]

Importantly for use as biological probes, QDs can absorb and emit light very efficiently, allowing highly sensitive detection, relative to conventionally used organic dyes and fluorescent proteins. QDs have very large molar extinction coefficients, on the order of $0.5\text{-}5 \times 10^6$ $M^{-1}cm^{-1}$,[15] roughly 10-50 times larger than those of organic dyes ($5\text{-}10 \times 10^4\ M^{-1}cm^{-1}$). Combined with the fact that QDs can have quantum efficiencies similar to that of organic dyes (up to 85%),[12] individual QDs have been found to be 10- 20 times brighter than organic dyes,[16,17] allowing highly sensitive detection of analytes in low concentration, which is particularly important for low copy number cancer markers. In addition, QDs are several thousand times more stable against photobleaching than organic dyes (Fig. 2A), and are thus well-suited for monitoring biological systems for long periods of time, which is important for developing robust sensors for cancer assays and for in vivo imaging.

A further advantage of QDs is that multicolor QD probes can be used to image and track multiple molecular targets simultaneously. Because cancer and many other diseases involve a large number of genes and proteins, this is certain to be one of the most powerful properties of QDs for medical applications. Multiplexing of QD signals is feasible because of the combination of broad absorption bands with narrow emission bands. Broad absorption bands allow

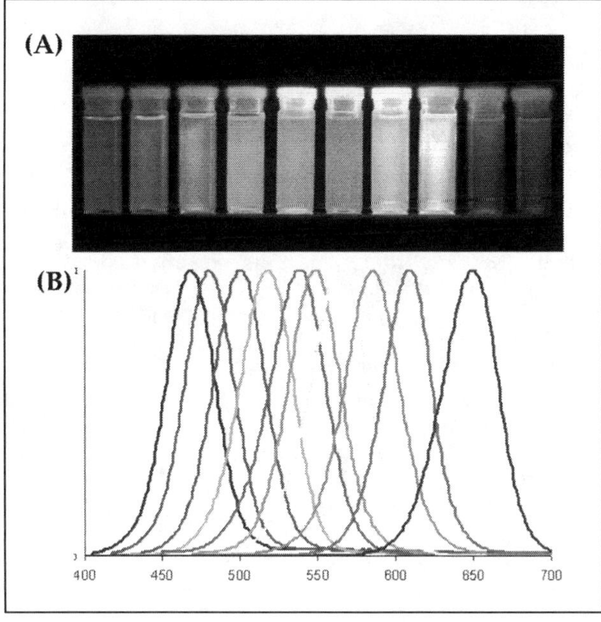

Figure 1. Size tunable emission of CdSe/ZnS quantum dots. A) Fluorescence image of a series QDs excited with a UV lamp. From left to right, the particle sizes vary from 2 to 6 nm in diameter. Reprinted with permission from Macmillian Publishers Ltd: Nature Biotechnology ©2001.[39] B) The corresponding fluorescence emission spectra.

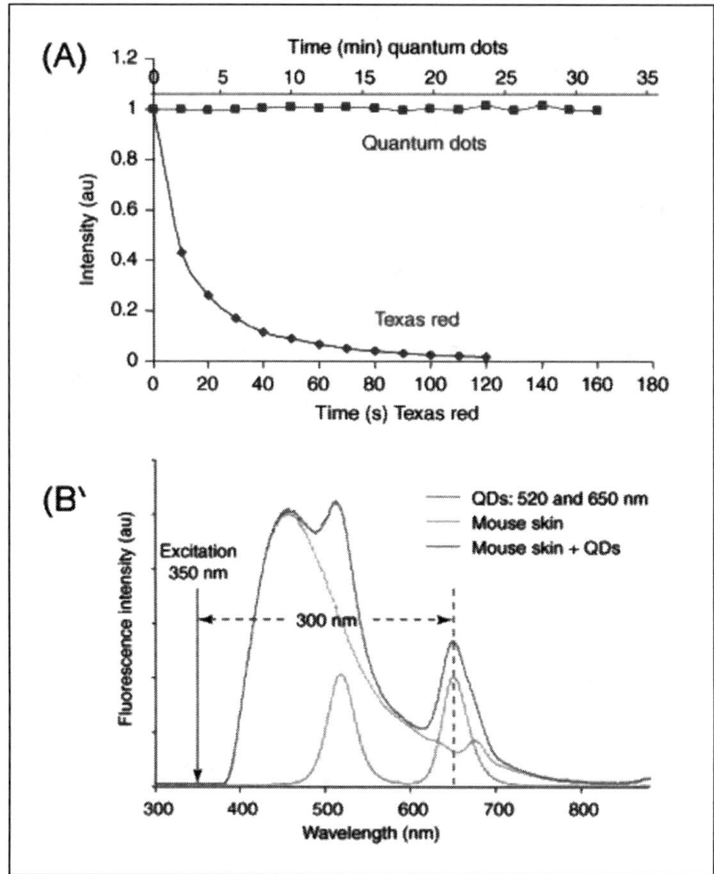

Figure 2. QDs' unique optical properties. A) Photostability comparison of QDs vs. organic dyes. Photobleaching curves showing that QDs are several thousand times more photostable than organic dyes (e.g., Texas red) under the same excitation conditions;[1] B) Stokes shift comparison. Comparison of mouse skin and QD emission spectra obtained under the same excitation conditions demonstrating that the QD signals can be shifted to a spectral region where autofluorescence is reduced.[1] Reprinted from Gao XH et al, Curr Opin Biotechnol 16:63-72; ©2005 with permission from Elsevier.[1]

multiple QDs to be excited with a single light source of short wavelength, simplifying instrumental design, increasing detection speed, and lowering cost. QD emission bands can be as narrow as 20 nm in the visible range, allowing for distinct signals to be detected simultaneously with very little crosstalk. In comparison, organic dyes and fluorescent proteins have narrow absorption bands and relatively wide emission bands, considerably increasing the difficulty of detecting well separated signals from distinct fluorophores.

Broad absorption bands are also useful for imaging of tissue sections and whole organisms in order to distinguish the QD signal from autofluorescent background signal (Fig. 2B). Biological tissue and fluids contain a variety of intrinsic fluorophores, particularly proteins and cofactors, yielding a background signal that decreases probe detection sensitivity. Intrinsic biological fluorescence is most intense in the blue to green spectral region, which is responsible for the faint greenish color of many cell and tissue micrographs. However QDs can be tuned to emit in spectral regions in which autofluorescence is minimized, such as longer wavelengths in the red or infrared spectra. Because of their broad absorption bands, QDs can still be efficiently excited by light hundreds of nanometers shorter than the emission wavelength, compared to organic dyes that require excitation close to the emission peak, burying the signal in autofluorescence. This can allow the sensitive detection of QDs over background autofluorescence in tissue biopsies and live organisms. Sensitivity can also be increased by using time-gated light detection, due to the fact that the excited state lifetimes of QDs (20-50 ns) are typically one order of magnitude longer than that of organic dyes. By delaying signal acquisition until background autofluorescence is decreased, QD fluorescence detection can be significantly increased.[18]

Synthesis and Bioconjugation

Research in probe development has focused on the synthesis, solubilization and bioconjugation of highly luminescent and stable QDs. Generally made from group II and VI elements (e.g., CdSe and CdTe) or group III and V elements (e.g., InP and InAs), recent advances have allowed the precise control of particle size, shape (dots, rods or tetrapods), and internal structure (core-shell, gradient alloy or homogeneous alloy).[5,19,20] In addition, QDs have been synthesized using both two-element systems (binary dots) and three-element systems (ternary alloy dots).

QDs can be prepared in a variety of media, from atomic deposition on solid-phases to colloidal synthesis in aqueous solution. However because the size-dependent properties of QDs are most pronounced when QDs are monodisperse in size, great strides have been made in the synthesis of highly homogeneous, highly crystalline QDs. The highest quality QDs are typically prepared at elevated temperatures in organic solvents, such as tri-n-octylphosphine oxide (TOPO) and hexadecylamine, both of which are high boiling bases containing long alkyl chains. These hydrophobic organic molecules not only serve as the reaction medium, but the basic moieties also coordinate with unsaturated metal atoms on the QD surface to prevent the formation of bulk semiconductors. As a result, the nanoparticles are capped with a monolayer of the organic ligands and are soluble only in hydrophobic solvents such as chloroform and hexane. The most commonly used and best understood QD system is a core of cadmium selenide (CdSe), coated with a shell of zinc sulfide (ZnS) to chemically and optically stabilize the core.

For biological applications, these hydrophobic dots must first be made water-soluble. Two general strategies have been developed to disperse QDs in aqueous biological buffers, as shown in (Fig. 3). In the first approach, the hydrophobic monolayer of ligands on the QD surface may be exchanged with hydrophilic ligands,[16] but this method tends to cause particle aggregation and decrease the fluorescent efficiency. As well, desorption of labile ligands from the QD surface increases potential toxicity due to exposure of toxic QD elements. Alternatively, the native hydrophobic ligands can be retained on the QD surface and rendered water-soluble through the adsorption of amphiphilic polymers that contain both a hydrophobic segment (mostly hydrocarbons) and a hydrophilic segment (such as polyethylene glycol [PEG] or multiple carboxylate groups). Several polymers have been reported, including octylamine-modified polyacrylic acid,[21] PEG-derivatized phospholipids,[22] block copolymers,[23] and amphiphilic polyanhydrides.[24] The hydrophobic domains strongly interact with alkyl chains of the ligands on the QD surface, whereas the hydrophilic groups face outwards and render the QDs water-soluble. Because the coordinating organic ligands (TOPO) are retained on the inner surface of QDs, the optical properties of QDs and the toxic elements of the core are shielded from the outside environment by a hydrocarbon bilayer. Indeed, after linking to PEG molecules, the polymer coated QDs are protected to such a degree that their optical properties does not change in a broad range of pH (1 to 14) and salt concentrations (0.01-1 M).[23] Parak and coworkers have also demonstrated that for polymer coated QDs the cytotoxicity is mainly due to the nanoparticle aggregation rather than the release of Cd ions.[24]

To achieve binding specificity or targeting abilities, polymer-coated QDs can be linked to bioaffinity ligands such as monoclonal antibodies, peptides, oligonucleotides or small-molecule inhibitors. As well, linking to polyethylene glycol (PEG) or similar ligands can allow improved biocompatibility and reduced nonspecific binding. Due to the large surface-area to volume ratio of QDs relative to their small molecule counterparts, single QDs can be conjugated to multiple molecules for multivalent presentation of affinity tags and multifunctionality. QD bioconjugation can be achieved using several approaches, including electrostatic adsorption,[26] covalent-bond formation,[16] or streptavidin-biotin linking.[27] Ideally, the molecular stoichiometry and orientation of the attached biomolecules could be manipulated to allow access to the active sites of all conjugated enzymes and antibodies, however this is very difficult in practice. Goldman et al first explored the use of a fusion protein as an adaptor for immunoglobulin G

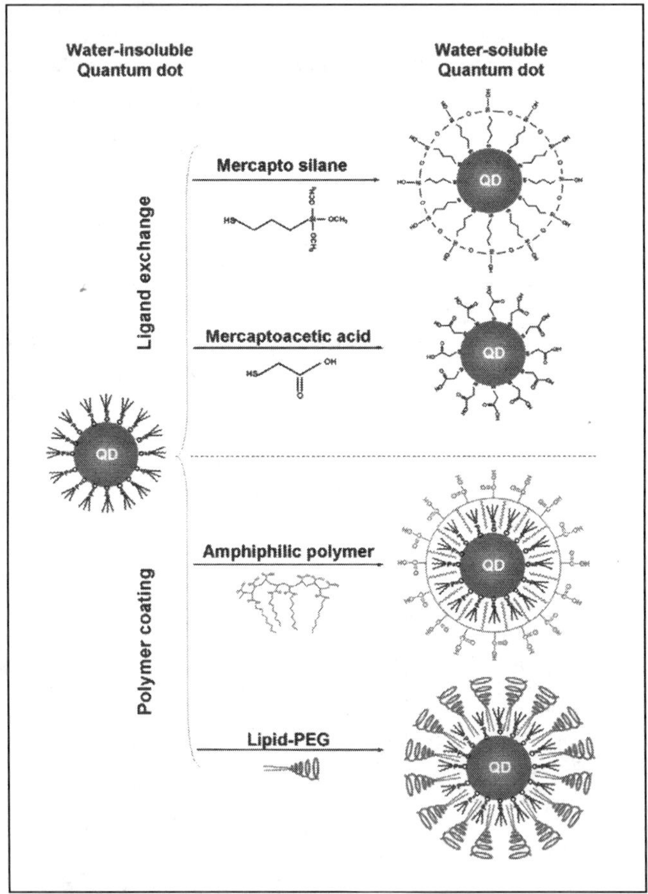

Figure 3. Diagram of two general strategies for phase transfer of TOPO-coated QD into aqueous solution. Ligands are drawn disproportionately large for detail, but the ligand-polymer coatings are usually only 1-2 nm in thickness. The top two panels illustrate the ligand exchange approach, where TOPO ligands are replaced by heterobifunctional ligands such as mercapto silanes or mercaptoacetic acid. This scheme can be used to generate hydrophilic QD with carboxylic acids or a shell of silica on the QD surfaces. The bottom two panels illustrate the polymer coating procedure, where the hydrophobic ligands are retained on the QD surface and rendered water soluble through micelle-like interactions with an amphiphilic polymer or lipids.

antibody coupling.[28] The adaptor protein has a protein G domain that binds to the antibody Fc region, and a positively charged leucine zipper domain for electrostatic interaction with anionic QDs. As a result, the Fc end of the antibody is connected to the QD surface, with the target-specific $F(ab')_2$ domain facing outwards. Surface engineering of nanoparticles is certain to be a greatly studied field in the near future.

Cancer Diagnostics with Quantum Dots

Bioconjugated QD probes have the potential to be useful for cancer diagnosis through many diverse approaches. Their bright and stable fluorescent light emission and multiplexing potential, combined with the intrinsic high spatial resolution and sensitivity of fluorescence imaging, have already demonstrated improvements in existing diagnostic assays. As well, new techniques have been developed framed on the unique properties of these nanoparticles.

In Vitro Diagnostic Assays

Screening of blood, urine, and other bodily fluids for the presence of cancer markers has become a commonly used diagnostic technique for cancer, yet it has been impeded by the lack of specific soluble markers, and sensitive means to detect them at low concentrations. The serum assay most commonly used for cancer diagnosis is the prostate specific antigen (PSA) screen for the detection of prostate cancer.[29] Although other biomarkers have been identified, including proteins, specific DNA or mRNA sequences, and circulating tumor cells, specific cancer diagnosis from serum samples alone may only be possible with a multiplexed approach to assess a large number of biomarkers.[30] QDs could not only serve as sensitive probes for biomarkers, but they could also allow the detection of hundreds to thousands of molecules simultaneously. Experimental groundwork has already begun to demonstrate the feasibility of these expectations, as QDs have found to be superior to conventional fluorescent probes in many clinical assay types.

Protein Biomarker Detection

The ability to screen for cancer in its earliest stages necessitates highly sensitive assays to detect biomarkers of carcinogenesis. The current gold standard for detecting low copy number protein is enzyme-linked immunosorbent assay (ELISA), which has a limit of detection in the pM range. Although these assays are used clinically, they are labor-intensive, time-consuming, prohibitive of multiplexing, and expensive. In this regard, the high sensitivity of QD detection could possibly increase the clinical relevance and routine use of diagnosis based on low-copy number proteins. QDs have been successfully used as substitutes for organic fluorophores and colorimetric reagents in a variety of immunoassays for the detection of specific proteins, yet they have not demonstrated an increase in sensitivity (100 pM).[28,30] Increasing the sensitivity of these probes may only be a matter of optimizing bioconjugation parameters and assay conditions, although the multiplexing capabilities of these probes have already been demonstrated. Goldman et al[32] simultaneously detected four toxins using four different QDs, emitting between 510 nm and 610 nm, in a sandwich immunoassay configuration with a single excitation source. Although there was spectral overlap of the emission peaks, deconvolution of the spectra revealed fluorescence contributions from all four toxins. This assay was far from quantitative, however, and it is apparent that fine-tuning of antibody cross-reactivity will be required to make multiplexed immunoassays useful. Similarly, Makrides et al demonstrated the ease of simultaneously detecting two proteins with two spectrally different QDs in a Western blot assay.[33]

Biosensors are a new class of probes developed for biomarker detection on a real-time or continuous basis in a complex mixture. Assays resulting from these new probes could be invaluable for protein detection for cancer diagnosis because of their high speed, ease of use, and low cost, enabling them to be used for quick point-of-care screening of cancer markers. QDs are ideal for biosensor applications due to their resistance to photobleaching, allowing for continuous monitoring of a signal. Fluorescence resonance energy transfer (FRET) has been the most prominent mechanism to render QDs switchable from a quenched "off" state to a fluorescent "on" state. FRET is the nonradiative energy transfer from an excited donor fluorophore to an acceptor. The acceptor can be any molecule (such as a dye or another nanoparticle) that absorbs radiation at the wavelength of the emission of the donor (the QD). Medintz et al used QDs conjugated to maltose binding proteins as an in situ biosensor for carbohydrate detection.[34] Adding a maltose derivative covalently bound to a FRET acceptor dye caused QD quenching (~60% efficiency), and fluorescence was restored upon addition of native maltose, which displaced the sugar-dye compound. QD biosensors have also been assembled that do not require binding and dissociation to modulate quenching and emission. The same group conjugated a donor QD to a photoresponsive dye that becomes an acceptor after exposure to UV light, and becomes FRET-inactive following white-light exposure, thus allowing light exposure to act as an on-off switch.[35] Before this work can be translated to a clinical tool, these probes must be optimized for higher detection sensitivity, which will require higher quenching efficiencies.

Nucleic Acid Biomarker Detection

Early detection and diagnosis of cancer could be greatly improved with genomic screening of individuals for hereditary predispositions to certain types of cancers, and by detecting mutated genes and other nucleic acid biomarkers for cancer in bodily fluids. The current gold standard for sensitive detection of nucleic acids is polymerase chain reaction (PCR) combined with a variety of molecular fluorophore assays, commonly resulting in a detection limit in the fM range. However, like ELISAs, the clinical utility of nucleic acid analysis for cancer diagnosis is precluded by its time and labor consumption and poor multiplexing capabilities. Many types of new technologies have been developed recently for the rapid and sensitive detection of nucleic acids, most notably RT-PCR and nanoparticle-based biobarcodes,[36] each of which have a limit of detection in the tens of molecules. However QDs could have an advantage in this already technologically crowded field because of their multiplexing potential. Gerion et al reported the detection of specific single nucleotide polymorphisms of the human p53 tumor suppressor gene using QDs in a microarray assay format,[37] although the level of sensitivity (2 nM) was far from matching current standards. Importantly, this work demonstrated the capacity to simultaneously detect two different DNA sequences using two different QDs.

Recently Zhang et al developed a QD biosensor for DNA, analogously to the aforementioned protein biosensor (Fig. 4A).[38] However in this case, the QDs acted as the donor, and an organic dye bound to the target DNA was the acceptor. Because QDs have broadband absorption compared to organic dyes, excitation of the QD at a short wavelength did not excite the dye, thereby allowing extremely low background signals. This allowed the highly sensitive and quantitative detection of as few as 50 DNA copies, and was sufficiently specific to differentiate single nucleotide differences. However this strategy is not ideal for high-throughput analysis of multiple biomarkers because sensitive detection required the analysis of single quantum dots, followed by statistical data analysis.

High-Throughput Multiplexing

Rather than using single QDs for identifying single biomarkers, it has been proposed that QDs of different colors can be combined into a larger structure, such as a microbead, to yield an "optical barcode." With the combination of 6 QD emission colors and 10 QD intensity levels for each color, one million different codes are theoretically possible. A vast assortment of biomarkers may be optically encoded by conjugation to these beads, opening the door to the multiplexed identification of many biomolecules for high-throughput screening of biological samples. Pioneering work was reported by Han et al in 2001, in which 1.2 μm polystyrene beads were encoded with three colors of QDs (red, green, and blue) and different intensity levels (Fig. 4B).[39] The beads were then conjugated to DNA, resulting in different nucleic acids being distinguished by their spectrally distinct optical codes. These encoded probes were incubated with their complementary DNA sequences, which were also labeled with a fluorescent dye as a target signal. The hybridized DNA was detected through colocalization of the target signal and the probe optical code, via single-bead spectroscopy,

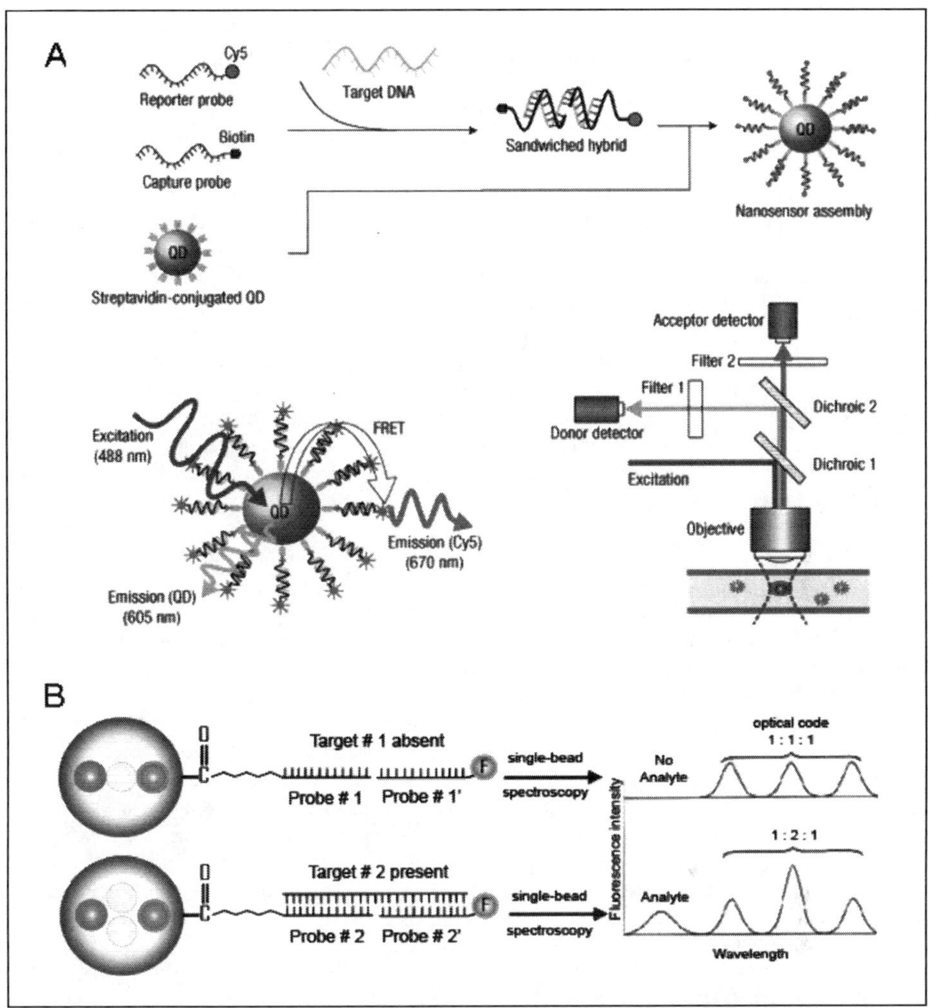

Figure 4. QD based biosensors and optical barcodes. A) Single QD DNA sensors: (top) conceptual scheme showing the formation of a nanosensor assembly in the presence of targets; (bottom left) fluorescence emission from Cy5 on illumination of QD caused by FRET between Cy5 acceptors and a QD donor; (bottom right) experimental setup. Reprinted with permission from Macmillan Publishers Ltd: Nature Materials ©2005.[38] B) Schematic illustration of DNA hybridization assays using QD-barcode beads. When the target molecule is in absence, only the QD barcode signal are detected by single bead spectroscopy or flow cytometry because hybridization doesn't occur. When the target molecule is in presence, it brings the barcode probe (probe #2) and reporter probe (probe #2') together, which results in a detection of both the barcode fluorescence and the reporter signal. The reporter signal not only indicates the presence or absence of the analyte, but also its abundance. Note that the reporter probe (probe # 1' & 2') can be labeled with either an organic fluorophore or a single QD (shown as a blue sphere). Reprinted with permission from Macmillian Publishers Ltd: Nature Biotechnology ©2001.[39]

using only one excitation source. The bead code identified the sequence, while the intensity of the target signal corresponded to the presence and abundance of the target DNA sequence. This uniformity and brightness of the QD encoded beads were substantially improved by Gao and Nie recently using mesoporous materials.[40,41]

The high-throughput potential of this technology was realized by combining it with flow cytometry. For example, DNA sequences from specific alleles of the human cytochrome P450 gene family were correctly identified by hybridization to encoded probes.[42] It's worth mentioning that the long excited state of QDs and the blinking effect do not interfere with bead decoding.[41] If three or more colors are used for microbead encoding, this identification would be considerably more difficult with organic dyes due to the fact that their emission peaks overlap, obscuring the distinct codes, and multiple excitation sources are required. Once encoded libraries have been developed for identification of nucleic acid sequences and proteins, solution-based multiplexing of QD-encoded beads could quickly produce a vast amount of gene and protein expression data. This could not only be used to discover new biomarkers for disease, but also open the door to simple and fast genotyping of patients and cancer classification for personalized medical treatment. Another approach to multiplexed gene analysis has been the use of planar chips, but bead-based multiplexing has the advantages of greater statistical analysis, faster assaying time, and the flexibility to add new probes at lower costs.[43]

Cellular Labeling

Pathological evaluation of biopsies of primary tumors and their distal metastases is the most important cancer diagnostic technique in practice. After microscopic examination of the tissue, the pathologist predicts a grade and stage of tumor progression, so that the cancer can be classified to give a prognosis and appropriate treatment regimen. However evaluation is based primarily on qualitative morphological assessment of the tissue sections, sometimes with fluorescent staining of the tissue for specific cancer biomarkers. This field is highly subjective, and diagnoses of identical tissue sections may vary between pathologists. A more objective and quantitative approach based on biomarker detection would increase diagnostic accuracy. Previous success has been made with colloidal gold and dye-doped silica nanoparticles, however, immunogold staining is essentially a single color assay, whereas dye-doped silica nanoparticles are limited by the unfavorable properties of organic fluorophores. In comparison, QDs would be better candidates for quantitative staining of tissues for biomarkers because of their ability to detect multiple analytes simultaneously and because they have already been proven to be outstanding probes for fluorescent detection of proteins and nucleic acids in cells.

Labeling of Fixed Cells and Tissues

The feasibility of using QD for biomarker detection in fixed cellular monolayers was first demonstrated by Bruchez and coworkers in 1998.[17] By labeling nuclear antigens with green silica-coated QD and F-actin filaments with red QD in fixed mouse fibroblasts, these two spatially distinct intracellular antigens were simultaneously detected. This article and others[16,21] have demonstrated that QDs are brighter and dramatically more photostable than organic fluorophores when used for cellular labeling. Many different cellular antigens in fixed cells and tissues have been labeled using QDs, including specific genomic sequences,[44,45] mRNA,[46] plasma membrane proteins,[21,47,48] cytoplasmic proteins,[17,21] and nuclear proteins,[16,21] and it is apparent that they can function as both primary and secondary antibody stains. In addition, high resolution microtubule filament imaging has been demonstrated using QDs (Fig. 5A),[21] and the fluorescence can be correlated directly to electron micrograph contrast due to the high electron density of QD.[49,50] It has now become clear that QDs are superior to organic dyes for fixed cell labeling. However the translation from fixed cell labeling to labeling of formaldehyde-fixed, paraffin-embedded tissue sections of tumor biopsies is not simple due to the high autofluorescence and the loss of antigen presentation associated with the embedding and fixation processes. Nonetheless, tissue section labeling with QDs has been successful for

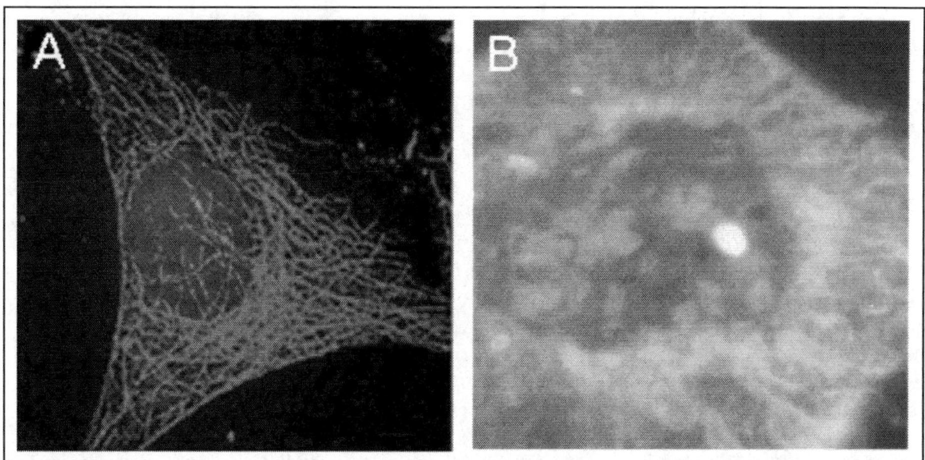

Figure 5. Molecular imaging of cells and tissues. A) Microtubules in NIH-3T3 cells labeled with red color QDs (reprinted by permission from Macmillian Publishers Ltd: Nature Biotechnology ©2003).[21] B) QD-Immunostaining of formalin-fixed, paraffin-embedded human prostate tumor specimens. Mutated p53 phosphoprotein over-expressed in the nuclei of androgen-independent prostate cancer cells is labeled with red color QDs. The Stokes shifted fluorescence signal is clearly distinguishable from the tissue autofluorescence. Reproduced from Smith et al, Expert Rev Mol Diagns 2006; 6:231-244 with permission of Future Drugs Ltd. A color version of this figure is available online at www.eurekah.com.

biomarker-specific staining of rat neural tissue,[51] human skin basal cell carcinomas,[47] and human tonsil tissue.[52] The recent advances in immunohistochemistry (IHC) for protein detection and fluorescence in situ hybridization (ISH) for nucleic acid detection using QD probes could revolutionize clinical diagnosis of biopsies because of the large number of biomarkers that could be simultaneously monitored (Fig. 5B).

Live Cell Imaging

In 1998, Chan et al demonstrated that QDs conjugated to a membrane-translocating protein, transferrin, could cause endocytosis of QDs by living cancer cells in culture.[16] The QDs retained their bright fluorescence in vivo and were not noticeably toxic, thus revealing that QDs could be used as intracellular labels for living cell studies. Most subsequent live cell studies with QDs have focused on labeling of plasma membrane proteins[53,54] and evaluating techniques for traversing the plasma membrane barrier,[55] and it is becoming evident that QDs will become powerful tools for unveiling cellular biology, and for optically "tagging" cells to determine lineage and distribution in multicellular organisms.[22] In this fast moving and exciting field, QDs have already been used to calculate plasma membrane protein diffusion coefficients[54] and observe a single erbB/HER receptor (a cancer biomarker) and its internalization after binding to epidermal growth factor (EGF).[53] As well, QD probes of living cells have prompted the discovery of a new filopodial transport mechanism.[53,56] While most of these studies have centered on biological discovery, a new clinically relevant assay for cancer diagnosis has already been developed from these living cell studies. Alivisatos and coworkers created a cell motility assay, in which the migration of cells over a substrate covered with silica-coated QD was measured in real time.[57] As the cells moved across their substrate, they endocytosed the QDs, causing an increase in fluorescence inside of the cells and a nonfluorescent "dark" path in their trails.[58] These phagokinetic tracks were used to accurately assess invasive potential of different cancer cell types, as motility of cells is strongly associated with their malignancy in vivo. This new assay could aid in the clinical classification of cancers with ambiguous subtypes and further separate subtypes into more discretely defined categories for better diagnosis.

In Vivo Imaging

Despite the large number of identified cancer biomarkers, targeted molecular imaging of cancer has yet to reach clinical practice, although it has been successful in animal models. The four major medical imaging modalities rely on signals that can transmit through thick tissue, using ultrasonic waves (ultrasound imaging), X-rays (computed X-ray tomography), gamma rays (positron emission tomography), or radio waves (magnetic resonance imaging). Image contrast from these techniques is generated from the differences in signal attenuation through different tissue types, which is largely a function of tissue structure and anatomy. Many tumor types can be identified purely based on their image contrast, and exogenous contrast agents are commonly intravenously infused in patients with tumors of poor contrast. However none of these acquired images can convey molecular information of the cancer like what is possible with quantitative in vitro assays and tissue biopsy evaluation. As well, detection of multiple markers is extremely difficult with these imaging techniques, and none of these modalities has innately high spatial resolution capable of detecting most very small, early stage tumors. Generating spatially accurate images of quantitative biomarker concentration would be a giant leap toward detection and diagnosis of cancers, especially for finding sites of metastasis.

Optical imaging, particularly fluorescence imaging, has high intrinsic spatial resolution (theoretically 200-400 nm), and has recently been used successfully in living animal models, yet is limited by the poor transmission of visible light through biological tissue. There is a near-infrared optical window in most biological tissue that is the key to deep-tissue optical imaging.[59] This is because Rayleigh scattering decreases with increasing wavelength, and because the major chromophores in mammals, hemoglobin and water, have local minima in absorption in this window. Few organic dyes are currently available that emit brightly in this spectral region, and they suffer from the same photobleaching problems as their visible counterparts, although this has not prevented their successful use as contrast agents for living organisms.[60] One of the greatest advantages of QDs for imaging in living tissue is that their emission wavelengths can be tuned throughout the near-infrared spectrum by adjusting their composition and size, resulting in photostable fluorophores that can be highly stable in biological buffers.[61] Visible QDs are more synthetically advanced than their near-infrared counterparts, which is why most of the living animal studies implementing QDs have used visible light emission. However even these have shown great promise, due to their ability to remain photostable and brightly emissive in living organisms.

Vascular Imaging

Quantum dots have been used to passively image the vascular systems of various animal models. In a report by Larson et al, intravenously injected QDs remained fluorescent and detectable when they circulated to capillaries in the adipose tissue and skin of a living mouse, as visualized fluorescently.[62] This report made use of two-photon excitation, in which near-infrared light is used to excite visible QDs, allowing for deeper penetration of excitation light, despite strong absorption and scattering of the emitted visible light. Lim et al intravenously injected near-infrared QDs to image the coronary vasculature of a rat heart.[63] The circulation lifetime of an injected molecule is dependent on the size of the molecule and its chemical properties. Small molecules, like organic dyes, are quickly eliminated from circulation minutes after injection because of renal filtration. QDs and other nanoparticles are too large to be cleared through the kidneys, and are primarily eliminated by nonspecific opsonization (a process of coating pathogenic organisms or particles so that they are more easily ingested by the macrophage system) by phagocytotic cells of the reticuloendothelial system (RES), mainly located in the spleen, liver, and lymph nodes. It was demonstrated by Ballou et al that the lifetime of QDs in the bloodstream of mice is significantly increased if the QDs are coated with PEG polymer chains,[64] an effect that has also been documented for other types of nanoparticles and small molecules. This effect is caused by a decreased rate of RES uptake, in part due to decreased nonspecific adsorption of the nanoparticle surface, and decreased antigenicity.[65] Recently PEG-coated QDs have been used to image the vasculature of subcutaneous tumors in mice. Stroh et al used two-photon microscopy to image the blood vessels within the microenvironment of the tumor.[66] Simultaneously, autofluorescence from collagen allowed high-resolution imaging of the extracellular matrix, and genetic modification to express green fluorescent protein (GFP) revealed perivascular cells. Stark contrast between cells, matrix, and the erratic, leaky vasculature was evident, pointing toward the use of fluorescence contrast imaging for the high resolution, noninvasive imaging and diagnosis of human tumors.

Lymph Node Tracking

The lymphatic system is another circulatory system that is of great interest for cancer diagnosis. Cancer staging, and therefore prognosis, is largely evaluated based on the number of lymph nodes involved in metastasis close to the primary tumor location, as determined from sentinel node biopsy and histological examination. It has been shown that QDs have an innate capacity to image sentinel lymph nodes, as first described by Lim et al in 2003.[63] Near-infrared QDs were intradermally injected into the paw of a mouse and the thigh of a pig. Dendritic cells nonspecifically phagocytosed the injected QD, and then migrated to sentinel lymph nodes that could then be fluorescently detected even 1 cm under the skin surface (Fig. 6A). Their results showed rapid uptake of QDs into lymph nodes, and clear imaging and delineation of involved sentinel nodes (which could then be excised). This work shows that QD probes could be used for real-time intra-operative optical imaging, providing an in situ visual guide so that a surgeon could locate and remove small lesions (e.g., metastatic tumors) quickly and accurately. The authors later demonstrated the ability to map esophageal and lung lymph nodes in pigs,[67,68] and also revealed preferential lymph nodes for drainage from the pleural space in rats.[69] Another interesting aspect of this research is that the QDs remained fluorescent after the biopsies were sectioned, embedded, stained, and frozen, allowing microscopic detection of the QDs postoperatively, and giving pathologists another visual aid in judging tissue morphology and cellular identity.

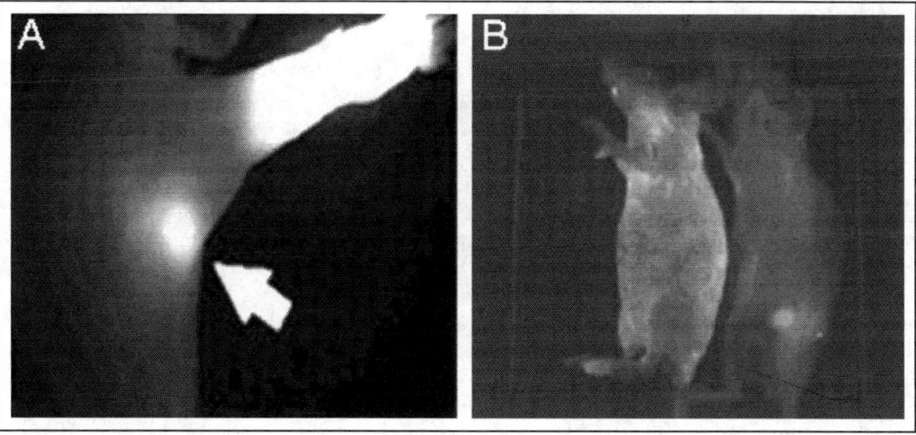

Figure 6. In vivo targeting and imaging with QDs. A) Near-infrared fluorescence of water-soluble type II QDs taken up by sentinel lymph nodes (reprinted by permission from Macmillian Publishers Ltd: Nature Biotechnology ©2004).[61] B) Molecular targeting and in vivo imaging of a prostate tumor in mouse using a QD-antibody conjugate (red). Reprinted from Gao XH et al, Nat Biotechnol 2004; 22(8):969-976, with permission from the author.[23] A color version of this figure is available online at www.eurekah.com.

Tumor Targeting and Imaging

Akerman et al first reported the use of QD-peptide conjugates to target tumor vasculatures, but the QD probes were not detected in living animals.[70] Nonetheless, in vitro histological results revealed that QDs homed to tumor vessels guided by the peptides, and were able to escape clearance by the RES. Most recently, a new class of multifunctional QD probe for simultaneous targeting and imaging of tumors in live animals has been reported by Gao et al.[23] This class of QD conjugate contains an amphiphilic triblock copolymer for in vivo protection, targeting ligands for tumor antigen recognition, and multiple PEG molecules for improved biocompatibility and circulation. Tissue section microscopy and whole animal spectral imaging allowed monitoring of in vivo behavior of QD probes, including their biodistribution, nonspecific uptake, cellular toxicity and pharmacokinetics. Under in vivo conditions, QD probes can be delivered to tumors either by a passive targeting mechanism or through an active targeting mechanism (Fig. 6B). In the passive mode, macromolecules and nanometer-sized particles are accumulated preferentially at tumor sites through an enhanced permeability and retention effect, which is a result of the permeable vasculature of the tumor and lack of effective lymphatic drainage. For active tumor targeting, antibody-conjugated QDs homed a prostate-specific membrane antigen (PSMA), which was previously identified as a cell-surface marker for both prostate epithelial cells and neovascular endothelial cells.

Toxicity and Clinical Potential

The potential toxic effects of semiconductor QDs have recently become a topic of considerable importance and discussion. Indeed, in vivo toxicity is likely to be a key factor in determining whether QD imaging probes would be approved by regulatory agencies for human clinical use. Recent work by Derfus et al[71] indicates that CdSe QDs are highly toxic to cultured cells under UV illumination for extended periods of time. This is not surprising because the energy of UV irradiation is close to that of a covalent chemical bond and dissolves the semiconductor particles in a process known as photolysis, releasing toxic cadmium ions into the culture medium. In the absence of UV irradiation, QDs with a stable polymer coating have been found to be essentially nontoxic to cells and animals.[22,23,56,62,64,66,67,72,73] Still, there is an urgent need to study the cellular toxicity and in vivo degradation mechanisms of QD probes. For polymer-encapsulated QDs, chemical or enzymatic degradation of the semiconductor cores is unlikely to occur. It is possible that the polymer-protected QDs might be cleared from the body by slow filtration and excretion out of the body. Although this should not impede the progress of cellular and solution-based assays using QDs, toxicity must be carefully examined before any human applications in medical imaging are considered.

Conclusion

QDs have already fulfilled some of their promises as a new class of molecular imaging agents. Through their versatile polymer coatings, QDs have also provided a "building block" to assemble multifunctional nanostructures and nanodevices. Multi-modality imaging probes could be created by integrating QDs with paramagnetic or superparamagentic agents. Indeed, researchers have recently attached QDs to Fe_2O_3 and FePt nanoparticles[74,75] and to paramagnetic gadolinium chelates (Gao et al, unpublished data). By correlating the deep imaging capabilities of magnetic resonance imaging (MRI) with ultrasensitive optical imaging, a surgeon could visually identify tiny tumors or other small lesions during an operation and remove the diseased cells and tissue completely. Medical imaging modalities such as MRI and PET can identify diseases noninvasively, but they do not provide a visual guide during surgery. The development of magnetic or radioactive QD probes could solve this problem.

Another desired multifunctional device would be the combination of a QD imaging agent with a therapeutic agent. Not only would this allow tracking of pharmacokinetics, but diseased tissue could be treated and monitored simultaneously and in real time. Surprisingly QDs may be innately multimodal in this fashion, as they have been shown to have potential activity as photodynamic therapy agents.[76,77] These combinations are only a few possible achievements for the future. Practical applications of these multifunctional nanodevices will not come without careful research, but the multidisciplinary nature of nanotechnology may expedite these goals by combining the great minds of many different fields. The success seen so far with QDs points toward the success of QDs in biological systems, and also predicts the success of other nanotechnologies for biomedical applications.

Acknowledgements

X. Gao acknowledge the University of Washington Bioengineering department for the startup fund and the Wallace H Coulter Foundation for a Translational Research Award. We also thank Profs. Lawrence True, Thomas Lawton, and Kimberly Allison in the Department of Pathology for fruitful discussion in molecular cancer research; and Dr. Andrew Wang at Ocean Nanotech Inc. for help on QD synthesis.

References

1. Gao XH, Yang L, Petros JA et al. In vivo molecular and cellular imaging with quantum dots. Curr Opin Biotechnol 2005; 16:63-72.
2. Jemal A, Murray T, Ward E et al. Cancer statistics, 2005. CA Cancer J Clin 2005; 55(1):0-30.
3. Jain RK, Stroh M. Zooming in and out with quantum dots. Nat Biotechnol 2004; 22(8):959-960.
4. Alivisatos AP. Perspectives on the physical chemistry of semiconductor nanocrystals. J Phys Chem 1996; 100(31):13226-13239.
5. Yin Y, Alivisatos AP. Colloidal nanocrystal synthesis and the organic-inorganic interface. Nature 2005; 437(7059):664-670.
6. Alivisatos AP. Semiconductor clusters, nanocrystals, and quantum dots. Science 1996; 271(5251):933-937.
7. Murphy CJ, Coffer JL. Quantum dots: A primer. Appl Spectrosc 2002; 56(1):16A-27A.
8. Sapra S, Sarma DD. Evolution of the electronic structure with size in II-VI semiconductor nanocrystals. Phys Rev B 2004; 69(12):125304.
9. Pietryga J, Schaller R, Werder D et al. Pushing the band gap envelope: Mid-infrared emitting colloidal PbSe quantum dots. J Am Chem Soc 2004; 126(38):11752-11753.
10. Zhong XH, Feng YY, Knoll W et al. Alloyed $Zn_xCd_{1-x}S$ nanocrystals with highly narrow luminescence spectral width. J Am Chem Soc 2003; 125(44):13559-13563.
11. Zhong XH, Han MY, Dong Z et al. Composition-tunable $Zn_xCd_{1-x}Se$ nanocrystals with high luminescence and stability. J Am Chem Soc 2003; 125(28):8589-8594.
12. Qu LH, Peng XG. Control of photoluminescence properties of CdSe nanocrystals in growth. J Am Chem Soc 2002; 124(9):2049-2055.
13. Kim S, Fisher B, Eisler HJ et al. Type-II quantum dots: CdTe/CdSe(core/shell) and CdSe/ZnTe(core/shell) heterostructures. J Am Chem Soc 2003; 125(38):11466-11467.
14. Smith A, Gao XH, Nie S. Quantum-dot nanocrystals for in-vivo molecular and cellular imaging. Photochem Photobiol 2004; 80(3):377-385.
15. Leatherdale C, Woo W, Mikulec F et al. On the absorption cross section of CdSe nanocrystal quantum dots. J Phys Chem B 2002; 106(31):7619-7622.
16. Chan WCW, Nie SM. Quantum dot bioconjugates for ultrasensitive nonisotopic detection. Science 1998; 281(5385):2016-2018.
17. Bruchez M, Moronne M, Gin P et al. Semiconductor nanocrystals as fluorescent biological labels. Science 1998; 281(5385):2013-2016.
18. Dahan M, Laurence T, Pinaud F et al. Time-gated biological imaging by use of colloidal quantum dots. Opt Lett 2001; 26(11):825-827.
19. Bailey RE, Nie SM. Alloyed semiconductor quantum dots: Tuning the optical properties without changing the particle size. J Am Chem Soc 2003; 125(23):7100-7106.
20. Yu WW, Wang YA, Peng XG. Formation and stability of size-, shape-, and structure-controlled CdTe nanocrystals: Ligand effects on monomers and nanocrystals. Chem Mater 2003; 15(22):4300-4308.
21. Wu XY, Liu HJ, Liu JQ et al. Immunofluorescent labeling of cancer marker Her2 and other cellular targets with semiconductor quantum dots. Nat Biotechnol 2003; 21(1):41-46.
22. Dubertret B, Skourides P, Norris DJ et al. In vivo imaging of quantum dots encapsulated in phospholipid micelles. Science 2002; 298(5599):1759-1762.

23. Gao XH, Cui YY, Levenson RM et al. In vivo cancer targeting and imaging with semiconductor quantum dots. Nat Biotechnol 2004; 22(8):969-976.
24. Kirchner C, Liedl T, Kudera S et al. Cytotoxicity of Colloidal CdSe and CdSe/ZnS Nanoparticles. Nano Lett 2005; 5(2):331-338.
25. Pellegrino T, Manna L, Kudera S et al. Hydrophobic nanocrystals coated with an amphiphilic polymer shell: A general route to water soluble nanocrystals. Nano Lett 2004; 4(4):703-707.
26. Mattoussi H, Mauro JM, Goldman ER et al. Self-assembly of CdSe-ZnS quantum dot bioconjugates using an engineered recombinant protein. J Am Chem Soc 2000; 122(49):12142-12150.
27. Goldman ER, Balighian ED, Mattoussi H et al. Avidin: A natural bridge for quantum dot-antibody conjugates. J Am Chem Soc 2002; 124(22):6378-6382.
28. Goldman ER, Anderson GP, Tran PT et al. Conjugation of luminescent quantum dots with antibodies using an engineered adaptor protein to provide new reagents for fluoroimmunoassays. Anal Chem 2002; 74(4):841-847.
29. Hernandez J, Thompson I. Prostate-specific antigen: A review of the validation of the most commonly used cancer biomarker. Cancer 2004; 101(5):894-904.
30. Goessl C. Noninvasive molecular detection of cancer - The bench and the bedside. Curr Med Chem 2003; 10(8):691-706.
31. Bakalova R, Zhelev Z, Ohba H et al. Quantum dot-based western blot technology for ultrasensitive detection of tracer proteins. J Am Chem Soc 2005; 127(26):9328-9329.
32. Goldman ER, Clapp AR, Anderson GP et al. Multiplexed toxin analysis using four colors of quantum dot fluororeagents. Anal Chem 2004; 76(3):684-688.
33. Makrides S, Gasbarro C, Bello J. Bioconjugation of quantum dot luminescent probes for Western blot analysis. BioTechniques 2005; 39(4):501-506.
34. Medintz IL, Clapp AR, Mattoussi H et al. Self-assembled nanoscale biosensors based on quantum dot FRET donors. Nat Mater 2003; 2(9):630-638.
35. Medintz IL, Trammell SA, Mattoussi H et al. Reversible modulation of quantum dot photoluminescence using a protein-bound photochromic fluorescence resonance energy transfer acceptor. J Am Chem Soc 2004; 126(1):30-31.
36. Penn SG, He L, Natan MJ. Nanoparticles for bioanalysis. Curr Opin Chem Biol 2003; 7(5):609-615.
37. Gerion D, Chen FQ, Kannan B et al. Room-temperature single-nucleotide polymorphism and multiallele DNA detection using fluorescent nanocrystals and microarrays. Anal Chem 2003; 75(18):4766-4772.
38. Zhang CY, Yeh HC, Kuroki MT et al. Single-quantum-dot-based DNA nanosensor. Nat Mater 2005; 4(11):826-831.
39. Han MY, Gao XH, Su JZ et al. Quantum-dot-tagged microbeads for multiplexed optical coding of biomolecules. Nat Biotechnol 2001; 19(7):631-635.
40. Gao XH, Nie S. Doping mesoporous materials with multicolor quantum dots. J Phys Chem B 2003; 107:11575-11578.
41. Gao XH, Nie SM. Quantum dot-encoded mesoporous beads with high brightness and uniformity: Rapid readout using flow cytometry. Anal Chem 2004; 76:2406-2410.
42. Xu HX, Sha MY, Wong EY et al. Multiplexed SNP genotyping using the Qbead (TM) system: A quantum dot-encoded microsphere-based assay. Nucleic Acids Res 2003; 31(8):e43.
43. Rosenthal SJ. Bar-coding biomolecules with fluorescent nanocrystals. Nat Biotechnol 2001; 19(7):621-622.
44. Pathak S, Choi SK, Arnheim N et al. Hydroxylated quantum dots as luminescent probes for in situ hybridization. J Am Chem Soc 2001; 123(17):4103-4104.
45. Xiao Y, Barker PE. Semiconductor nanocrystal probes for human metaphase chromosomes. Nucleic Acids Res 2004; 32(3):e28.
46. Matsuno A, Itoh J, Takekoshi S et al. Three-dimensional imaging of the intracellular localization of growth hormone and prolactin and their mRNA using nanocrystal (Quantum dot) and confocal laser scanning microscopy techniques. J Histochem Cytochem 2005; 53(7):833-838.
47. Sukhanova A, Devy M, Venteo L et al. Biocompatible fluorescent nanocrystals for immunolabeling of membrane proteins and cells. Anal Biochem 2004; 324(1):60-67.
48. Howarth M, Takao K, Hayashi Y et al. Targeting quantum dots to surface proteins in living cells with biotin ligase. Proc Nat Acad Sci USA 2005; 102(21):7583-7588.
49. Giepmans BNG, Deerinck TJ, Smarr BL et al. Correlated light and electron microscopic imaging of multiple endogenous proteins using quantum dots. Nature Methods 2005; 2(10):743-749.
50. Nisman R, Dellaire G, Ren Y et al. Application of quantum dots as probes for correlative fluorescence, conventional, and energy-filtered transmission electron microscopy. J Histochem Cytochem 2004; 52(1):13-18.

51. Ness JM, Akhtar RS, Latham CB et al. Combined tyramide signal amplification and quantum dots for sensitive and photostable immunofluorescence detection. J Histochem Cytochem 2003; 51(8):981-987.

52. Sukhanova A, Venteo L, Devy J et al. Highly stable fluorescent nanocrystals as a novel class of labels for immunohistochemical analysis of paraffin-embedded tissue sections. Lab Investig 2002; 82(9):1259-1261.

53. Lidke DS, Nagy P, Heintzmann R et al. Quantum dot ligands provide new insights into erbB/ HER receptor-mediated signal transduction. Nat Biotechnol 2004; 22(2):198-203.

54. Dahan M, Levi S, Luccardini C et al. Diffusion dynamics of glycine receptors revealed by single-quantum dot tracking. Science 2003; 302(5644):442-445.

55. Derfus AM, Chan WCW, Bhatia SN. Intracellular delivery of quantum dots for live cell labeling and organelle tracking. Adv Mater 2004; 16(12):961-966.

56. Lidke DS, Lidke KA, Rieger B et al. Reaching out for signals: Filopodia sense EGF and respond by directed retrograde transport of activated receptors. J Cell Biol 2005; 170(4):619-626.

57. Parak WJ, Boudreau R, Le Gros M et al. Cell motility and metastatic potential studies based on quantum dot imaging of phagokinetic tracks. Adv Mater 2002; 14(12):882-885.

58. Pellegrino T, Parak W, Boudreau R et al. Quantum dot-based cell motility assay. Diferentiation 2003; 71(9-10):542-548.

59. Weissleder R. A clearer vision for in vivo imaging. Nat Biotechnol 2001; 19(4):316-317.

60. Frangioni JV. In vivo near-infrared fluorescence imaging. Curr Opin Chem Biol 2003; 7(5):626-634.

61. Kim S, Lim YT, Soltesz EG et al. Near-infrared fluorescent type II quantum dots for sentinel lymph node mapping. Nat Biotechnol 2004; 22(1):93-97.

62. Larson DR, Zipfel WR, Williams RM et al. Water-soluble quantum dots for multiphoton fluorescence imaging in vivo. Science 2003; 300(5624):1434-1436.

63. Lim YT, Kim S, Nakayama A et al. Selection of quantum dot wavelengths for biomedical assays and imaging. Molecular Imaging 2003; 2(1):50-64.

64. Ballou B, Lagerholm BC, Ernst LA et al. Noninvasive imaging of quantum dots in mice. Bioconjug Chem 2004; 15(1):79-86.

65. Roberts M, Bentley M, Harris J. Chemistry for peptide and protein PEGylation. Adv Drug Deliv Rev 2002; 54(4):459-476.

66. Stroh M, Zimmer JP, Duda DG et al. Quantum dots spectrally distinguish multiple species within the tumor milieu in vivo. Nat Med 2005; 11(6):678-682.

67. Parungo C, Ohnishi S, Kim S et al. Intraoperative identification of esophageal sentinel lymph nodes with near-infrared fluorescence imaging. J Thorac Cardiovasc Surg 2005; 129(4):844-850.

68. Soltesz E, Kim S, Laurence R et al. Intraoperative sentinel lymph node mapping of the lung using near-infrared fluorescent quantum dots. Ann Thorac Surg 2005; 79(1):269-277.

69. Parungo C, Colson Y, Kim S et al. Sentinel lymph node mapping of the pleural space. Chest 2005; 127(5):1799-1804.

70. Akerman ME, Chan WCW, Laakkonen P et al. Nanocrystal targeting in vivo. Proc Natl Acad Sci USA 2002; 99(20):12617-12621.

71. Derfus AM, Chan WCW, Bhatia SN. Probing the cytotoxicity of semiconductor quantum dots. Nano Lett 2004; 4(1):11-18.

72. Jaiswal JK, Mattoussi H, Mauro JM et al. Long-term multiple color imaging of live cells using quantum dot bioconjugates. Nat Biotechnol 2003; 21(1):47-51.

73. Voura E, Jaiswal J, Mattoussi H et al. Tracking metastatic tumor cell extravasation with quantum dot nanocrystals and fluorescence emission-scanning microscopy. Nat Med 2004; 10(9):993-998.

74. Wang DS, He JB, Rosenzweig N et al. Superparamagnetic Fe2O3 beads-CdSe/ZnS quantum dots core-shell nanocomposite particles for cell separation. Nano Lett 2004; 4(3):409-413.

75. Gu HW, Zheng RK, Zhang XX et al. Facile one-pot synthesis of bifunctional heterodimers of nanoparticles: A conjugate of quantum dot and magnetic nanoparticles. J Am Chem Soc 2004; 126(18):5664-5665.

76. Samia ACS, Chen X, Burda C. Semiconductor quantum dots for photodynamic therapy. J Am Chem Soc 2003; 125(51):15736-15737.

77. Bakalova R, Ohba H, Zhelev Z et al. Quantum dot anti-CD conjugates: Are they potential photosensitizers or potentiators of classical photosensitizing agents in photodynamic therapy of cancer? Nano Lett 2004; 4(9):1567-1573.

CHAPTER 6

Carbon Nanostructures as a New High-Performance Platform for MR Molecular Imaging

Keith B. Hartman and Lon J. Wilson*

Abstract

Over the last several years, great interest has developed in the potential use of carbon nanostructures (C_{60} fullerenes and nanotubes) in medicine. In some cases, medical agents derived from these materials have demonstrated greater efficacy than existing clinical agents in many imaging and therapeutic applications. This chapter provides an overall review of the application of these materials in the area of magnetic resonance imaging (MRI), with an emphasis on their future applications in targeted MR molecular imaging for the early detection of cancer and other life-threatening diseases.

Introduction

Widely considered the birth of nanotechnology, the discovery of C_{60}, a new form of elemental carbon, by Curl, Kroto, and Smalley in 1985[1] marshaled in a new wave of research into nanoscale science. These new nanomaterials can be engineered to produce properties unattainable at the larger macro or smaller quantum levels, and they have been the stimulus behind a myriad of proposed nanotechnology applications, including energy storage,[2,3] electronics,[4,5] catalysis,[6] and space travel.[7,8] Most recently, great strides have been made in the biotechnology sector for nanoscale materials whose proposed uses include: quantum dots as fluorescent-imaging agents,[9] C_{60} as drug scaffolds,[10] gold nanoshells as cancer therapeutics,[11] and metallofullerenes as radiotracers and radiopharmaceuticals.[12] In this chapter, we discuss one of the more promising new areas of nanobiotechnology, carbon nanomaterials engineered as high-performance magnetic resonance imaging (MRI) contrast agents.

A fundamental aspect of clinical medicine is the imaging and diagnosis of a given malady. Before clinicians can begin treatment, they must first identify the area of concern (cancerous tumor, arterial blockage, ligament tear, etc.) and ascertain the extent of the ailment in order to devise treatment strategies. To this end, a variety of different imaging techniques have been developed, many within the last 10 years. One of the key techniques is MRI, for which the 2003 Nobel Prize in Medicine was awarded to Paul Lauterbur and Peter Mansfield for their development of the technique.

MRI is based on nuclear magnetic resonance, which measures the relaxation rate of water proton spins exposed to a magnetic field. About 35% of the 60 million annual MRI procedures

*Corresponding Author: Lon J. Wilson—Department of Chemistry, Rice University, P.O. Box 1892, Houston, Texas 77251-1892, U.S.A. Email: durango@rice.edu

Bio-Applications of Nanoparticles, edited by Warren C.W. Chan. ©2007 Landes Bioscience and Springer Science+Business Media.

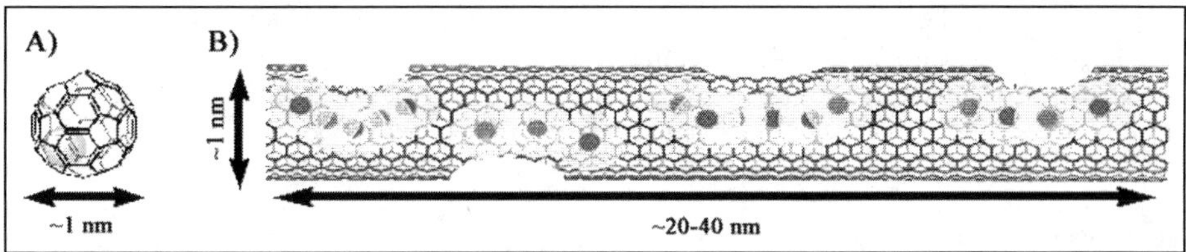

Figure 1. The carbon nanostructure-based MRI contrast agents: (A) Gd@C_{60} gadofullerene and (B) Gd^{3+}_n@US-tube gadonanotube.

utilize a chemical contrast agent (CA) to enhance the sensitivity of the MR image. Most clinical CAs are Gd^{3+}-based due to the ion's 7 unpaired f-electrons (the greatest number of unpaired electrons for any element or ion) and their symmetrical distribution within the ion. Dozens of such CAs have been approved by the FDA, each consisting of a chelated Gd^{3+} compound that circulates in the blood until elimination via the kidney.

Current imaging methods are able to distinguish between neighboring tissues of different composition and anomalies within these tissues. However, when a tumor cannot be detected because of limitations of current imaging technology (limited to the detection of a large mass of tumorigenic cells), they can become metastatic and thereafter much more difficult to treat. To this end, current medical imaging research strives to image the most fundamental unit of life: the single cell. Known as molecular imaging, researchers seek to develop "smart CAs" that are molecularly targeted to diseased cells of interest (tumor cells, arterial plaque cells, etc.) so as to accumulate within targeted cells. This can lead to the detection of disease (via some abnormal cellular biochemistry) at its earliest stage when relatively few cells are present. Such capabilities would allow for earlier diagnosis of cancer or arterial blockage, leading to an increase in successful therapies and the preservation of more lives. Pursuant to these goals, current clinical MRI CAs have important limitations that engineered nanostructures should be able to overcome. Current clinical CAs are generally not targeted to sites of disease, are not readily internalized by cells, and have efficacies much too low to realize molecular imaging.

Engineered carbon nanostructure CAs based on C_{60} fullerene and shortened single-walled carbon nanotubes (SWNTs), as shown in Figure 1, have the potential to become clinical molecular imaging CAs for the following reasons:

- Recent research has demonstrated that this new class of CAs far outperforms conventional CAs, with relaxivities (efficacies) up to 100 times larger than current clinical CAs when internally loaded with Gd^{3+} ions.
- The exterior carbon sheaths of these nanostructures can be functionalized for targeting and biocompatibility without compromising the performance of the sequestered Gd^{3+} ions within. For example, a carbon-based nanostructure CA could be linked to an antibody or peptide that targets a specific cancer cell type.
- Because of their lipophilicities and very small size (1 nm diameters), these carbon nanostructures can readily cross cell membranes and accumulate within cells.

With their extremely high relaxivites and targeting potential, cell-internalized carbon nanostructure CAs are providing one of the first real opportunities to image single cells or small groups of cells, a feat that would revolutionize MRI in medicine.

A Primer in Magnetic Resonance Imaging (MRI)

Derived from the principles of nuclear magnetic resonance (NMR), clinical MRI uses a strong homogenous magnetic field to align the proton spins of water so as to measure the time it takes them to equilibrate, a process known as relaxation.[13] Differences in relaxation times are perceived in the MRI image as varying shades of white and gray. Although water has a very slow relaxation time and is ubiquitous in living organisms, complex instrumentation and mathematical algorithms are able to filter the data and coarsely differentiate between neighboring tissues which have slightly different relaxation times.

Because it is difficult to distinguish between similar tissue (e.g., a tumor within fat tissue), CAs are often used to enhance images. CAs operate by decreasing the relaxation time of the local water protons near the agent, thus allowing for better differentiation between areas of interest. Relaxation occurs via two mechanisms: inner-sphere, in which the water protons are relaxed by unpaired electron spins (spin-lattice relaxation, r_1), and outer-sphere, in which the relaxation is propagated through the lattice field (spin-spin relaxation, r_2). An r_1-agent produces a positively-enhanced image (the image brightens), and it is usually a paramagnetic ion (Gd^{3+}, Mn^{2+}). An r_2-agent produces a negatively-enhanced image (the image darkens), and it is typically a superparamagnetic species (e.g., nanocrystalline Fe_3O_4). All things being equal, clinicians prefer r_1 agents as it is easier to distinguish features in a brightened image opposed to a darkened image.

Relaxivity is relatively simple to measure, although it requires sophisticated instrumentation. About 0.2 ml of sample is placed in an NMR tube and positioned in a relaxometer. Temperatures can be varied to determine characteristics of the system, however, the most useful medical information is acquired at 37°C (body temperature). A uniform magnetic field aligns the proton spins of the sample and a pulse inverts the magnetization vector. The time required for the magnetization vector to randomize is known as the relaxation time (T_1). Fortunately, water protons relax slowly, on the order of several seconds. Upon the addition of a CA, the relaxation time decreases to tens of milliseconds. Because the relaxation time is concentration dependent, mathematical adjustments are made to yield "relaxivity", a standard by which all contrast agents are compared quantitatively for efficacy. The greater the relaxivity, the more efficacious a given contrast agent; units are $mM^{-1}s^{-1}$. Typical relaxivities for clinical agents at 37°C and clinical-field strengths (~60 MHz or 1.5 Tesla) are approximately 4 $mM^{-1}s^{-1}$.[14,15]

Fullerene(C_{60})-Based Contrast Agents

Endohedrals are a unique class of fullerenes that have a single metal atom (or in some cases, up to three[16,17]) entrapped inside the fullerene carbon cage. These materials are commonly referred to as metallofullerenes.[18] Most of the lanthanide ions have been captured inside fullerene cages of 60, 70, 72, or 82 carbon atoms. These metallofullerenes are generated during fullerene synthesis by using carbon rods soaked in a solution of the appropriate metal salt before carbon-arc ablation. While several lanthanides give appreciable relaxivities, Gd^{3+} endohedrals, as expected, demonstrate the best performance and will be exclusively discussed here as 'gadofullerenes'.

Gadofullerenes were initially attractive as possible CAs for two reasons: the fullerene cage acts as a perfect chelate, preventing Gd^{3+} leakage in vivo (clinical CAs demetallate in vivo to a minor extent), and initial measurements demonstrated enhanced relaxivity as compared to traditional CAs.[19,20]

Initially, the C_{60}-based gadofullerenes were written-off as low-yield, inert polymeric materials, however, technological innovations overcame these limitations and now a wide-variety of C_{60}-functionalized gadofullerenes can be produced in large quantities.[21] Two of the most studied water-soluble species are $Gd@C_{60}[C(COOH)_2]_{10}$ and $Gd@C_{60}(OH)_x$, which have relaxivities ranging from 20 $mM^{-1}s^{-1}$ to 100 $mM^{-1}s^{-1}$ at 60 MHz and 37 °C, respectively, or approximately 5-20 times greater than current clinical agents (~4 $mM^{-1}s^{-1}$).[19,21]

Initial studies exploring the proton relaxation processes of the gadofullerenes revealed that they behave unlike any classical contrast agent, making them especially interesting candidates for study. Because it is entrapped inside a carbon cage, the Gd^{3+} ion relaxes water protons via an outer-sphere mechanism since it has no direct contact with water molecules. ^{17}O relaxation studies indicate that the outer-sphere relaxation mechanism is ten times that of chelated Gd^{3+} compounds without inner-sphere water molecules (e.g., $[Gd(TETA)]^-$ where H_4TETA = 1,4,8,11-tetraazacyclotetradecane-1,4,8,11 acetic acid), and it is believed that the C_{60} carbon cage (formally a C_{60}^{3-} cage) itself plays an important role in the large relaxivity of the gadofullerenes.[21] The surface of a gadofullerene is paramagnetic and approximately 200 $Å^2$, resulting in a large outer-sphere relaxivity surmised to be caused by the simultaneous relaxation of many water protons on the cage surface.

A powerful tool for studying the factors influencing MRI contrast agent behavior is the nuclear magnetic relaxation dispersion (NMRD) profile, which measures the relaxivity as a function of magnetic field (0.01-400 MHz). Classic Solomon, Bloembergen, and Morgen (SBM) theory of relaxation[22-26] failed to model the NMRD profile of the gadofullerenes. However, using algorithms originally developed by Bertini and coworkers,[27] the NMRD behavior of gadofullerenes was successfully modeled as a slowly-tumbling structure in solution.[21] A typical Gd^{3+} chelate's rotational tumbling time is in the picosecond regime, whereas the gadofullerenes tumble on the nanosecond scale. Cryo-TEM and dynamic light scattering (DLS) studies concluded that the slower-tumbling gadofullerenes exist as large aggregates, estimated to be about 100-200 nm in size, depending on the metallofullerene species in question.[28]

In addition to determining the relaxivity maxima, evaluating the NMRD profile at several temperatures yields insight into the two separate processes that affect proton relaxivity, namely the proton exchange rate and the molecular reorientation (or tumbling) rate which have opposing temperature dependencies.[28] For the most part, the $Gd@C_{60}(OH)_x$ gadofullerenes did not exhibit widely-varying relaxivities as a function of temperature, however the r_1 relaxivity of $Gd@C_{60}[C(COOH)_2]_{10}$ decreased with increasing temperature, indicating that slow proton exchange is not a limiting factor in relaxivity. This reaffirmed that slowly-tumbling aggregates are the source of the large relaxivities displayed by the gadofullerenes. The influence of pH on r_1 also demonstrated a remarkable dependency, with decreasing pHs resulting in increasing relaxivity. Relaxivites increased up to a factor of three as pH was lowered. Lowering the pH causes the aggregate size to increase, which slows the tumbling time to produce higher relaxivities until the gadofullerenes precipitate from solution around pH ~3. In the case of $Gd@C_{60}[C(COOH)_2]_{10}$, DLS data determined the aggregate size to range from 70 to 700 nm as pH was lowered from 9 to 4.

Finally, high salt concentration breaks up gadofullerene aggregates, drastically reducing the relaxivity since the smaller aggregates tumble faster.[29] Interestingly, a 10 mM phosphate solution is far more effective than a 150 mM NaCl solution at breaking up the gadofullerene aggregates, indicating that this phenomenon is not solely linked to ionic strength. It is believed that the phosphate species are able to intercalate and hydrogen bond to the malonate or hydroxyl groups facilitating disaggregation of the gadofullerenes. DLS studies supported this disaggregate model.

Another interesting fullerene-based MRI CA of note is the tri-gadolinium nitride endohedral metallofullerene $(Gd_3N@C_{80})$.[30] Synthesized by substituting nitrogen gas for argon during the laser ablation process, three Gd^{3+} ions are bound to a central nitride ion to form a four-atom cluster that is encapsulated by a C_{80} cage. Further study is underway with these exciting new compounds, but the production of derivatives for study is hampered by the relative poor chemical reactivity of the $M_3N@C_{80}$ metallofullerenes as compared to their $Gd@C_{60}$ brethren.[31]

Nanotube-Based Contrast Agents

Gadofullerenes have shown great promise as MRI CAs, but synthetic difficulties (low yields and problematic purification steps) coupled with prohibitively high synthetic cost, reduce the likelihood that such CA materials will become clinical agents. However, a related new Gd^{3+}-based nanomaterial which employs ultra-short single-walled carbon nanotubes (US-tubes) has been recently reported (Fig. 1B).[32] Not only does this CA platform outperform the gadofullerenes with respect to relaxivity (Fig. 2), but the synthetic limitations that the gadofullerenes present are no longer an issue since Gd^{3+}-ion loading occurs post US-tube synthesis. Thus, the synthesis of the $Gd^{3+}_n@$US-tube CAs is limited only by the ability to produce single-walled carbon nanotubes (SWNTs).

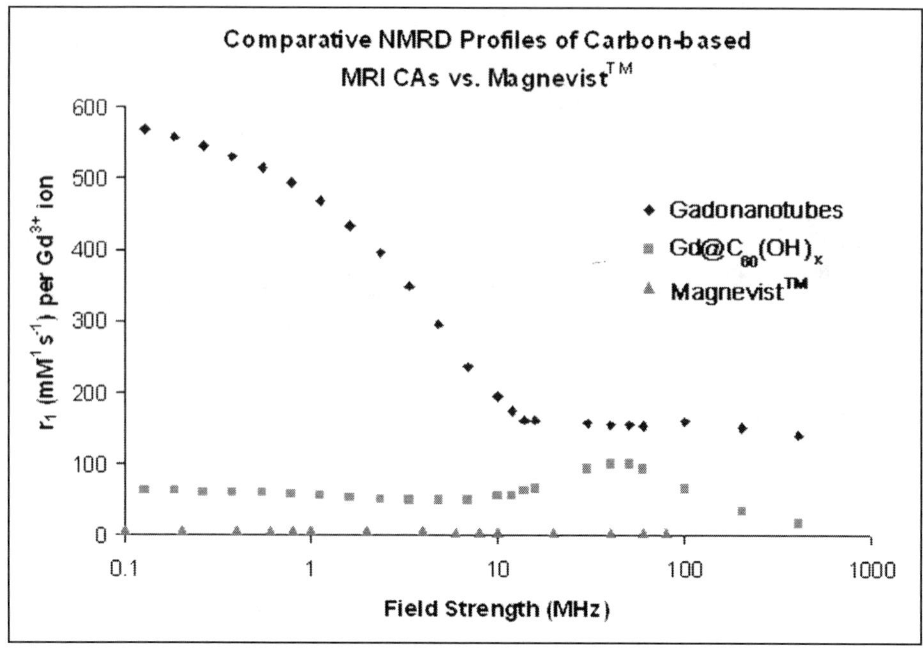

Figure 2. Comparative Nuclear Magnetic Resonance Dispersion (NMRD) profiles of the gadonanotubes, $Gd@C_{60}(OH)_x$, and a common clinical agent, Magnevist™.

To prepare the $Gd^{3+}_n@US$-tube CAs, full-length nanotubes are first cut chemically via fluorination and pyrolysis under inert atmosphere, resulting in US-tubes with an average length of 40 nm.[33] The carbon sheath of the US-tubes can then be functionalized externally for water solubility and biocompatibility, while medically-interesting agents can be loaded internally. One consequence of cutting SWNTs by fluorination is the creation of defects in the sidewalls of US-tubes. This, in turn, serendipitously facilitates filling by Gd^{3+} ions which accumulate at the defects sites, with the nanotube acting as a template for Gd^{3+} loading. Interestingly, aqueous Gd^{3+} ions do not uniformly accumulate down the length of the US-tube, but rather exist in small 2-5 nm clusters of up to 10 Gd^{3+} ions sitting in the defect site. The $Gd^{3+}_n@US$-tube gadonanotubes can then be suspended in a surfactant solution, 1% sodium dodecylbenzene sulfate (SDBS), for relaxivity measurements and electron microscope imaging.

The Gd^{3+}_n-clusters in the $Gd^{3+}_n@US$-tubes are clearly distinguished in both cryo-TEM and high-resolution TEM images since the electron-dense Gd^{3+}-clusters appear as darkly-contrasted areas (Fig. 3).[32] The high-resolution TEM was equipped with EDAX, an elemental analysis attachment, which confirmed that the dark areas tested positive for Gd, while the lighter areas did not (Fig. 3A). It is also worth noting that the cryo-TEM sample (Fig. 3B) was flash-frozen from surfactant-suspended gadonanotubes to give individual-appearing $Gd^{3+}_n@US$-tubes, whereas in the high-resolution TEM sample (Fig. 3A), water (without surfactant) was removed by evaporation, which produced gadonanotube aggregation. For this reason, the cryo-TEM image is a more accurate reflection of the structure of the suspended gadonanotube material in solution.

So what is the functional consequence of the small Gd^{3+}_n clusters within the gadonanotubes? By filling the small surface defect sites of US-tubes with only a few ions of Gd^{3+}, a superparamagnetic material is created. Superconducting quantum interference device (SQUID) measurements show that not only are the gadonanotubes superparamgnetic, but that the system exists in a spin-glass state.[32] This is characteristic of uncommon magnetically-frustrated

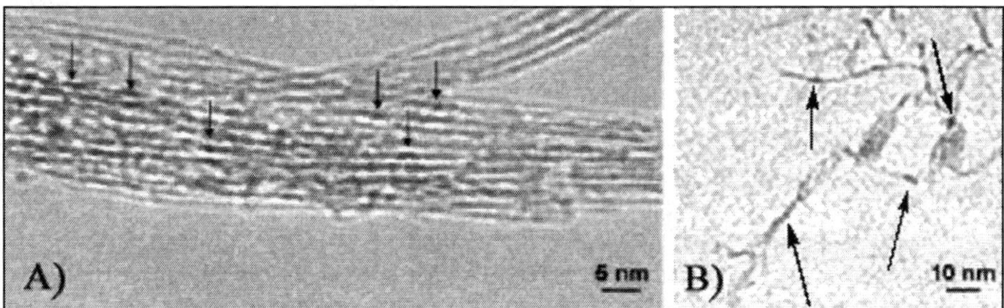

Figure 3. An HRTEM image of aggregated gadonanotubes (A). The arrows indicate dark-contrast spots which tested positive for Gd^{3+} as measured by EDAX; lightly-contrasted areas tested negative for Gd^{3+}. A Cryo-TEM image (B) establishes the presence of Gd^{3+}_n clusters (dark spots indicate an electron-dense center and are designated by the arrows). [Sitharaman B, Kissell KR, Hartman KB, et al. Superparamagnetic gadonanotubes are high-performance MRI contrast agents. Chem Commun (Cambridge, U.K.) 2005; (31):3915-3917.] A and B reproduced by permission of the Royal Society of Chemistry.

environments, confirming the existence of nanoscalar superparamagnetic clusters, acting independently of one another. Thus, the gadonanotubes are the first superparamagnetic Gd^{3+}-based system which results in never-before-seen performance in an MRI CA.

Comparing the NMRD profile of the gadonanotubes to that of the gadofullerenes, several contrasting features can be seen (Fig. 2). Most striking is the magnitude of the gadonanotube r_1 values, an astounding 174 $mM^{-1}s^{-1}$ per Gd^{3+} at 60 MHz (clinical field strengths), compared to ~20-40 $mM^{-1}s^{-1}$ for gadofullerenes and 4 $mM^{-1}s^{-1}$ for Magnevist™, a common clinical MRI CA. Also, the shape of the curve for the gadonanotubes is quite different from that of the gadofullerenes and traditional clinical CAs. The shape of the gadonanoube curve cannot be modeled with either SBM theory or with a slowly-tumbling aggregate algorithm. This indicates that there are features of the gadonanotube system that have never been observed for an MRI CA. We currently speculate that this agent has significant inner-sphere relaxation because of the Gd^{3+} ions, as well as a significant outer-sphere contribution from the superparamagnetic character of the Gd^{3+}_n -ion cluster. However, much work remains to be done to better understand this unique system.

A major difference between the gadonanotubes and the gadofullerenes is that aggregation does not seem to play as important a role in gadonanotube relaxation, as it does for the gadofullerenes. While pH does strongly affect the relaxivity of gadonanotubes, preliminary DLS data indicates that this is not due to an aggregation effect since the hydrodynamic radius of the gadonanotubes is fairly independent of pH.

Due to their exceptionally high performance and previously unobserved behavior, gadonanotube MRI CAs appear poised to make an important impact on MRI. At current clinical field strengths (20-60 MHz), gadonanotubes are roughly 40 times better CAs than their current clinical counterparts (170 $mM^{-1}s^{-1}$ vs. 4.0 $mM^{-1}s^{-1}$). Furthermore, at higher fields the relaxivity of the gadonanotubes stays fairly constant, in contrast to clinical CAs whose relaxivities decline as the magnetic field is increased. With the trend toward higher-field MRI instrumentation, this behavior is a significant advantage for gadonanotube-based CAs in the future. Perhaps the biggest impact may actually lie in the opposite direction, specifically, millitesla imaging. The astonishing relaxivities of the gadonanotubes at low fields, 635 $mM^{-1}s^{-1}$ for gadonanotubes vs. 7.0 $mM^{-1}s^{-1}$ for Magnevist™ (Fig. 2), will undoubtedly encourage the emerging area of millitesla imaging which requires greater CA relaxivities to overcome instrumentation signal-to-noise issues.

Molecular Targeting of Carbon Nanostructures

While gadofullerenes and gadonanotubes clearly demonstrate superior relaxivity properties, they still must be solubilized (without the aid of surfactant) and targeted to specific areas of interest in the body if they are to realize their full potential. An obvious advantage of these engineered carbon-based nanostructures over their inorganic nanostructure counterparts (e.g., iron oxide nanocrystals) is the potential for the facile derivatization of their carbon exterior. Gadofullerenes and gadonanotubes can be functionalized for biocompatibility, water-solubility, and molecular targeting. Such functionalization can link the carbon nanostructure and its cargo to a variety of different targeting moieties that may include peptides, steroids, antibodies, or viral vectors.

While seemingly a simple process in design, the strong innate bundling of nanotubes to one another is a property often overlooked. Due to the π-conjugated electronic structure of carbon nanotubes, the π-π interaction between different tubes results in aggregated bundles with van der Waals energies of approximately 0.5 eV per nanometer of tube contact.[34] A single, full-length nanotube can be isolated with ultrasonication and polymer or surfactant coating.[35,36] It is believed that the full-length tubes are long enough to permit structural bending, eventually allowing for them to be peeled free from the bundle. However, this strategy is not possible with US-tubes because their small length:width ratio (relative to full-length SWNTs) results in rigid-rod behavior, preventing separation.

In order to obtain single-molecule nanotube "capsules" derived from US-tubes, the tubes must first be separated from one another and quickly derivatized to prevent reaggregation. The π-π system responsible for their extreme bundling can also be exploited to overcome the aggregation. The π-conjugated system of a nanotube has been shown to accept and stabilize a large number of electrons (1 electron per 10 carbon atoms).[37,38] Taking advantage of this behavior, we have chemically reduced US-tubes by an amended Birch reduction method (Na°/THF reduction), creating single US-tubes salt complexes in tetrahydrofuran, which are stable for well over 10 days. The extent of exfoliation can be determined by spin-coating the samples onto a mica surface and measuring their height, and therefore, bundle thickness, by atomic force microscopy (AFM). AFM measurements of the reduced US-tubes yield z-heights (tube diameters) of 0.5-1.5 nm, far smaller than for bundled US-tubes which have z-heights in excess of 7 nm. Since a typical HiPco-produced, single-wall nanotube has an average diameter of 1.0 nm (although it may vary from 0.5 nm to 2.0 nm[39]) an AFM height measurement of ~1.0 nm indicates that complete exfoliation of the nanotubes is achieved by this method.

Individualized US-tubes are then easily functionalized via the Bingel reaction.[40,41] Briefly, a bromomalonate is deprotonated by a strong base, creating a nucleophile that readily attacks the fullerene or nanotube conjugated system, resulting in formation of a cyclopropyl group on the surface. The Bingel group is both chemically and biologically stable and easily allows for subsequent derivatization of esters or amides to produce any desired targeting

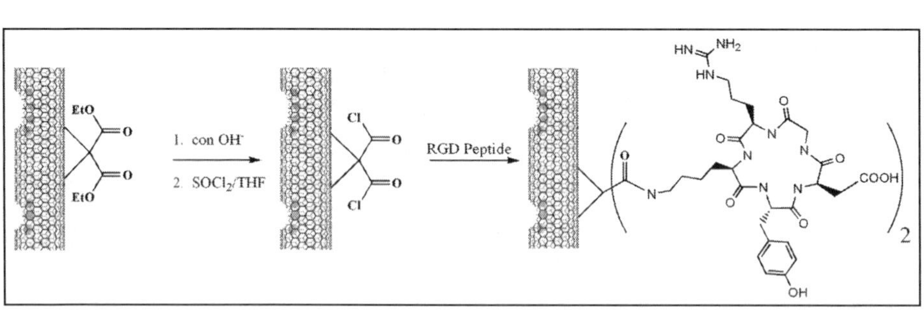

Figure 4. A reaction scheme for attaching the RGD peptide to a gadonanotube.

moiety as shown schematically in Figure 4 for the case of the RGD targeting peptide, which specifically targets the $\alpha_v\beta_3$ integrin receptor upregulated in breast cancers.[42] Several analytical techniques, including NMR can confirm covalent attachment of the malonate group, and X-ray photoelectron spectroscopy (XPS), an elemental analysis technique, has been used to establish that approximately 5 malonate groups per nanometer of US-tube can be attached. AFM indicated that once functionalized in this manner, US-tubes do not flocculate and reform bundles, but exist as individual functionalized US-tubes. The attainment of "single-molecule" nanotube building block of the nature shown in Figure 4 is an important milestone for the continued development of carbon nanotube biotechnology.

Fullerene-Antibody Conjugates

Using similar Bingel functionalization strategies, the first fullerene(C_{60})-antibody immunoconjugates have been recently produced.[43] Antibodies are a particularly promising class of cellular-targeting agents.[44,45] They are large, specialized proteins (MW >100,000 Daltons) that contain antigen binding sites which preferentially bind to a specific antigen that is over-expressed on a given cell type, allowing for cell-specific targeting. Recently, our group engineered the first fullerene(C_{60})-antibody conjugate using a murine antibody that specifically targets melanoma cancer cells.[43]

To produce the fullerene(C_{60})-antibody immunoconjugate, two unique Bingel-derivatized fullerenes were investigated—one that can form a covalent disulfide bond with the antibody (C_{60}-SPDP) and one that cannot form covalent bonds with the antibody (C_{60}-Ser).[43] These C_{60} derivatives are shown in Figure 5. Amazingly, the results of this study indicated that fullerene(C_{60})-antibody covalent bond conjugation may not be requisite for immunoconjugate formation. In fact, the C_{60}-Ser derivative loaded into the antibody to a greater extent (on average, 42 C_{60}'s derivatives per antibody) than the C_{60}-SPDP derivative (on average, 15 C_{60}'s derivatives per antibody). These C_{60}:antibody ratios were obtained from the UV-Vis spectroscopic signatures of the C_{60} derivatives and Biorad protein assays for the antibody. TEM images of the C_{60}-Ser immunoconjugates and the antibody-only control are shown in Figure 6. From the image, a contrast between the unloaded antibody (Fig. 6A) with the C_{60}-SPDP-loaded antibody (Fig. 6B) is observed, where the loaded antibody appears to have tripled in size and exhibits small dark clusters (~5 nm) believed to be small fullerene aggregates inside the antibody. Enzyme-linked immunoabsorbant assay

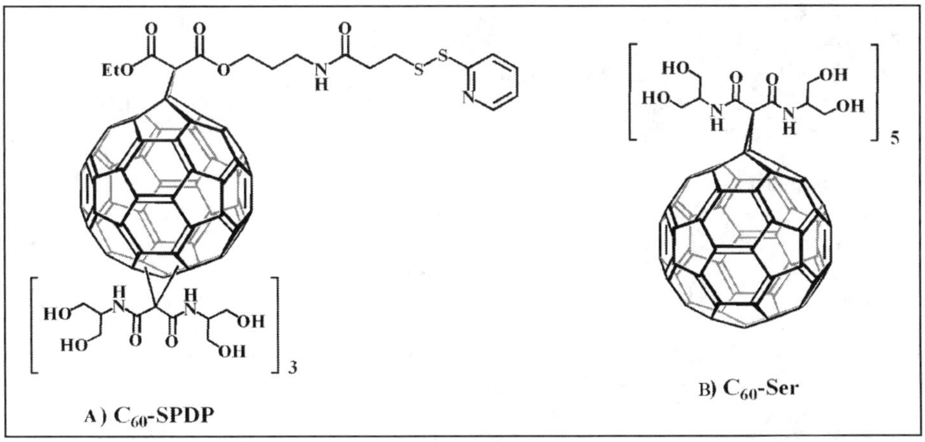

Figure 5. The water-soluble, Bingel-derivatized fullerene (C_{60}) derivatives. The C_{60}-SPDP derivative (A) was designed to form a covalent attachment through the disulfide linker, whereas the C_{60}-Ser derivative (B) cannot form a covalent attachment with the antibody.

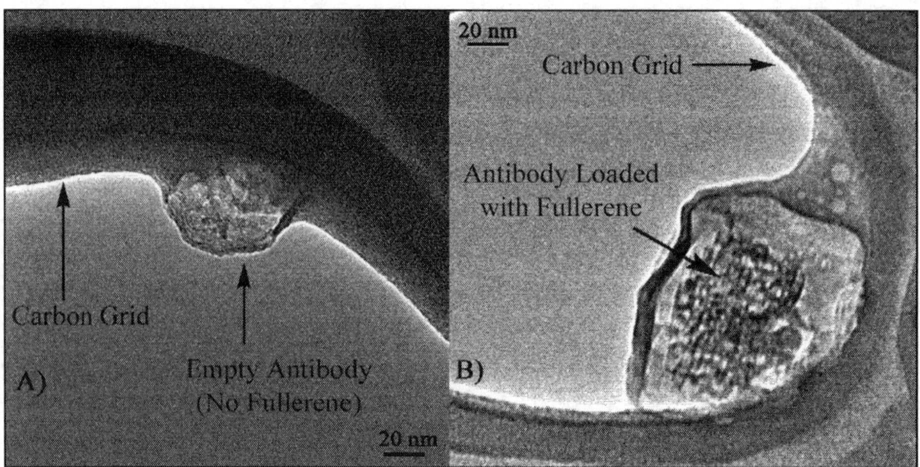

Figure 6. TEM images of an unloaded melanoma antibody (A) and a C_{60}-Ser loaded melanoma antibody (B), at identical magnification (scale bar = 20 nm). The fullerene derivative is readily taken up by the antibody, resulting in increased size as can easily be seen. The dark contrast of the C_{60}-Ser antibody conjugate is likely due to small C_{60} aggregates adsorbed in the antibody interior.

(ELISA) determined that the C_{60}-Ser-antibody is still 70 times more specific than a generic murine antibody. For the C_{60}-SPDP -antibody derivative, there is essentially no loss in specificity and binding efficiency as compared to an unconjugated antibody. This data suggests that the C_{60} moieties are internalized by the antibody and that the antibody can be highly loaded with a payload of ~15 fullerene nanostructures and still retain its biological function. It is believed that no chemical attachment is needed to link the fullerene to the antibody since the water-soluble, lipophillic fullerene derivative, C_{60}-Ser, is readily absorbed into the antibody interior.

Since gadofullerenes are very similar materials to C_{60}, it is logical to assume that they will also load into an antibody. Experiments are currently underway in our laboratory to determine proton relaxivity properties of gadofullerene-antibody conjugates and to attempt targeted, MR imaging of melanoma cells dispersed in a field of normal cells. Perhaps the greatest advantage of carbon-based nanostructures for molecular targeting is the ability of these materials to cross cell membranes,[46,47] resulting in the accumulation of deliverable payloads inside cells, whether for molecular imaging or guided therapy.

Closing Remarks

Clearly the future of carbon-based nanostructures in MRI is bright. In only a few short years, these materials have been shown to dramatically outperform classical MRI CAs in vitro, and they have the potential to revolutionize the field of medical MR imaging. They could compliment (or even replace) current MRI CAs because of their superior relaxivities or they may encourage the development of new techniques of the future, such as millitesla imaging. Clearly, the most exciting frontier for these carbon nanostructures is their potential for overcoming many of the limitations of classical MRI CAs and thus, to produce new agents for molecular imaging. In particular, they are high-performance nanoscalar MRI probes which are intrinsically intracellular and chemically very versatile. A worthy long-term goal is their development as a new universal platform, based on single-molecule US-tubes as disease-targeted carbon nanocapsules, for the containment and delivery of an array of diagnostic and therapeutic agents in medicine.

Acknowledgements

The research at Rice University was supported by the Robert A. Welch Foundation (Grant C-0627), the U.S. National Institutes of Health (Grant NIH R01 EB000703), the NASA Johnson Space Center and University of Texas Health Science Center (Grant NNJ05HE75A). KBH is supported by a Rice University Center for Biological and Environmental Nanotechnology Fellowship (EEC-0118007).

References

1. Kroto HW, Heath JR, O'Brien SC et al. C_{60}: Buckminsterfullerene. Nature (London, United Kingdom) 1985; 318(6042):162-163.
2. Arico AS, Bruce P, Scrosati B et al. Nanostructured materials for advanced energy conversion and storage devices. Nature Materials 2005; 4(5):366-377.
3. Gomez-Romero P, Cuentas-Gallegos K, Lira-Cantu M et al. Hybrid nanocomposite materials for energy storage and conversion applications. J Mat Sci 2005; 40(6):1423-1428.
4. Seminario JM. Molecular electronics. Approaching reality. Nature Materials 2005; 4(2):111-113.
5. Balandin AA. Nanophononics: Phonon engineering in nanostructures and nanodevices. Journal of Nanoscience and Nanotechnology 2005; 5(7):1015-1022.
6. Johnson BFG. Nanoparticles in Catalysis. Topics in Catalysis 2003; 24(1-4):147-159.
7. Arepalli S, Fireman H, Huffman C et al. Carbon-nanotube-based electrochemical double-layer capacitor technologies for spaceflight applications. J of Mat 2005; 57(12):26-31.
8. Sharma SK, Kumar R, Dolia SN et al. Magnetic nanoparticles for space applications. Materials Research Society Symposium Proceedings 2005; 851:481-486.
9. Alexson D, Chen H, Cho M, et al. Semiconductor nanostructures in biological applications. J Phys: Condensed Matter 2005; 17(26):R637-R656.
10. Zakharian TY, Seryshev A, Sitharaman B et al. A Fullerene-paclitaxel chemotherapeutic: Synthesis, characterization, and study of biological activity in tissue culture. J Am Chem Soc 2005; 127(36):12508-12509.
11. Hirsch LR, Stafford RJ, Bankson JA et al. Nanoshell-mediated near-infrared thermal therapy of tumors under magnetic resonance guidance. PNAS 2003; 100(23):13549-13554.
12. Cagle DW, Kennel SJ, Mirzadeh S et al. In vivo studies of fullerene-based materials using endohedral metallofullerene radiotracers. PNAS 1999; 96(9):5182-5187.
13. Merbach AE, Toth E, eds. The Chemistry of Contrast Agents in Medical Magnetic Resonance Imaging. Chichester: John Wiley and Sons, 2001.
14. Muller RN, Vander Elst L, Rinck PA et al. The importance of nuclear magnetic relaxation dispersion (NMRD) profiles in MRI contrast media development. Inv Rad 1988; 23(Suppl 1):S229-231.
15. Lauffer RB. Paramagnetic metal complexes as water proton relaxation agents for NMR imaging: Theory and design. Chem Rev 1987; 87:901-927.
16. Stevenson S, Fowler PW, Heine T et al. A stable nonclassical metallofullerene family. Nature (London) 2000; 408(6811):427-428.
17. Stevenson S, Rice G, Glass T et al. Small-bandgap endohedral metallofullerenes in high yield and purity. Nature (London) 1999; 401(6748):55-57.
18. Cioslowski J, Fleischmann ED. Endohedral complexes: Atoms and ions inside the carbon sixty-atom molecule (C_{60}
19. Bolskar RD, Benedetto AF, Husebo LO et al. First soluble $M@C_{60}$ derivatives provide enhanced access to metallofullerenes and permit in vivo evaluation of $Gd@C_{60}[C(COOH)_2]_{10}$ as a MRI contrast agent. J Am Chem Soc 2003; 125(18):5471-5478.
20. Mikawa M, Kato H, Okumura M et al. Paramagnetic water-soluble metallofullerenes having the highest relaxivity for MRI contrast agents. Bioconj Chem 2001; 12(4):510-514.
21. Toth E, Bolskar RD, Borel A et al. Water-soluble gadofullerenes: Toward high-relaxivity, pH-responsive MRI contrast agents. J Am Chem Soc 2005; 127(2):799-805.
22. Bloembergen N, Morgan LO. Proton relaxation times in paramagnetic solutions effects of electron spin relaxation. J Chem Phys 1961; 34:842-850.
23. Bloembergen N. Proton relaxation times in paramagnetic solutions. J Chem Phys 1957; 27:572-573.
24. Solomon I, Bloembergen N. Nuclear magnetic interactions in the HF molecule. J Chem Phys 1956; 25:261-266.
25. Solomon I. Relaxation processes in a system of two spins. Phys Rev 1955; 99:559-565.
26. Bloembergen N, Purcell EM, Pound RV. Relaxation effects in nuclear magnetic resonance absorption. Phys Rev 1948; 73:679-712.

27. Bertini I, Galas O, Luchinat C et al. A computer program for the calculation of paramagnetic enhancements of nuclear-relaxation rates in slowly rotating systems. J Mag Res A 1995; 113(2):151-158.
28. Sitharaman B, Bolskar RD, Rusakova I et al. $Gd@C_{60}[C(COOH)_2]_{10}$ and $Gd@C_{60}(OH)_x$: Nanoscale aggregation studies of two metallofullerene MRI contrast agents in aqueous solution. Nano Lett 2004; 4(12):2373-2378.
29. Laus S, Sitharaman B, Tóth É et al. Destroying gadofullerene aggregates by salt addition in aqueous solution of $Gd@C_{60}(OH)_x$ and $Gd@C_{60}[C(COOH_2)]_{10}$. J Am Chem Soc 2005; 127(26):9368-9369.
30. Stevenson S, Phillips JP, Reid JE et al. Pyramidalization of Gd_3N inside a C_{80} cage. Chem Commun (Cambridge, United Kingdom) 2004; (24):2814-2815.
31. Stevenson S, Stephen RR, Amos TM et al. Synthesis and purification of a metallic nitride fullerene bisadduct: Exploring the reactivity of $Gd_3N@C_{80}$. J Am Chem Soc 2005; 127(37):12776-12777.
32. Sitharaman B, Kissell KR, Hartman KB et al. Superparamagnetic gadonanotubes are high-performance MRI contrast agents. Chem Commun (Cambridge, United Kingdom) 2005; (31):3915-3917.
33. Gu Z, Peng H, Hauge RH et al. Cutting single-wall carbon nanotubes through fluorination. Nano Lett 2002; 2(9):1009-1013.
34. Girifalco LA, Hodak M, Lee RS. Carbon nanotubes, buckyballs, ropes, and a universal graphitic potential. Phys Rev B 2000; 62(19):13104-13110.
35. Bachilo SM, Strano MS, Kittrell C et al. Structure-assigned optical spectra of single-walled carbon nanotubes. Science (Washington, DC, United States) 2002; 298(5602):2361-2366.
36. Moore VC, Strano MS, Haroz EH et al. Individually suspended single-walled carbon nanotubes in various surfactants. Nano Lett 2003; 3(10):1379-1382.
37. Penicaud A, Poulin P, Derre A et al. Spontaneous dissolution of a single-wall carbon nanotube salt. J Am Chem Soc 2005; 127(1):8-9.
38. Sauvajol JL, Bendiab N, Anglaret E et al. Phonons in alkali-doped single-wall carbon nanotube bundles. Comptes Rendus Physique 2003; 4(9):1035-1045.
39. Kim UJ, Gutierrez HR, Kim JP et al. Effect of the tube diameter distribution on the high-temperature structural modification of bundled single-walled carbon nanotubes. J Phys Chem B 2005; 109(49):23358-23365.
40. Worsley KA, Moonoosawmy KR, Kruse P. Long-range periodicity in carbon nanotube sidewall functionalization. Nano Lett 2004; 4(8):1541-1546.
41. Coleman KS, Bailey SR, Fogden S et al. Functionalization of single-walled carbon nanotubes via the bingel reaction. J Am Chem Soc 2003; 125(29):8722-8723.
42. Chen X, Conti PS, Moats RA. In vivo near-infrared fluorescence imaging of integrin $\alpha_v\beta_3$ in brain tumor xenografts. Can Res 2004; 64(21):8009-8014.
43. Ashcroft JM, Tsyboulski DA, Hartman KB et al. Fullerene (C_{60}) immunoconjugates: Interaction of water-soluble C_{60} derivatives with the murine anti-gp240 melanoma antibody. Chem Commun (Cambridge, United Kingdom) 2006; 3004-3006.
44. Goodwin DA. Strategies for antibody targeting. Antibody, Immunoconjugates, and Radiopharmaceuticals. 1991:4(4):427-434.
45. Courtenay-Luck NS, Epenetos AA. Targeting of monoclonal antibodies to tumors. Current Opinion in Immunology 1990; 2(6):880-883.
46. Pantarotto D, Briand JP, Prato M et al. Translocation of bioactive peptides across cell membranes by carbon nanotubes. Chem Commun (Cambridge, United Kingdom) 2004; (1):16-17.
47. Kam NWS, Jessop TC, Wender PA et al. Nanotube molecular transporters: Internalization of carbon nanotube-protein conjugates into mammalian cells. J Am Chem Soc 2004; 126(22):6850-6851.

CHAPTER 7

Magnetic Nanoparticle Assisted Molecular MR Imaging

Young-wook Jun, Jung-tak Jang and Jinwoo Cheon*

Abstract

Magnetic nanoparticles exhibit unique nanoscale properties of superparamagnetism and have the potential to be utilized as excellent probes for magnetic resonance imaging (MRI). Especially, clinically benign iron oxide nanoparticles provide good MR probing capability and some of them are currently available for clinical applications. However, limited magnetic property and inability to escape from reticuloendothelial system (RES) of the currently used nanoparticles impede their further advancements and therefore it is necessary to develop advanced magnetic nanoparticle probes for next-generation molecular MR imaging. In this chapter, we overview recent progresses on the development of magnetic nanoparticle probes for molecular MR imaging. Utilization of these nanoparticle probes for both in vitro and in vivo molecular MR imaging will be described.

Introduction

Nanoparticles produced in biological system can be utilized as key biofunctional materials for protection, navigation, and sensing.[1-3] For example, magnetotatic bacterium possesses a chain of about 20 magnetite crystals on a nanometer scale diameter.[3] They selectively collect iron ions from the seawater into organic hollow spheres and subsequent reduction of iron ion results in formation of magnetite nanoparticles.[3] This bacterium utilizes the chained magnetic nanoparticles as a magnetic compass to orient itself along the Earth's magnetic field and can navigate to its desired destination.[3]

On the other hand, researchers have explored new ways to chemically synthesize nanoparticles for their applications in biomedical sciences.[4-9] When these are used in biological system, in addition to their comparable small size to biological molecules and functional units, unique nanoscale property of inorganic nanoparticles enables them to be utilized as probes and vectors for biomedical diagnosis and therapy. In particular, magnetic nanoparticles with superparamagnetism are emerging as next-generation molecular imaging probes for observing and tracking molecular events via magnetic resonance imaging (MRI). MRI is one of the most powerful medical diagnostic tools due to its noninvasive nature and multi-dimensional tomographic capabilities coupled with high spatial resolution.[10] Although MRI lags in sensitivity when compared with other tools,[11] such weakness can be significantly improved by using magnetic nanoparticle based contrast agents.[12,13] Under an applied magnetic field, these nanoparticles are magnetized and generate induced magnetic fields, which can perturb the magnetic relaxation processes of the protons in water molecules that surround the magnetic nanoparticles.

*Corresponding Author: Jinwoo Cheon—Department of Chemistry and Nano-Medical National Core Research Center, Yonsei University, Seoul 120-749, Korea. Email: jcheon@yonsei.ac.kr

Bio-Applications of Nanoparticles, edited by Warren C.W. Chan. ©2007 Landes Bioscience and Springer Science+Business Media.

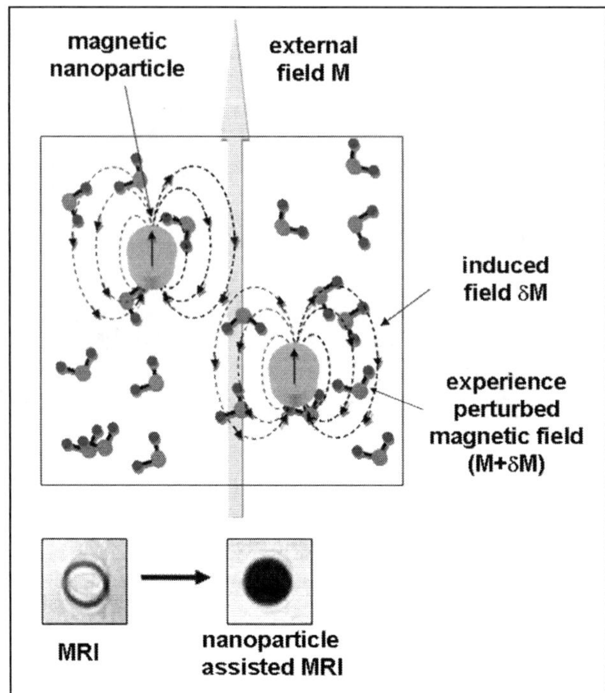

Figure 1. MR contrast effects of magnetic nanoparticles. Under an applied magnetic field (M), magnetic nanoparticles are magnetized and generate an induced magnetic field (δM), which perturbs the magnetic relaxation processes of the proton in water molecules, which is reflected as dark MR contrast.

For example, this phenomena lead to the shortening of the spin-spin relaxation time (T2) of the proton, which results in the darkening of MR images (Fig. 1).

According to the outer sphere spin-spin relaxation formula of solvent protons by solute magnetic particles, spin-spin relaxation time of the proton is

$$\frac{1}{T2} = \left(\frac{32\pi\,N_A\,[M]}{405000\;rD}\right)\gamma_I^2\mu^2\{6.5j_2(\omega_s,\pi)+1.5j_1(0,\pi)\}$$

where, γ_I is gyromagnetic ratio of protons in water, M is the molarity of magnetic nanoparticles, r is their radius, N_A is Avogadro's' number, μ is the magnetic moment of the nanoparticle, ω_s and ω_I are the respective Larmor angular precession frequencies of the solute electronic and water proton magnetic moments, the functions $j_n(\omega, \tau)$ are spectral density functions, and τ (= r^2/D) is the time scale of fluctuations in the particle-water proton magnetic dipolar interaction arising from the relative diffusive motion (D) of a particle and water molecules.[14] Therefore, the shortening of T2 is achieved by increasing the magnetic moment of the nanoparticles. Recently, J. Cheon, J. Suh, and coworkers experimentally demonstrated such magnetic moment effects on T2 by elucidating the correlated nanoscale effects of iron oxide nanoparticles between size, magnetism, and T2 relaxivity.[15] Figure 2a shows transmission electron microscopic images of highly monodispersed iron oxide nanoparticles with the size of 4, 6, 9, and 12 nm respectively. These magnetic nanoparticles exhibit size-dependent magnetic moments and as the nanoparticle size is increased from 4 to 6, 9, and 12 nm, the mass magnetization value at 1.5 T changes from 25 to 43, 80, and 102 emu/(g Fe) respectively (Fig. 2e). Such a trend is clearly reflected in the T2-weigthed MR images. The 1/T2 relaxivity gradually increases from

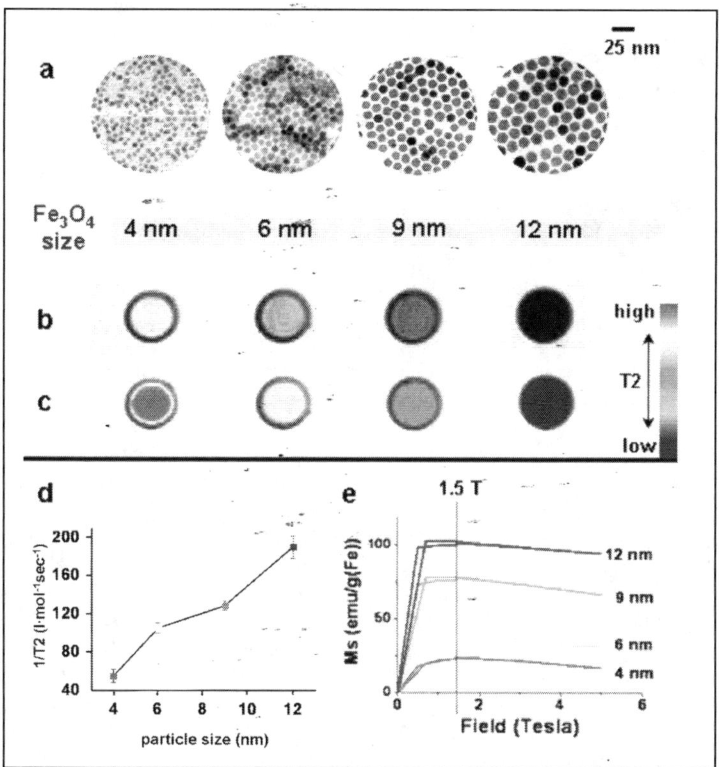

Figure 2. Nanoscale size effects of iron oxide nanoparticles on magnetism and induced magnetic resonance (MR) signals. a) TEM images of Fe_3O_4 nanoparticles of 4 to 6, 9, and 12 nm. b) Size dependent T2-weighted MR images of iron oxide nanoparticles in aqueous solution at 1.5 Tesla. c) Size dependent changes from red to blue in color-coded MR images based on T2 values. d) Graph of 1/T2 relaxivity value versus iron oxide nanoparticle size. e) Magnetization of iron oxide nanoparticles measured by a SQUID magnetometer. (From ref. 15, with permission).

56 to 106, 130, and to 190 $l \cdot mol^{-1} sec^{-1}$, which is imaged by the gradual change of the MR contrast from white to black through gray (Figs. 2b-d).

Such effect of magnetic nanoparticles on MR contrast provides them the ability to report on various biological events. For example, magnetic nanoparticles of larger than 30 nm have been used for phagocytosis imaging.[16,17] When phagocytes uptake magnetic nanoparticles, they are imaged as dark contrast. But tumor cells without phagocytic ability are imaged as white contrast. By utilizing such effect, liver metastasis,[18,19] spleen,[18] and lymph node detection[20] have been performed.

On the other hand, when the smaller nanoparticles (e.g., 10 nm) can easily escape from phagocytes, magnetic nanoparticles, upon conjugation with a target specific biomolecule, can detect target tissues through molecular interactions between nanoparticle-biomolecule conjugates and molecular markers expressed from target tissues.[21,22] Various types of clinically benign iron oxide based magnetic nanoparticles (e.g., superparamagnetic iron oxide (SPIO)) have been explored, and the imaging of infarct,[23-25] angiogenesis,[26] apoptosis,[27] gene expression,[28,29] and cancer[15,30-33] have been reported. However, MR signal sensitivity and specificity of nanoparticle probes to the target tissue are still unsatisfactory for clinical applications and further efforts are needed to make them better. In this section, we briefly review recently developed bio-compatible magnetic nanoparticles and their utilization in molecular MR imaging.

Recent Developments in the Synthesis of Magnetic Nanoparticle Probes

For their successful utilization as molecular MR contrast agents, magnetic nanoparticles must fulfill the following requirements : i) uniform and high superparamagnetic moment, ii) high colloidal stability under physiological conditions (e.g., high salt concentration and pH changes), iii) the ability to escape from the reticuloendothelial system (RES), iv) low toxicity and biocompatibility, and v) possession of functionality to be linked to biologically active species (e.g., nucleic acid, proteins). Since these properties are highly related to their size, stoichiometry, and surface structures, various types of iron oxide nanoparticles have been developed.

Silica- or Dextran- Coated Iron Oxide Contrast Agents

For conventional MR contrast agents, iron oxide nanoparticles are synthesized through the precipitation of iron oxide in an aqueous solution containing ferrous salt by adding an alkaline solution.[34] Such iron oxide nanoparticles are usually insoluble as-is and a coating material is required for them to be soluble in aqueous media. In early attempts to make them water soluble, silica was used as a coating material.[35] The size of the core magnetic iron oxide can vary between 4 to 10 nm and the total particle size varies from 10 nm to 1 μm including coating materials. Since these nanoparticles have a broad size distribution, further size sorting procedures including differential centrifugation and dialysis are required. One of the representative silica-coated iron oxide contrast agent is AMI-121 (generic name : Ferumoxsil), which is now commercially available as Lumirem® (Guerbet) and Gastromark® (Advance Magnetics). The core is approximately ~10 nm sized polycrystalline iron oxide and the hydrodynamic size is approximately ~300 nm. This agent is delivered orally and used for abdomen MR imaging.[36]

Although the silica coated iron oxide nanoparticles are reasonably stable in aqueous media, they tend to aggregate in blood and therefore are inadequete for blood injection. To enhance colloidal stability of iron oxide nanoparticles, another type of coating agent, dextran or carbodextran, has been developed.[37] Since dextran possesses high colloidal stability against harsh physiological condition, dextran coated iron oxide nanoparticles can have good stability. Dextran coated iron oxide nanoparticles are prepared through a coprecipitation method from aqueous solution containing ferrous salt and dextran by adding an alkaline solution.[34] There are three representative dextran-coated iron oxide nanoparticles: AMI-25 (Feridex®(Berlex Lab.) and Endorem® (Guerbet, SHU 555A Resovist® Schering, and AMI-227 Combidex® Advanced Magnetics and Sinerem® Guerbet). AMI-25 is composed of ~5 nm iron oxide core and dextran coating materials. The total size is in between 80 and 150 nm with T2 relaxivity of ~98.3 l·mmol^{-1}s^{-1}.[38] SHU 555A has an ~4.2 nm iron oxide core coated with carbodextran with total size of ~62 nm, Resovist has a higher T2 relaxivity value of 151.0 l·mmol^{-1}s^{-1} [39] and has no known side effect after fast intravenous injection.[40] These magnetic contrast agents are generally trapped and accumulated by the reticuloendothelial cells in the liver with a short blood half life time of less than 10 min. and therefore used for liver imaging.[17] Compared to these two iron oxide contrast agents, AMI-227 has similar iron oxide core size (~5 nm) but a smaller overall size of 20-40 nm. Although AMI-227 has lower T2 contrast effects (T2 relaxivity of ~53 l·mmol^{-1}s^{-1}), its smaller size provides a much higher blood half life time of ~24 h, which enables MR angiography and lymph node detection.[41]

Smaller dextran-coated iron oxide nanoparticles including monocrystalline iron oxide (MION) and its derivative, cross-linked iron oxide (CLIO) are composed of an ~2.8 nm core iron oxide and dextran shell with the total size of 10-30 nm.[42-44] Since these nanoparticles are relatively small in size and have a long blood half life time and their surface can be readily linked with biologically active molecules, they are useful for in vivo molecular MR imaging of biological targets.[42]

Nanoparticles Grown in Organic/Biological Templates: Magnetoferritin, Magnetodendrimers, and Magnetoliposomes

Ferritin is a well-known iron storage protein used to sequester and store iron, which is comprised of a ~6 nm hydrated iron oxide, ferrihydrite ($5Fe_2O_3 \cdot 9H_2O$), core and polypeptide apoferritin shell.[45] Ferritin has been used as an efficient synthesizer for other magnetic materials.[46-48] For example, magnetically less useful ferrihydrite can be replaced by iron sulfide or magnetite nanoparticles (Fig. 3).[46] Researchers have also utilized magnetoferritin as contrast agents for MR imaging. Magnetoferritin possesses a reasonably high T2 relaxation value of 157 $l \cdot mmol^{-1}s^{-1}$.[49] Although magnetoferritin is expected to have high biocompatibility and colloidal stability in the blood when considering that they mimic naturally-occurring ferritin, the results are contradictory. These magnetoferritin particles are rapidly cleared from the blood circulation (blood half life time of less than 10 min) by the reticuloendothelial system in the liver, spleen, and lymph nodes.[49] Therefore, magnetoferritins are only suitable for liver, spleen, and lymph node detection rather than molecular imaging.

The unique pore structures and multiple functional end-groups of dendrimers make them useful as host materials in drug and gene delivery. Similarly, dendrimers can efficiently deliver magnetic nanoparticles to cells. Bulte, Frank, and coworkers have demonstrated carboxy-terminated dendrimer (G = 4.5) coated iron oxide contrast agents.[50,51] Typically, magnetodendrimers are synthesized through the pH controlled reaction of a ferrous salt and a trimethylamine oxide oxidant in a methanol/water mixture containing polyamidoamine dendrimers (Fig. 4a). The core size of the magnetodendrimers is 7-8 nm and they tend to aggregate to oligomers with the size of 20-30 nm (Fig. 4b). Magnetodendrimers show enhanced magnetic property (saturation magnetism: ~94 emu/g Fe) and a high T2 relaxivity of 200-406 $l \cdot mol^{-1}s^{-1}$,[50] compared to those of dextran coated MION. Since dendrimers can be efficiently transfected to cells without any transfection agents, these magnetodendrimers can be used as labels for cellular MR imaging and trafficking.[50]

Similarly, liposomes which are also widely used for drug and gene delivery can be good coating materials to solubilize iron oxide nanoparticles. Liposomes have bilayer assembly of surfactant molecules with a hydrophilic head and a hydrophobic tail. As shown in (Fig. 4c), the hydrophilic ends of the inner layer surfactants encapsulate the iron oxide nanoparticles and the hydrophilic head of the outer layer surfactant make them soluble in water. Bulte, Frank, and coworkers have reported that such magnetoliposome can be utilized for bone marrow MR contrast agents.[52] The iron oxide core size of the magnetoliposome is ~16 nm and the entire size is ~40 nm (Fig. 4d) with a T2 relaxivity of ~240 $l \cdot mmol^{-1}s^{-1}$.

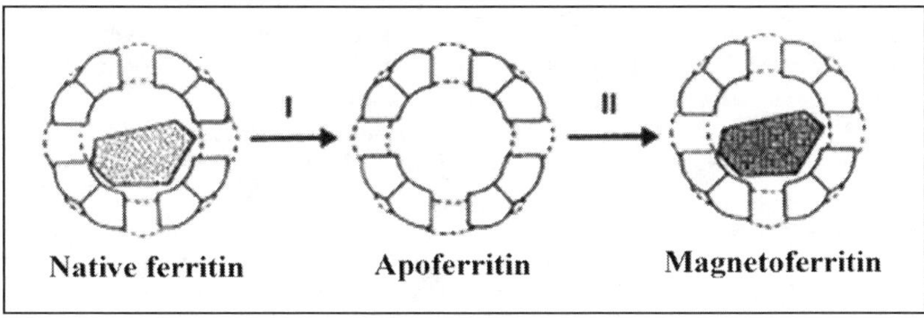

Figure 3. Synthetic scheme of magnetoferritin. Removal of ferrihydrite from native ferritin produces apoferritin and subsequent formation of magnetite nanoparticles inside apoferritin results in the formation of magnetoferritin. (From ref. 46, with permission).

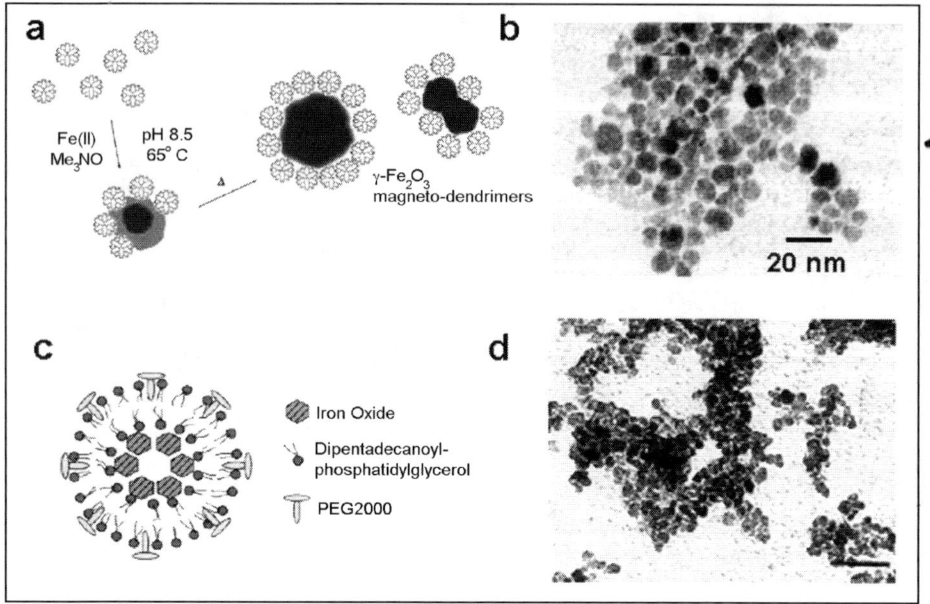

Figure 4. Schematics and TEM images of (a, b) magnetodendrimers and (c, d) magnetoliposomes. (From refs. 51, 52, with permission).

Phase Transfer Strategy: Nonhydrolytically Synthesized High Quality Iron Oxide Nanoparticles

With the exception of MION and CLIO, previously developed iron oxide MR contrast agents undergo rapid uptake by the reticuloendothelial systems (RES) and therefore are effective for liver, spleen, and lymph node detection. Therefore, researchers have encountered difficulties when they are utilized for molecular MR imaging.[16-20] For the success of molecular imaging, it is necessary to have high performance magnetic nanoparticle systems which exhibit excellent magnetic properties, the ability to escape from the RES, and the possession of active functionality that can be linked with biologically active molecules.[12] Since magnetic properties of nanoparticles depend highly on the materials properties such as size, shape, stoichiometry, and crystallinity,[15,53,54] the ability to control such properties is critical. However, conventional water-phase protocols, which have been widely adopted for superparamagnetic iron oxide (SPIO) contrast agents, generally lack precise size-control and monodispersity, have poor crystallinity, and have nonstoichiometric composition.[34] In contrast, nonhydrolytic high temperature growth methods allow one to have size-controllability, high single crystallinity, and good stoichiometry.[15,54,56,57] For example, the nanoparticle size can be easily controlled from 4 nm to ~20 nm with a very narrow size distribution (σ < 8 %) by controlling the growth conditions.[15,54] One difficulty that must be overcome prior to their utilization as MR contrast agents is obtaining water-solubility since nonhydrolytically synthesized iron oxide nanoparticles are soluble only in organic media. Various surface modification methods have been developed including bifunctional ligand,[15,33,58] micellular,[59,60] polymer,[61-63] and siloxane-linking procedures.[64,65] For example, nonhydrolytically synthesized nanoparticles can be transferred to aqueous media by overcoating the nanoparticles with polyethyleneglycol-(PEG)-ylated phospholipid micelles. Such a micellular coating strategy has been demonstrated in the case of quantum dots,[66] but Bao and coworkers have successfully extended this strategy to make water-soluble iron oxide nanoparticles (Fig. 5).[60] The PEGylated nanoparticles can be further linked to cellular transfection Tat peptides and utilized for MR cellular labeling.[60]

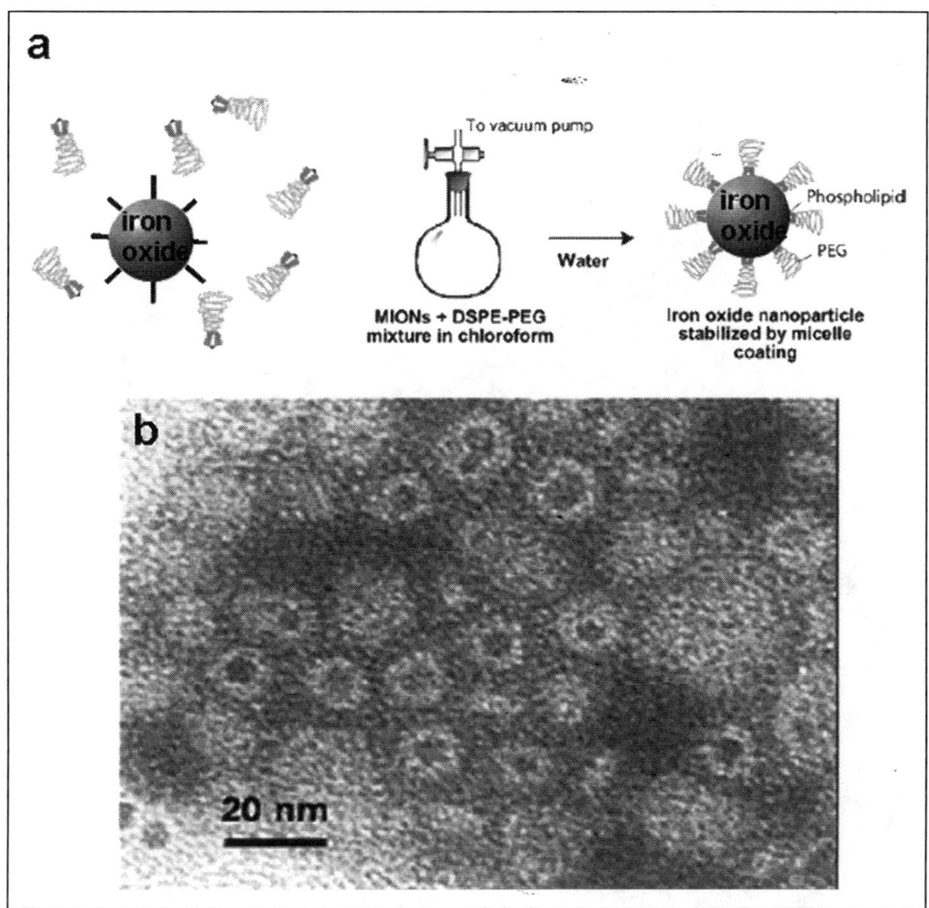

Figure 5. a) Synthetic scheme and (b) TEM image of polyethylene glycol (PEG)-ylated iron oxide nanoparticles. (From ref. 60, with permission).

Bawendi and coworkers have proposed another approach to transfer iron oxide nanoparticles from organic to aqueous media by coating them with polymeric phosphine oxide ligands.[67] These polymeric ligands can tightly bind to the iron oxide nanoparticle surface through multidentate bondings (Fig. 6).

It is well-known that the siloxane linkage to a metal oxide surface is efficient and strong. Zhang and coworkers have successfully applied this strategy for the synthesis of water-soluble iron oxide nanoparticles.[68] Refluxing toluene solution containing triethoxysilyl-terminated PEG ligands and nonhydrolytically synthesized iron oxide nanoparticles provides iron oxide nanoparticles with high colloidal stability in aqueous media (Fig. 7).

The major advantage of these nonhydrolytic synthesized iron oxide nanoparticles, as mentioned above, is the precise size control with high monodispersity. Cheon, Suh, and coworkers have demonstrated such advantages for the synthesis of iron oxide MR contrast agents.[15,33,58] As shown in transmission electron microscopic image (Fig. 8), nanoparticles obtained are ~9 nm with narrow size distribution ($\sigma < 8$ %). HR-TEM and X-ray analyses show that nanoparticles are single crystalline stoichiometric Fe_3O_4. Then, water-soluble iron oxide nanoparticles (WSIO) are obtained by introducing 2,3-dimercaptosuccinic acid (DMSA) ligand onto the nanoparticle

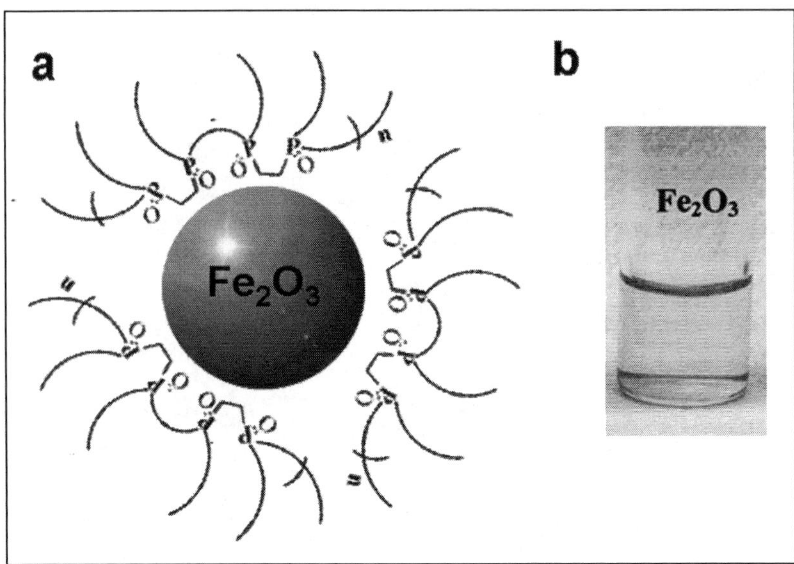

Figure 6. Multi-dentate phosphine oxide ligand approach for the synthesis of iron oxide nanoparticles. a) The phosphine oxide functional groups bind to the surface of iron oxide and exposed PEG groups to make them water-soluble. b) Iron oxide nanoparticles dissolved in water. (From ref. 67, with permission).

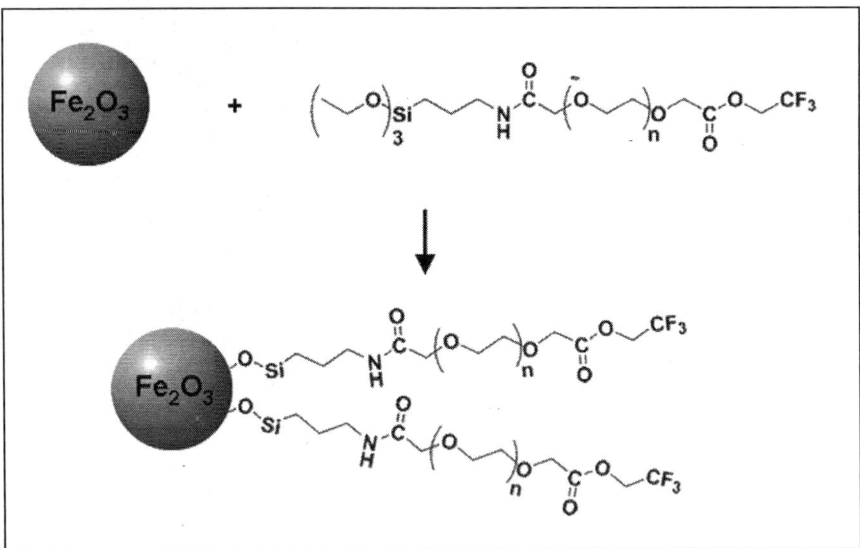

Figure 7. Siloxane-polyethyleneglycol (PEG) coated iron oxide nanoparticles. Silanization of terminal ethoxysilane group of PEG ligand on top of iron oxide nanoparticles induces the formation of PEG coated iron oxide nanoparticles. (From ref. 68, with permission)

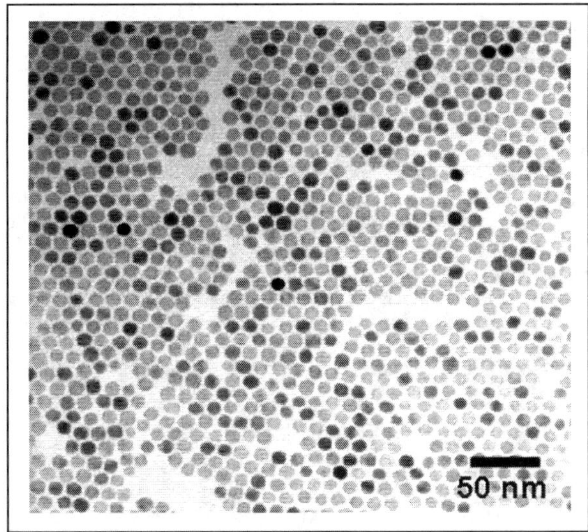

Figure 8. TEM image of ~9 nm Fe_3O_4 nanoparticles.

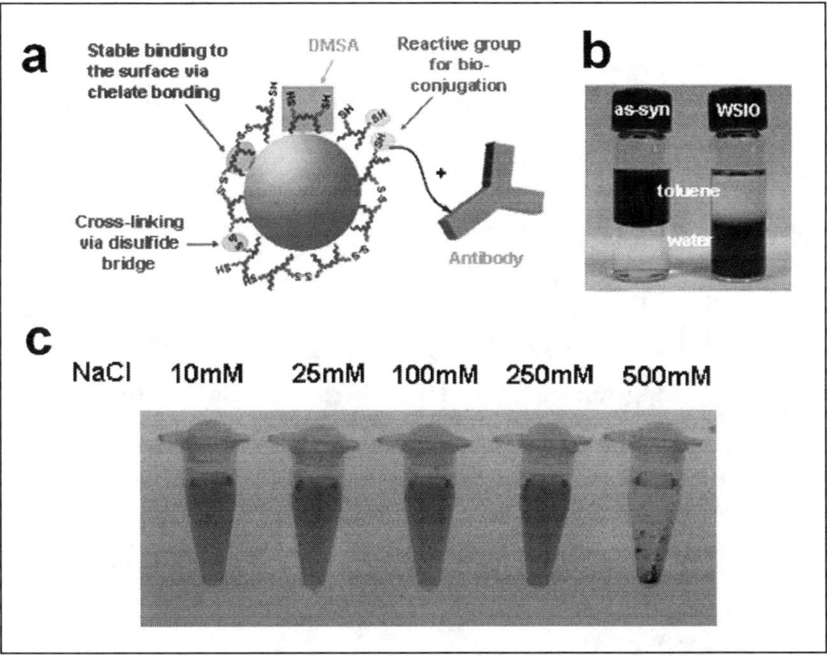

Figure 9. a) Schematic of 2,3-dimercaptosuccinic acid (DMSA)-coated iron oxide nanoparticles. The carboxylic ends of DMSA bind to the surface iron oxide nanoparticles and they are further stabilized through interligand disulfide cross-linkages. Remaining free thiol can be used for further conjugation for biomolecules such as antibody. b) Solubility test of as-synthesized and DMSA-coated iron oxide nanoparticles. c) Stability test of water-soluble iron oxide nanoparticles in various concentrations of NaCl solution. (From ref. 15, with permission).

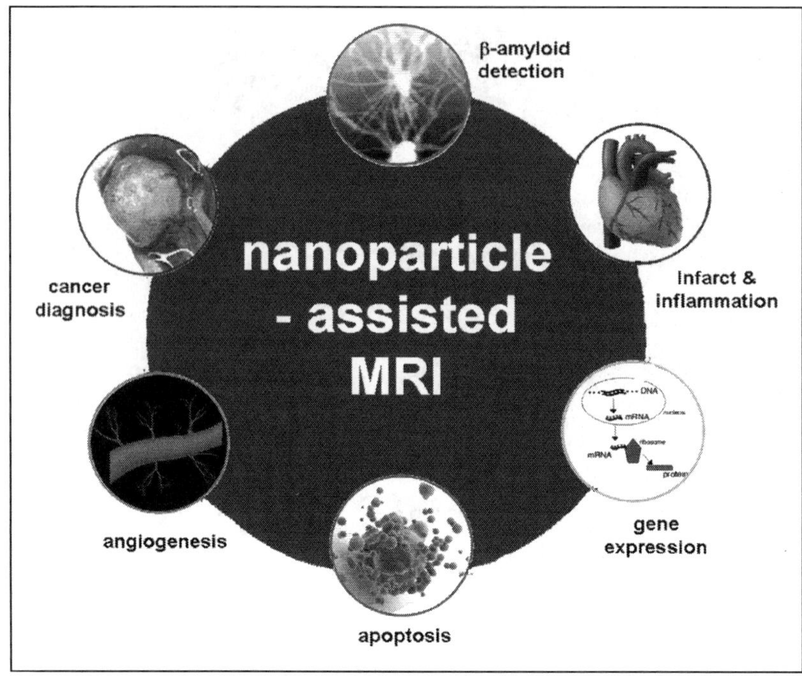

Figure 10. Nanoparticle assisted molecular MR imaging of biological systems.

surface. This ligand endows them high water-phase stability through (i) carboxylate chelate bonding to iron and (ii) disulfide cross-linkages between the ligands (Fig. 9a).[15] Furthermore, the remaining free thiol group of the ligand can be used for the attachment of target-specific biomolecules. Obtained Fe_3O_4 nanoparticles with the DMSA ligand are fairly stable in water and phosphate-buffered saline (PBS) up to NaCl concentration of 250 mM without any aggregation. These nanoparticles are utilized as MR probes, upon conjugation with cancer targeting antibody, not only for the detection of in vitro detection of cancer cells but also in vivo imaging of cancer implanted in a mouse.[15,33]

Molecular MR Imaging Utilizing Iron Oxide Nanoparticle Probes

When iron oxide nanoparticles are conjugated with biologically active materials (e.g., antibody), the resulting iron oxide-biomolecule conjugates possess dual-functionalities of both the MR contrast enhancers and the molecular recognition capability. These conjugates act as molecular imaging probes which can efficiently report on various molecular/biological events occurring in region-of-interest targets. Molecular MR imaging studies utilizing such iron oxide-biomolecule conjugates include the imaging of inflammation,[69,70] infarct,[23-25] angiogenesis,[26] apoptosis,[27] gene expression,[28,29] β-amyloid plaques,[71] and cancer.[15,30-33]

Infarct and Inflammation

For imaging of infarcts and inflammations, monocrystalline iron oxide (MION) nanoparticles are conjugated with specific antibodies through electrostatic interactions or covalent linkages by the reaction of potassium periodate-activated surface hydroxyl group with lysine residues of antibodies. For example, $R_{11}D_{10}$ antimyosin Fab was electrostatically conjugated to hydroxyl groups on the MION surfaces for cardiac infarct imaging.[24] Since infarcted cardiac cells have increased permeability to myocardial cell membranes compared with normal cells, iron oxide-antimyosin Fab conjugates can be efficiently transported into damaged cells and recognize

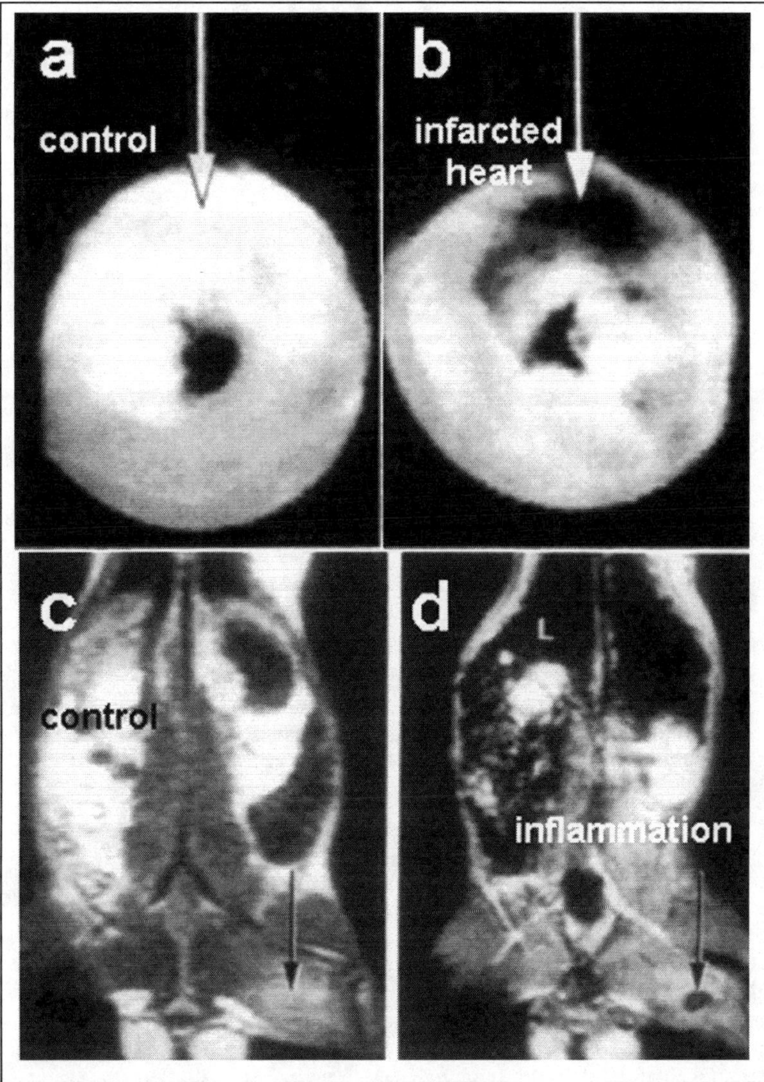

Figure 11. MR detection of (a,b) cardiac infarct and (c,d) inflammation of a mouse. After intravenous injection of iron oxide-antimyosin antibody, dark MR contrast in the infarcted area is observed. Similarly, inflammation is not imaged at the control mouse but inflammation site is clearly shown as dark contrast. (From refs. 24 and 69, with permission).

myosin. Figure 11 shows T2-weighted MR images of a mouse with cardiac infarct after injection of MION-$R_{11}D_{10}$ antimyosin Fab conjugates. The infarcted region is clearly observed as dark MR images, while no contrast effect was seen when unconjugated MIONs were administered. Such targeting effect of MION-$R_{11}D_{10}$ antimyosin Fab conjugates were evaluated through ex vivo immunohistological analyses through Prussian blue staining. Weissleder and coworkers further extended this strategy for the detection of inflammation by conjugating MIONs with polyclonal human immunoglobulin G. MION-IgG conjugates consistently detected the area of inflammation in T2-weighted spin-echo MR images (Fig. 11c,d), which was also further confirmed by histological Prussian blue staining study.[69]

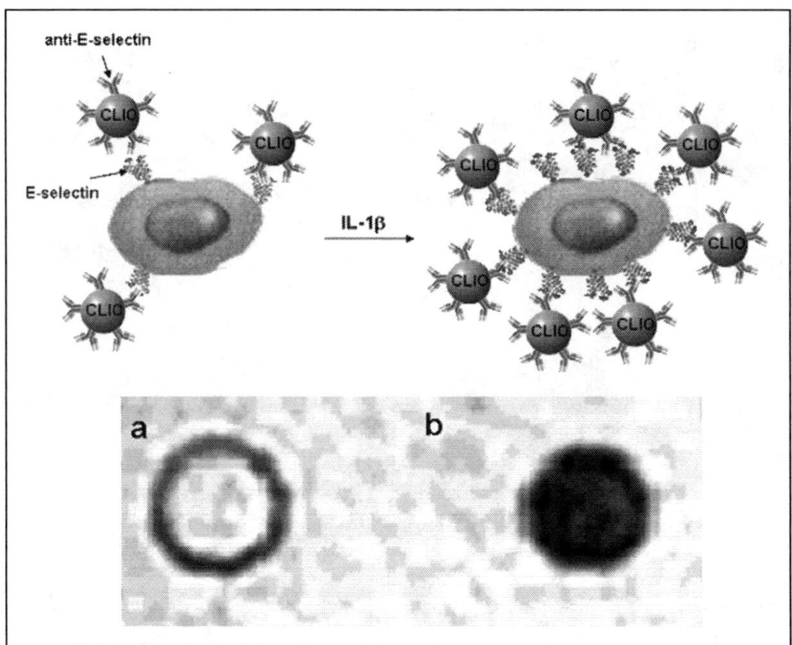

Figure 12. In vitro MR detection of E-selectin stimulated by interleukin-1β. a) Without interleukin-1β, white MR image is obtained from HUVEC only treated with CLIO-anti E-selectin antibody. b) In contrast, after interleukin-1β and CLIO-anti E-selectin is sequentially dosed to HUVEC, E-selectin expression is clearly imaged as dark MR image. (From ref. 26, with permission).

Angiogenesis

Angiogenesis is a fundamental growth process of new blood vessels for development, reproduction, and wound repair. This process is also related to the progression of tumor growth. Therefore, the development of anti-angiogenic agents can be a potential pathway to efficient cancer treatment. Imaging of angiogenesis is also related to the cancer diagnosis and the evaluation of anti-cancer agents. Several molecular markers are involved in angiogenesis: vascular endothelial growth factor (VEGF), fibroblast growth factors (FGG), platelet-derived endothelial cell growth factor (PD-EDGF), Tie-2 receptor, integrin, and E-selectin.[72] Among the various angiogenesis markers, VEGF, Tie 2 receptor, and integrin have been extensively studied.[73-75] A. Bogdanov and coworkers reported that E-selectin expression in human endothelial cells can be imaged by using cross-linked iron oxide (CLIO) - monoclonal anti-human E-selectin antibody conjugate MR contrast agent.[26] When only CLIO-antibody conjugates are treated to endothelial cells with a low E-selectin expression level, MR contrast effect is hardly detected (Fig. 12a). In contrast, interleukin-1β, which stimulates the expression of E-selectin, is treated to the cells and then nanoparticle-antibody is dosed to the cells, a significant MR contrast effect is shown (Fig. 12b).

Apoptosis

Apoptosis is an active process of programmed self-destruction of cells. In the early stages of apoptosis, the redistribution of phosphatidylserine in the cell membrane occurs and the detection of such processes can be an indicator of the programmed cell death. Representative binding proteins to phosphatidylserine are annexin V and synaptotagmin I. Although imaging of apoptosis using these antibodies has been already performed through radio-isotope techniques,[76] the spatial resolution is only ~1-3 mm and needs improvement. Brindle and coworkers have

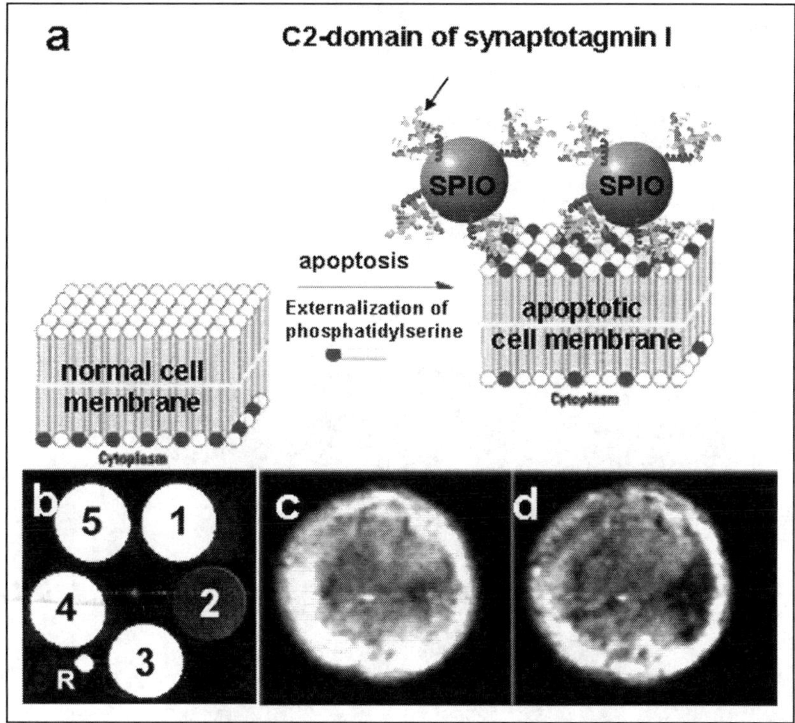

Figure 13. a) MR Imaging of apoptosis using SPIO-C2 domain of synaptotagmin I. b) T2-weighted MR images of (b 1) water, (b 2) SPIO-C2 conjugate, (b 3) SPIO-BSA control conjugate, (b 4) SPIO only treated apoptotic cells, and (b 5) SPIO-C2 conjugate treated normal cells. In vivo MR images of tumors implanted in a mouse (c) before and (d) after drug treatment. (From ref. 27, with permission).

shown that conjugates of the SPIO and C2 domain of synaptotagmin I (C2-SPIO) can detect apoptotic cells through MRI with ~0.1 mm resolution.[27] While there are no MR signals for nonspecific SPIO (BSA-SPIO) (Fig. 13 b3), SPIO only treated apoptotic cells (Fig. 13 b4), and C2-SPIO treated normal cells (Fig. 13 b5), C2-SPIO treated apoptotic cells (Fig. 13 b2) clearly show dark MR contrast with a significant change in T2 values ($\Delta T2$ = ~90%). Further extension of this strategy to an in vivo animal study was also successful. When C2-SPIO was intravenously injected to drug-treated tumor-bearing mice, the nanoparticle conjugates are able to detect apoptotic regions with a significant MR signal change (Figs. 13, c,d).

Gene Expression

Gene expression is also an emerging field in the biomedical sciences and the imaging of such gene expression processes is of importance. Although several approaches to detect in vivo gene expression have been performed through optical[77,78] and radioisotope imaging techniques,[79] there are limitations (i.e.: low-penetration depth of light for optical imaging and poor spatial resolution of radioisotope imaging). Weissleder and coworkers demonstrated that MR detection of transgene expression of engineered transferrin receptor (ETR) in tumors is possible by using MION-transferrin (MION-Tf) contrast agents.[29] When MION-Tf is treated to the cells with various ETR expression levels, a gradual decrease in T2 is observed as the ETR expression levels of cells are increased due to proportional binding of MION-Tf conjugates to the expressed ETR. They also determined whether ETR expression can be detected in in vivo

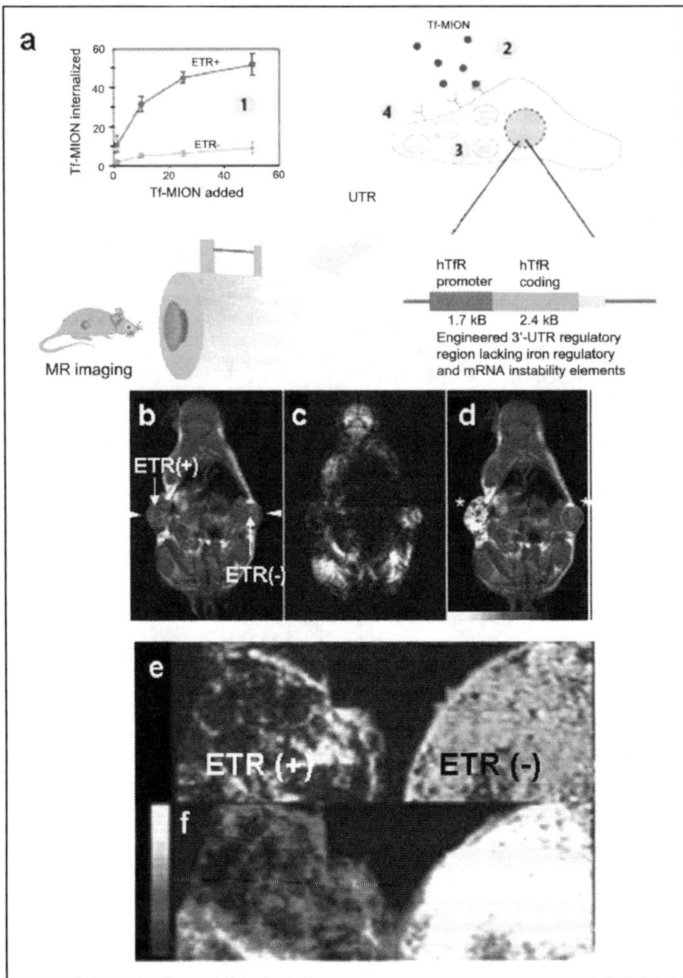

Figure 14. In vivo and ex vivo imaging of engineered transferrin receptor (ETR) expression in tumors. (a) Schematic of MR imaging, (b,c) in vivo imaging of ETR (+) and ETR (-) tumor implanted mouse before (b) and after (c,d) the injection of MION-transferrin. Ex vivo imaging of excised ETR(+) and ETR(-) tumors (e) and their color maps based on T2 (f). (From ref. 29, with permission).

live mice with ETR positive tumors and ETR negative tumors. The results show that the MR contrast effect is only observed for ETR positive tumors (Fig. 14b-d). Ex vivo MR imaging of excised tumors shows more dramatic differences between these two tumors (Fig. 14e,f).

Cancer Imaging

Noninvasive detection of cancer in its early stages is of great interest since early detection of cancer can significantly increase the survival rate of patients. With conventional MRI technique, current detectable size of cancer is roughly ~1 cm^3. If nanoparticle contrast agents can specifically recognize cancer cells through molecular interaction, selective enhancement of the MR signal of cancer cells can provide one with a way to clearly distinguish cancer from normal tissues. Tiefenauer and coworkers reported the detection of cancers is possible through such molecular recognition of superparamagnetic iron oxide (SPIO) nanoparticle-antibody conjugates.[32] Conjugation of poly(glutamic acid-lysine-tyrosine) coated nanoparticles with anti

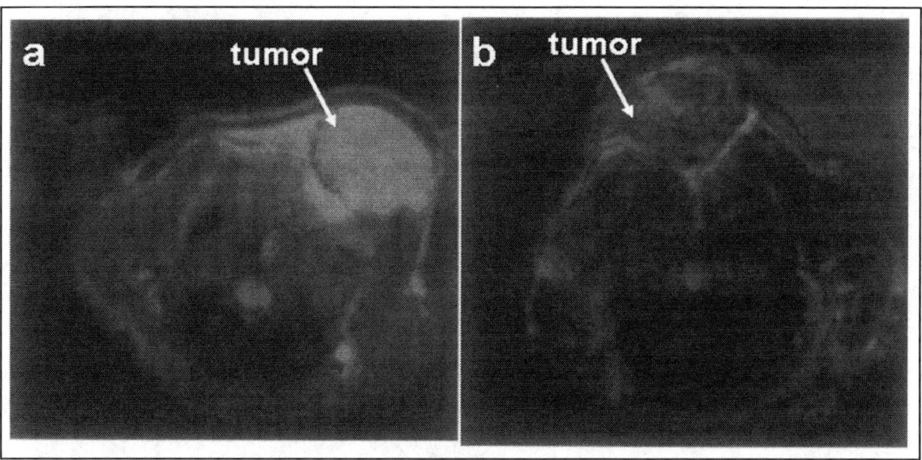

Figure 15. MR detection of carcinoembryonic antigen (CEA) overexpressed tumors by using iron oxide-CEA antibody conjugates. MR images of before (a) and after (b) injection of SPIO-CEA antibody. (From ref. 32, with permission).

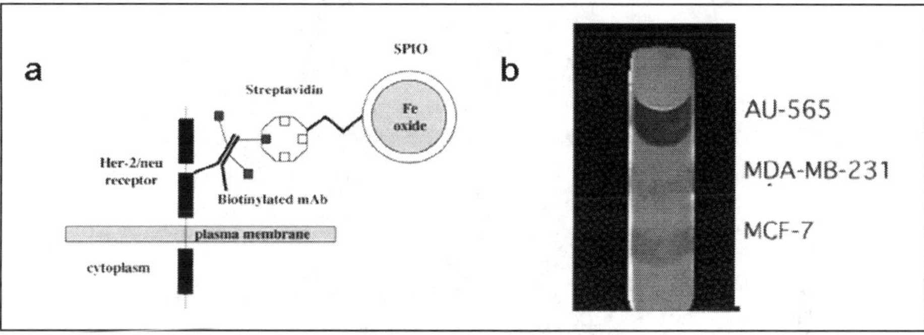

Figure 16. In vitro MR detection of HER2/neu overexpressed cancer cells by using avidin coated SPIO and biotinylated Herceptin. (a) Schematic and (b) MR images of SPIO-avidin conjugate treated cells (AU-565, MDA-MB-231, MCF-7). (From ref. 31, with permission).

carcinoembryonic antigen (CEA) antibody is performed through a conventional sulfo-MBS cross-linking method. In the T2-weighted MR images, dark contrast is imaged at the CEA-expressed tumors, although contrast difference is not highly pronounced (Fig. 15). Artemov and coworkers utilized another approach to detect cancer cells,[31] using the avidin-biotin recognition. Avidin conjugated SPIO can efficiently detect biotinylated cancer specific antibodies which bind to the cancer cells (Fig. 16a). Their in vitro fluorescence-assisted cell sorting analyses and MR imaging confirm cancer detection (Fig. 16b). Au-565 cells with high expression of HER2/neu cancer markers are imaged as dark MR contrast, while no contrast effect is obtained from MDA-MB-231 cells with low HER2/neu expression.

Recently, Cheon, Suh, and coworkers have shown that highly efficient cancer targeting can be achieved using high quality, small sized water-soluble iron oxide (WSIO) nanoparticle-antibody conjugates.[33] The WSIO nanoparticles have high magnetic momentum (~100 emu/g Fe) and small hydrodynamic size (~9 nm), which are advantageous for both in vitro and in vivo cancer imaging. When these nanoparticles are conjugated with Herceptin, they successfully detect cancer cells (SK-BR-3) as dark MR image (Fig. 17b) through molecular

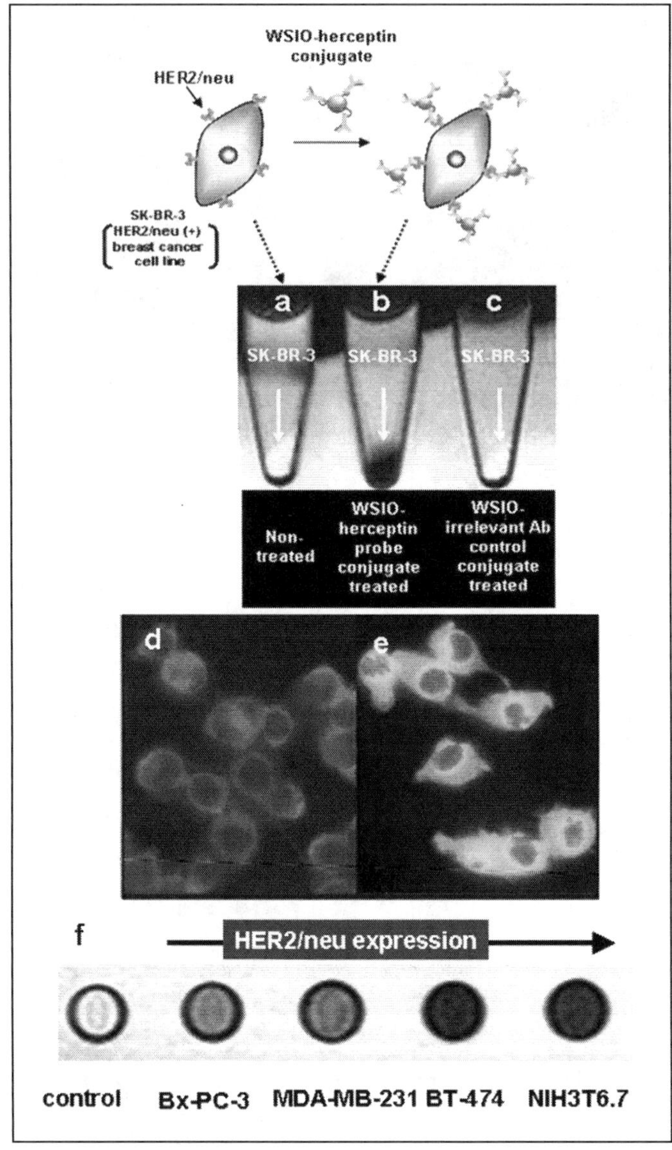

Figure 17. In vitro cancer detection using water-soluble iron oxide (WSIO)-Herceptin conjugates. MR images of (a) nontreated, (b) WSIO-Herceptin treated, (c) WSIO-irrelevant antibody treated breast cancer cells (SK-BR-3). (d) MR images of WSIO-Herceptin conjugate treated cell lines with increasing expression levels of HER2/*neu* receptors: Bx-PC-3, MDA-MB-231, BT-474, and NIH3T6.7 cell lines. Control conjugates are treated to Bx-PC-3 cell lines. (From ref. 33, with permission).

interaction between nanoparticle surface-bound Herceptin and HER2/neu cancer markers, compared to nontreated (Fig. 17a) and WSIO-irrelevant conjugate treated cells (Fig. 17c). This MRI result is also confirmed by an optical technique where vivid green fluorescence from the fluorescein (FITC) is clearly observed only for FITC-WSIO-Herceptin probe conjugate

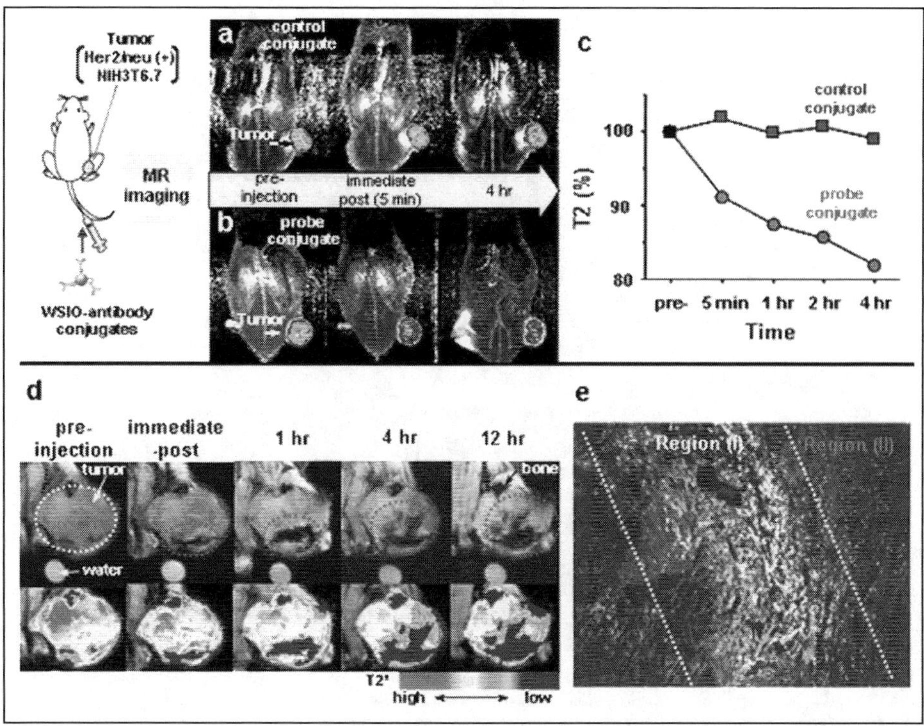

Figure 18. In vivo MRI of cancer targeting events of WSIO-Herceptin conjugates. Color maps of T2-weighted MR images of cancer cell implanted (NIH3T6.7) mice at different temporal points (preinjection, immediate post, 4 hr) after the intravenous injection of (a) WSIO-irrelevant antibody control conjugates and (b) WSIO-Herceptin probe conjugates. c) Plot of T2 values versus time after the injection of WSIO-antibody conjugates in (a) and (b) samples. d) T2*-weighted MR images of cancer cell implanted (NIH3T6.7) mouse at 9.4 Tesla and their color maps at different temporal points after probe conjugate injection. Tumor area is circled with white dotted lines. e) Fluorescence immunohistochemical analyses of an excised tumor slice. Endothelial vessels were stained with Rhodamine-anti-CD31 (red fluorescence) and probe conjugates were stained with FITC-anti-human IgG (green fluorescence). (From ref. 33, with permission). Color version available at Eurekah.com.

treated cells (Figs 17d, e). Furthermore, WSIO nanoparticles enable the detection of various cell lines with different levels of HER2/neu cancer marker expression: Bx-PC-3, MDA-MB-231, BT-474, and NIH3T6.7 cell lines, which are arranged in the order of increasing HER2/*neu* expression level. T2-weighted MR signals of the cell lines treated with WSIO-Herceptin probe conjugates become darker as the expression level of the HER2/*neu* receptors is increased (Fig. 17f). It is noteworthy that the MR signal intensities of the cell lines treated with WSIO-Herceptin probe conjugates show a marked difference from that of control conjugates indicating excellent specific binding efficiency of the probe conjugates.

These magnetic probes are successfully extended to the in vivo detection of cancer cells implanted in mouse.[33] When these WSIO-Herceptin conjugates are intravenously injected to a mouse, they successfully reach and recognize HER2/neu receptors overexpressed from cancer cells which results in a significant MR contrast effect in the tumor sites with the ~20 % decrease in T2 value compared to the control experiments (Fig. 18a-c). In high resolution MR images of WSIO-Herceptin conjugate treated mouse measured at 9.4 T MRI, a dark MR image initially appears near the bottom region of the tumor and then gradually grows and

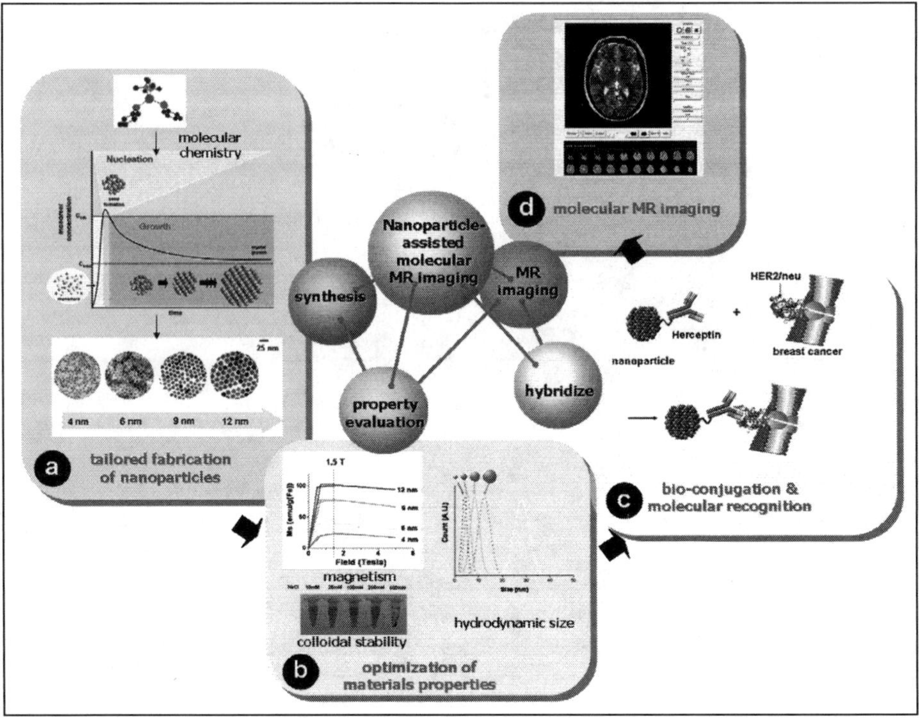

Figure 19. Nanoparticle-assisted molecular MR imaging. a) Modern molecular chemistry approach enables the tailored synthesis of high quality magnetic nanoparticles with desired sizes and monodispersity. b) Then, evaluation and optimization of materials properties such as magnetism, hydrodynamic size, and colloidal and biostability are important. c) When these nanoparticles are conjugated to biomolecules, the resulting nanoparticle-biomolecule conjugates possess the capabilities of both MR contrast effects and molecular recognition of target biosystems. d) This will enable molecular MR imaging which can report various biological events such as pinpointing cancer diagnosis, cell migration, cell signaling, and genetic developments, with high sensitivity and specificity.

spreads to the central and upper region of the tumor as time elapses (Fig. 18d). They found that such time-dependent MR signal change is revealing the heterogeneous pattern of the intratumoral vasculatures, where the bottom side of the tumor has well-developed vascular structures.

Outlook

Although there has been much progress in the development of magnetic nanoparticle contrast agents for molecular MR imaging in the past few years, their successful utilization is limited to in vitro systems except for a few in vivo cases. The main difficulties lie in their poor MR contrast effects and limited stability under in vivo condition. The MR signal enhancing effect of conventional iron oxide-based nanoparticles is unsatisfactory compared to other diagnostic tools such as fluorescence and PET and needs to be improved. Therefore, it is important to develop new types of magnetic nanoparticle contrast agents which can significantly improve contrast effects. Since nanoparticles with higher magnetization values provide stronger MR contrast effects, the development of novel nanoparticles with superior magnetism is the first prerequisite. Concurrently, since the MR contrast effects of nanoparticles are strongly correlated

Table 1. Various silica- or dextran- coated iron oxide contrast agents

Agent	*AMI-121[36] Lumiren, Gastromark	*AMI-25[38] Feridex Endorem	*SHU 555A[39] Resovist	*AMI-227[41] Combidex Sinerem	MION[42-44], CLIO
Iron oxide core size	~10 nm	5~6 nm	~4.2 nm	4~6 nm	~2.8 nm
Total size	~300 nm	80~150 nm	~62 nm	20~40 nm	10~30 nm
Coating material	silica	dextran	carbodextran	dextran	dextran
Magnetization	N/A	78 emu/g	N/A	69.8 emu/g	60~68 emu/g
T2 relaxitivity	72 l·mol^{-1}s^{-1}	98 l·mol^{-1}s^{-1}	151 l·mol^{-1}s^{-1}	53 l·mol^{-1}s^{-1}	~69 l·mol^{-1}s^{-1}
Blood half life time	<5 mm	~ 6 mm	3 mm	>24 hr	10 hr

*commercialized

to materials characteristics in terms of their size, shape, composition, single crystallinity, and magnetism, it is important to have a good nanoparticle model system which can clearly describe the relationship between nanoscale materials characteristics and MR contrast effects (Fig. 19a).

The next required step is to impart high colloidal stability and biocompatibility to the magnetic nanoparticles. As described in previous sections, various coating materials have been developed for such applications, but it is still necessary to develop general and more reliable protocols for tailoring nanoparticle surfaces with desired coating materials. Since a smaller overall size is advantageous for escaping the reticuloendothelial system, the coating materials should be as small as possible while possessing high colloidal stability without any aggregation under physiological conditions (Fig. 19b).

The toxicity of magnetic nanoparticles is also a very important issue that needs to be resolved prior to clinical utilization. Although iron oxide nanoparticles have been regarded as clinically benign materials, potential cytotoxicity arising from their size, shape, and coating materials should also be examined along with systematic guidelines for the nano-toxicity evaluation of newly developed novel nanoparticles.

Once novel nanoparticles with highly enhanced magnetism, small size, and high colloidal and bio-stability are developed, significant improvements in MR detection sensitivity and target specificity are expected. (Figs. 19 c,d) This can bring huge advances in current cancer diagnosis and biomedical imaging fields. For example, highly enhanced MR contrast of a biological target through the molecular recognition of such nanoparticle contrast agents will promise in vivo diagnosis of early staged cancer with sub-millimeter dimension. In addition, many unrevealed biological processes such as in vivo pathways of cell evolutions, cell differentiations, cell-to-cell interactions, molecular signaling pathways can be precisely monitored at the molecular level by using next-generation nanoparticle contrast agents.

Acknowledgement

We would like to thank J.-M. Oh (KBSI) for TEM, Dr. H. C. Kim (KBSI) for SQUID, and Dr. Y. J. Kim for HVEM (JEM-ARM1300S) analyses. This work is supported in part by National Research Laboratory (M10600000255), NCI center for Cancer Nanotechnology Excellence, NCRC (R15-2004-024-02002-0), and 2nd stage of BK21.

References

1. Belcher AM, Wu XH, Christensen RJ et al. Control of crystal phase switching and orientation by soluble mollusc-shell proteins. Nature 1996; 381:56-58.
2. Aizenberg J, Tkachenko A, Weiner S et al. Calcitic microlenses as part of the photoreceptor system in brittlestars. Nature 2001; 412:819-822.
3. Bazylinski DA, Frankel RB. Magnetosome formation in prokaryotes. Nat Rev Microbiol 2004; 2:217-230.
4. Weissleder R, Mahmood U. Molecular imaging. Radiology 2001; 219:316-333.
5. Michalet X, Pinaud FF, Bentolila LA et al. Quantum dots for live cells, in vivo imaging, and diagnostics. Science 2005; 307:538-544.
6. Bruchez Jr M, Moronne M, Gin P et al. Semiconductor nanocrystals as fluorescent biological labels. Science 1998; 281:2013-2016.
7. Chan WCW, Nie S. Quantum dot bioconjugates for ultrasensitive nonisotopic detection. Science 1998; 281:2016-2018.
8. LaConte L, Nitin N, Bao G. Magnetic nanoparticle probes. Nanotoday 2005; 8:32-38.
9. Wu X, Liu H, Liu J et al. Immunofluorescent labeling of cancer marker Her2 and other cellular targets with semiconductor quantum dots. Nat Biotechnol 2003; 21:41-46.
10. Mitchell DG. MRI Principles. 1st ed. Philadelphia: W. B. Saunders Company, 1999.
11. Massoud TF, Gambhir SS. Molecular imaging in living subjects: Seeing fundamental biological processes in a new light. Gene Dev 2003; 17:545-580.
12. Bulte JWM, Kraitchman DL. Iron oxide MR contrast agents for molecular and cellular imaging. NMR Biomed 2004; 17:484-499.
13. Wang YXJ, Hussain SM, Krestin GP. Superparamagnetic iron oxide contrast agents: Physicochemical characteristics and applications in MR imaging. Eur Radiol 2001; 11:2319-2331.
14. Koenig SH, Kellar KE. Theory of 1/T1 and 1/T2 NMRD profiles of solutions of magnetic nanoparticles. Magn Reson Med 1995; 34:227-233.
15. Jun YW, Huh YM, Choi JS et al. Nanoscale size effect of magnetic nanocrystals and their utilization for cancer diagnosis via magnetic resonance imaging. J Am Chem Soc 2005; 127:5732-5733.
16. Hamm B, Staks T, Taupitz M et al. Contrast-enhanced MR imaging of liver and spleen: First experience in humans with a new superparamagnetic iron oxide. J Magn Reson Imaging 1994; 4:659-668.
17. Reimer P, Tombach B. Hepatic MRI with SPIO: Detection and characterization of focal liver lesions. Eur Radiol 1998; 8:1198-1204.
18. McLachlan SJ, Morris MR, Lucas MA et al. J Magn Reson Imaging 1994; 4:301-307.
19. Weissleder R. Radiology 1994; 193:593-595.
20. Bengele HH, Palmacci S, Rogers J et al. Magn Reson Imaging 1994; 12:433-442.
21. Suwa T, Ozawa S, Ueda M et al. Magnetic resonance imaging of esophageal squamous cell carcinoma using magnetite particles coated with anti-epidermal growth factor receptor antibody. Int J Cancer 1998; 75:626-634.
22. Kresse M, Wagner S, Pfefferer D et al. Targeting of ultrasmall superparamagnetic iron oxide (USPIO) particles to tumor cells in vivo by using transferrin receptor pathways. Magn Reson Med 1998; 40:236-242.
23. Krieg FM, Andres RY, Winterhalter KH. Superparamagnetically labelled neutrophils as potential abscess-specific contrast agent for MRI. Magn Reson Imaging 1995; 13:393-400.
24. Weissleder R, Lee AS, Khaw BA et al. Antimyosin-labeled monocrystalline iron oxide allows detection of myocardial infarct: MR antibody imaging. Radiology 1992; 182:381-385.
25. Kraitchman DL, Heldman AW, Atalar E et al. In vivo magnetic resonance imaging of mesenchymal stem cells in myocardial infarction. Circulation 2003; 107:2290-2293.
26. Kang HW, Josephson L, Petrovsky A et al. Magnetic resonance imaging of inducible E-selectin expression in human endothelial cell culture. Bioconj Chem 2002; 13:122-127.
27. Zhao M, Beauregard DA, Loizou L et al. Noninvasive detection of apoptosis using magnetic resonance imaging and a targeted contrast agent. Nat Med 2001; 7:1241-1244.
28. Högemann D, Josephson L, Weissleder R et al. Improvement of MRI probes to allow efficient detection of gene expression. Bioconj Chem 2000; 11:941-946.
29. Weissleder R, Moore A, Mahmood U et al. In vivo magnetic resonance imaging of transgene expression. Nat Med 2000; 6:351-355.
30. Tiefenauer LX, Kühne G, Andres RY. Antibody-magnetite nanoparticles: In vitro characterization of a potential tumor-specific contrast agent for magnetic resonance imaging. Bioconj Chem 1993; 4:347-352.

31. Artemov D, Mori N, Okollie B et al. MR molecular imaging of the Her-2/neu receptor in breast cancer cells using targeted iron oxide nanoparticles. Magn Reson Med 2003; 49:403-408.
32. Tiefenauer LX, Tschirky A, Iwhne G et al. In vivo evaluation of magnetite nanoparticles for use as a tumor contrast agent in MRI. Magn Reson Imaging 1996; 14:391-402.
33. Huh YM, Jun YW, Song HT et al. In vivo magnetic resonance detection of cancer by using multifunctional magnetic nanocrystals. J Am Chem Soc 2005; 127:12387-12391.
34. Sjögren CE, Johansson C, Naevestad A et al. Crystal size and properties of superparamagnetic iron oxide (SPIO) particles. Magn Reson Imaging 1997; 15:55-67.
35. Hahn PF, Stark DD, Lewis JM et al. First clinical trial of a new superparamagnetic iron oxide for use as an oral gastrointestinal contrast agent in MR imaging. Radiology 1990; 175:695-700.
36. Bach-Gansmo T. Ferrimagnetic susceptibility contrast agents. Acta Radiol 1993; 387:1-30.
37. Weissleder R, Elizondo G, Josephson L et al. Experimental lymph node metastases: Enhanced detection with MR lymphography. Radiology 1989; 171:835-839.
38. Weissleder R, Stark DD, Engelstad BL et al. Superparamagnetic iron oxide: Pharmacokinetics and toxicity. Am J Roentgenol 1989; 152:167-173.
39. Reimer P, Rummeny EJ, Daldrup HE et al. Clinical results with Resovist: A phase 2 clinical trial. Radiology 1995; 195:489-496.
40. Grubnic S, Padhani AR, Revell PB et al. Comparative efficacy of and sequence choice for two oral contrast agents used during MR imaging. Am J Roentgenol 1999; 173:173-178.
41. Bartolozzi C, Lencioni R, Donati F et al. Abdominal MR: Liver and pancreas. Eur Radiol 1999; 9:1496-1512.
42. Moore A, Marecos E, Bogdanov A et al. Tumoral distribution of long-circulating dextran-coated iron oxide nanoparticles in a rodent model. Radiology 2000; 214:568-574.
43. Bengele HH, Palmacci S, Rogers J et al. Biodistribution of an ultrasmall superparamagnetic iron oxide colloid, BMS 180549, by different routes of administration. Magn Reson Imaging 1994; 12:433-442.
44. Shen T, Weissleder R, Papisov M et al. Monocrystalline iron oxide nanocompounds (MION): Physicochemical properties. Magn Reson Med 1993; 29:599-604.
45. Ford GC, Harrison PM, Rice DW et al. Ferritin: Design and formation of an iron-storage molecule. Philos Trans R London Ser B 1984; 304:551-565.
46. Meldrum FC, Heywood BR, Mann S. Magnetoferritin: In vitro synthesis of a novel magnetic protein. Science 1992; 257:522-523.
47. Bulte JWM, Douglas T, Mann S et al. Magnetoferritin: Characterization of a novel superparamagnetic MR contrast agent. J Magn Reson Imaging 1994; 4:497-505.
48. Meldrum FC, Wade VJ, Nimmo DL et al. Synthesis of inorganic nanophase materials in supramolecular protein cages. Nature 1991; 349:684-687.
49. Bulte JWM, Douglas T, Mann S et al. Initial assessment of magnetoferritin biokinetics and proton relaxation enhancement in rats. Acad Radiol 1995; 2:871-878.
50. Bulte JWM, Douglas T, Witwer B et al. Magnetodendrimers allow endosomal magnetic labeling and in vivo tracking of stem cells. Nat Biotech 2001; 19:1141-1147.
51. Strable E, Bulte JWM, Moskowitz B et al. Synthesis and characterization of soluble iron oxide-dendrimer composites. Chem Mater 2001; 13:2201-2209.
52. Bulte JWM, De Cuyper M, Despres D et al. Short- vs. long-circulating magnetoliposomes as bone marrow-seeking MR contrast agents. J Magn Reson Imaging 1999; 9:329-335.
53. Cheon J, Kang NJ, Lee SM et al. Shape evolution of single-crystalline iron oxide nanocrystals. J Am Chem Soc 2004; 126:1950-1951.
54. Sun S, Zeng H. Size-controlled synthesis of magnetite nanoparticles. J Am Chem Soc 2002; 124:8204-8205.
55. Redl FX, Black CT, Papaefthymiou GC et al. Magnetic, electronic, and structural characterization of nonstoichiometric iron oxides at the nanoscale. J Am Chem Soc 2004; 126:14583-14599.
56. Park J, An K, Hwang Y et al. Ultra-large-scale syntheses of monodisperse nanocrystals. Nat Mater 2004; 3:891-895.
57. Jana NR, Chen Y, Peng X. Size- and shape-controlled magnetic (Cr, Mn, Fe, Co, Ni) oxide nanocrystals via a simple and general approach. Chem Mater 2004; 16:3931-3935.
58. Song HT, Choi JS, Huh YM et al. Surface modulation of magnetic nanocrystals in the development of highly efficient magnetic resonance probes for intracellular labeling. J Am Chem Soc 2005; 127:9992-9993.
59. Pileni MP. The role of soft colloidal templates in controlling the size and shape of inorganic nanocrystals. Nat Mater 2003; 2:145-150.
60. Nitin N, LaConte LEW, Zurkiya O et al. Functionalization and peptide-based delivery of magnetic nanoparticles as an intracellular MRI contrast agent. J Biol Inorg Chem 2004; 9:706-712.

61. Yee C, Kataby G, Ulman A et al. Self-assembled monolayers of alkanesulfonic and -phosphonic acids on amorphous iron oxide nanoparticles. Langmuir 1999; 15:7111-7115.
62. Harris LA, Goff JD, Carmichael AY et al. Magnetite nanoparticle dispersions stabilized with triblock copolymers. Chem Mater 2003; 15:1367-1377.
63. Burke NAD, Stover HDH, Dawson FD. Magnetic nanocomposites: Preparation and characterization of polymer-coated iron nanoparticles. Chem Mater 2002; 14:4752-4761.
64. Lu Y, Yin Y, Mayers BT et al. Modifying the surface properties of superparamagnetic iron oxide nanoparticles through a sol-gel approach. Nano Lett 2002; 2:183-186.
65. Santra S, Tapec R, Theodoropoulou N et al. Synthesis and characterization of silica-coated iron oxide nanoparticles in microemulsion: The effect of nonionic surfactants. Langmuir 2001; 17:2900-2906.
66. Dubertret B, Skourides P, Norris DJ et al. In vivo imaging of quantum dots encapsulated in phospholipid micelles. Science 2002; 298:1759-1762.
67. Kim SW, Kim S, Tracy JB et al. Phosphine oxide polymer for water-soluble nanoparticles. J Am Chem Soc 2005; 127:4556-4557.
68. Kohler N, Fryxell GE, Zhang M. A bifunctional poly(ethylene glycol) silane immobilized on metallic oxide-based nanoparticles for conjugation with cell targeting agents. J Am Chem Soc 2004; 126:7206-7211.
69. Weissleder R, Lee AS, Fischman AJ et al. Polyclonal human immunoglobulin G labeled with polymeric iron oxide: Antibody MR imaging. Radiology 1991; 181:245-249.
70. Schmitz SA, Taupitz M, Wagner S et al. Magnetic resonance imaging of atherosclerotic plaques using superparamagnetic iron oxide particles. J Magn Reson Imaging 2001; 14:355-361.
71. Wadghiri YZ, Sigurdsson EM, Sadowski M et al. Detection of Alzheimer's amyloid in Transgenic mice using magnetic resonance microimaging. Magn Reson Med 2003; 50:293-302.
72. Risau W. Mechanisms of angiogenesis. Nature 1997; 386:671-674.
73. Veikkola T, Karkkainen M, Claesson-Welsh L et al. Regulation of angiogenesis via vascular endothelial growth factor receptors. Cancer Res 2000; 60:203-212.
74. Kim I, Kim HG, So JN et al. Angiopoietin-1 regulates endothelial cell survival through the phosphatidylinositol 3'-kinase/Akt signal transduction pathway. Circ Res 2000; 86:24-29.
75. Saeed M, Wendland MF, Engelbrecht M et al. Value of blood pool contrast agents in magnetic resonance angiography of the pelvis and lower extremities. Eur Radiol 1998; 8:1047-1053.
76. Lahorte CM, Vanderheyden JL, Steinmetz N et al. Apoptosis-detecting radioligands: Current state of the art and future perspectives. Eur J Nucl Med Mol Imaging 2004; 31:887-919.
77. Shah K, Weissleder R. Molecular optical imaging: Applications leading to the development of present day therapeutics. NeuroRx 2005; 2:215-225.
78. Gross S, Piwnica-Worms D. Spying on cancer: Molecular imaging in vivo with genetically encoded reporters. Cancer Cell 2005; 7:5-15.
79. Haberkorn U, Altmann A. Functional genomics and radioisotope-based imaging procedures. Ann Med 2003; 35:370-379.

CHAPTER 8

Patterning Metallic Nanoparticles by DNA Scaffolds

Rahul Chhabra, Jaswinder Sharma, Yan Liu and Hao Yan*

Abstract

In order to miniaturize nanoelectronic circuitry and architectures, organizing the nanoelectronic components in deliberately designed complex patterns is desired. Bio-mimetic self-assembly is a possible approach to achieve this goal. Bio-macromolecules can serve as scaffolds to template the nanoelectronic components into patterns with precise periodicity and complexity. In this review, we will summarize the progress in organizing metallic nanoparticles templated by DNA scaffolds into one and two dimensional architectures.

Introduction

Recent advances in scientific ventures have realized the potential of using self-assembly of biomolecules as templates to organize matter in the nanometer scale and thus an era of 'nanobiotechnology' has begun. DNA (deoxyribonucleic acid) came across as a promising scaffolding molecule with its remarkable physical and structural properties.[1] DNA nanostructures can serve as scaffolds for the assembly of nanoelectronic components into patterns with precise periodicity and complexity. In this review, we will summarize the progresses in organizing metallic nanoparticles by DNA scaffolds into one and two dimensional architectures and will comment on the future aspects of the research area

DNA: As a Polymer

Long admired as storage and processing carrier of genetic information, DNA also has properties of a polymer that can be used as a structural material (Fig. 1). It has four unique aromatic nitrogenous bases, namely Adenosine (A), Guanine (G), Thymine (T) and Cytosine (C) as recognition tags, which account for its addressability. In solution, it exists in a right handed B-form double helix where 'A' binds with 'T' and 'G' binds with 'C'. Molecular recognition properties of DNA lie in the specific interactions between the bases; thus, DNA can be programmed with specific sequences for nanoassembly purposes. DNA double helix has dimensions in the nanometer scale, the B-form helix completes one turn every 10.5 base pairs that span ~3.4 nm, and a diameter ~2 nm. On nanometer scale, double stranded DNA (dsDNA) exhibits stiffness with a persistence length of ~50 nm, and single stranded DNA (ssDNA) is relatively flexible with a persistence length of <1 nm. A combination of ssDNA and dsDNA in a structure provides tailored rigidity and flexibility. DNA can be synthesized using automated phosphoramidite chemistry, where a large pool of possible modifications enhances its use as a

*Corresponding Author: Hao Yan—Department of Chemistry and Biochemistry, and Center for Single Molecular Biophysics, The Biodesign Institute Arizona State University, Tempe, AZ 85287, U.S.A. Email: hao.yan@asu.edu

Bio-Applications of Nanoparticles, edited by Warren C.W. Chan. ©2007 Landes Bioscience and Springer Science+Business Media.

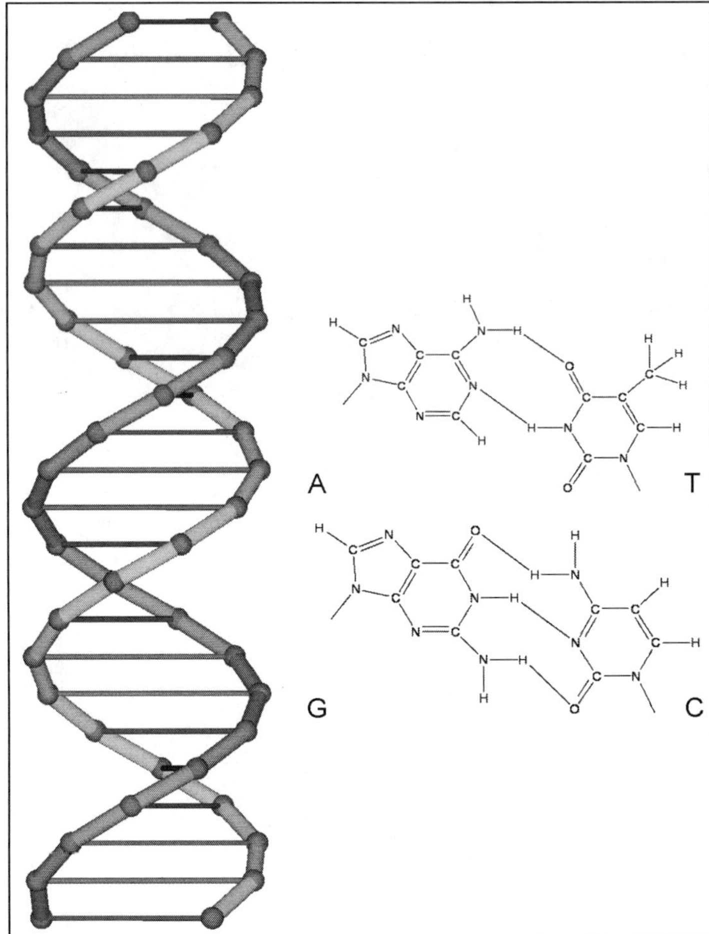

Figure 1. A cartoon showing the double helix DNA and hydrogen bonding pattern of A pairs with T and G pairs with C.

versatile material. A myriad of DNA-modifying enzymes is available, including site-selective DNA cleavage, ligation, labeling, amplifications etc,[2] which can manipulate DNA for various applications. Local structure and intermolecular interactions can be predicted from the sequences of DNA. Conclusively, DNA is a suitable polymer with material like properties.

Use of DNA as a structural motif has been realized by Seeman in his ingenius vision of DNA self-assembly. He proposed some basic rules that synthetic DNA oligonucleotides can be molded into rigid branched building blocks, which can further self-assemble into complex periodic structures in two or three dimensional space.[3] Since then, there have been innumerable reports exploiting DNA as structural building blocks. Besides generating structures with various complexities, it can act as a scaffold, showing regular periodicities with easily tailored dimensions.

Structural DNA nanotechnology has seen development of a rich toolbox, including different logo blocks, such as double-crossover molecule (DX), triple crossover molecule (TX), 2-, 4-, 6-helix bundles and cross shaped tiles (see Fig. 2), etc.[4] These tools exhibit different structural and mechanical properties conferring combined rigidity and flexibility on nanoarchitectures

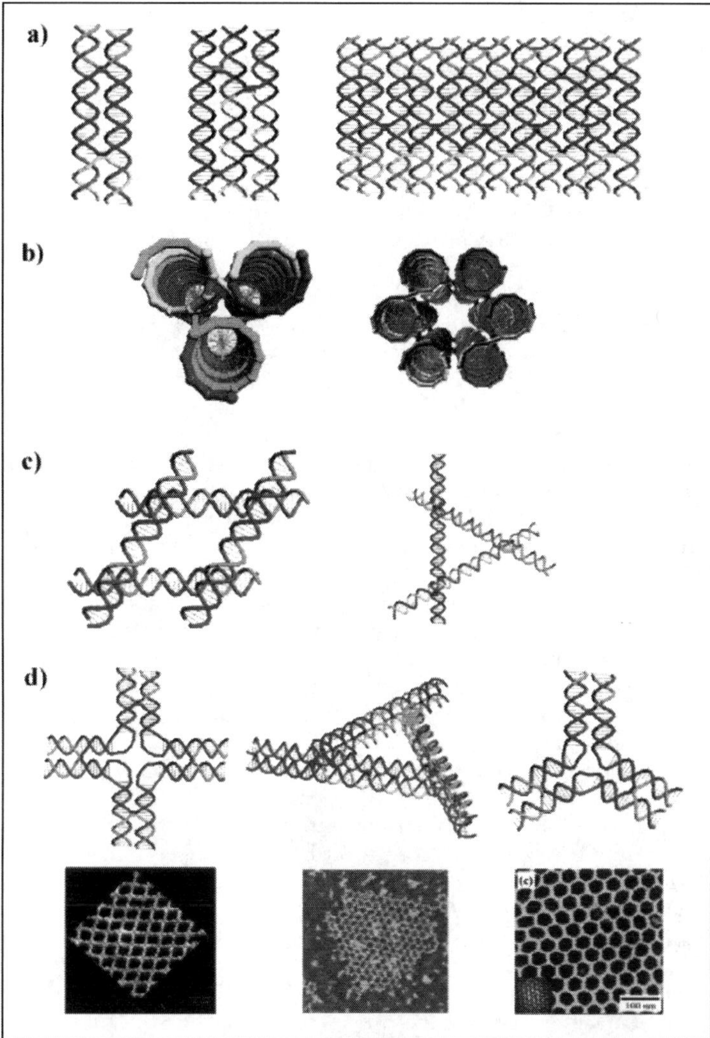

Figure 2. Examples of DNA tiles. a) left: a DX DNA tile; middle: a TX DNA tile; right: 12-helix DNA tile. b) left: a three helix bundle DNA tile; right: a six helix bundle DNA tile. c) left: a parallelogram DNA tile; right: a DNA triangle. d) upper left: a cross shaped tile; upper middle: a triangular DNA tile; upper right: a 3 point star DNA tile; bottom: Atomic force microscopy (AFM) images showing the self-assembly of the tiles forming 2D periodic lattices with square cavities. Reprinted with permission from: Lin C et al. Chem Phys Chem 2006; 7:1641-47.[4]

and allow the assembly of various 2D and 3D structures with precisely defined periodicities and complexities. A detailed review on DNA nanostructures can be found elsewhere.[5,6] As DNA can be chemically functionalized, Seeman visioned the use of DNA scaffold to template nano-scale components (proteins and nanoelectronics, such as metal nanoparticles and semiconductor nanocrystals) in a bottom-up approach.

Metal nanoparticles exhibit interesting optical and optoelectronic properties depending upon their sizes, shapes and interparticle-distances in the nanometer scale. Much research has been done in conjugation of DNA with gold nanoparticles (AuNPs) due to well studied surface chemistry and stability of gold-thiol bond. This review will be focusing on the self-assembly of AuNPs on DNA scaffolds. Organizing AuNPs on DNA templates usually utilized two forces; (1) nonspecific interactions, such as electrostatic interactions between positively charged AuNPs and negatively charged phosphate backbone of the DNA and (2) sequence specific interactions dictated by Watson-Crick base pairing rules.

Assembling AuNPs Using DNA Template—An Electrostatic Approach

DNA in physiological conditions behaves as a negatively charged molecule due to its phosphate backbone. Any positively charged ligand can bind to it nonspecifically through electrostatic interactions.

Sastry and coworkers[9] realized the potential of DNA as a template to generate metal nanowires by organizing metal nanoparticles along the DNA. They modified AuNPs with the amino acid lysine and used synthetic DNA and calf thymus DNA as a template. Lysine behaves as a positively charged ligand at physiological pH ~7.0. As synthetic DNA was custom made, the length of metal nanowires can be controlled by varying the length of DNA template (Fig. 3a).

Warner et al[10] used the positively charged ligand thiocholine to modify the surface of AuNPs and λ-DNA *Hind*III was used as a scaffold to pattern AuNPs on the surface. Patterns like linear chain-like arrays, ribbon shaped structures and branched architectures were observed, depending on the concentrations of λ-DNA and AuNPs, and the incubation time. The two dimensional ribbon and branched structures observed were due to cross-linking of DNA templates by polyvalent AuNPs (Fig. 3b).

Ohtani and coworkers[11] reported aniline-capped AuNPs (AN-AuNPs) assembled on the DNA templates. Positive charged AN-AuNPs were prepared by reducing $HAuCl_4$ using aniline as a reducing agent. Upon mixing with λ-DNA, highly ordered and long-range self-assembly of AuNPs on λ-DNA was formed. These may be due to the electrostatic interactions of the positively charged AN-AuNPs and the negatively charged phosphate backbone of λ-DNA, and also the hydrogen bonding interactions between amino group of aniline and the exposed nitrogen and oxygen atoms of the bases in the major or minor grooves of the DNA.

Jaeger and coworkers[12] have used tecto-RNA molecules to generate a large number of complex structures. The tecto-RNA molecules form tectosquares (TS), which can self-assemble into nanostructures of different shapes and geometrics with a superb programmability. Two tectosquares TS1 and TS2 were formed first, which were then self-assembled into a RNA ladder with periodic square cavities. Positively charged thiocholine capped AuNPs were organized onto the TS ladder based on electrostatic interactions and recognition properties due to shape and size of the scaffold (Fig. 3c).

Self-Assembly mediated by intercalation of small molecules into the double helix of DNA is another approach to organize metallic nanoparticles on DNA scaffolds. Willner and coworkers[13] exploited the use of intercalation strategy to generate metallic nanowires of AuNPs templated by DNA. AuNPs (1.4 nm) functionalized with single N-hydroxysuccinide were covalently linked to amino psoralen, an aromatic DNA intercalator. Modified AuNPs were then intercalated with a synthetic dsDNA. AuNPs-DNA assembly, irradiated with λ >360 nm, lead to cycloaddition of the psoralen with the thymine residues, rendering covalent attachment of the psoralen to the DNA template and thus generating a patterned array of AuNPs (Fig. 3d).

Self-assembly mediated by coulombic interactions has shown a vast potential to organize AuNPs in linear arrays. This approach is limited to simple designs due to the lack of control on the assembly process. Strategies using stronger and more specific binding interactions are required to arrange metal nanoparticles in deliberately designed complex assemblies.

Self-Assembly of AuNPs on DNA Scaffold Using Covalent Au-DNA Conjugates

Mirkin et al[14] and Alivisatos et al,[15] concurrently, succeeded in covalently conjugating DNA with AuNPs. Mirkin exploited the gold-thiol chemistry whereas Alivisatos utilized chemical

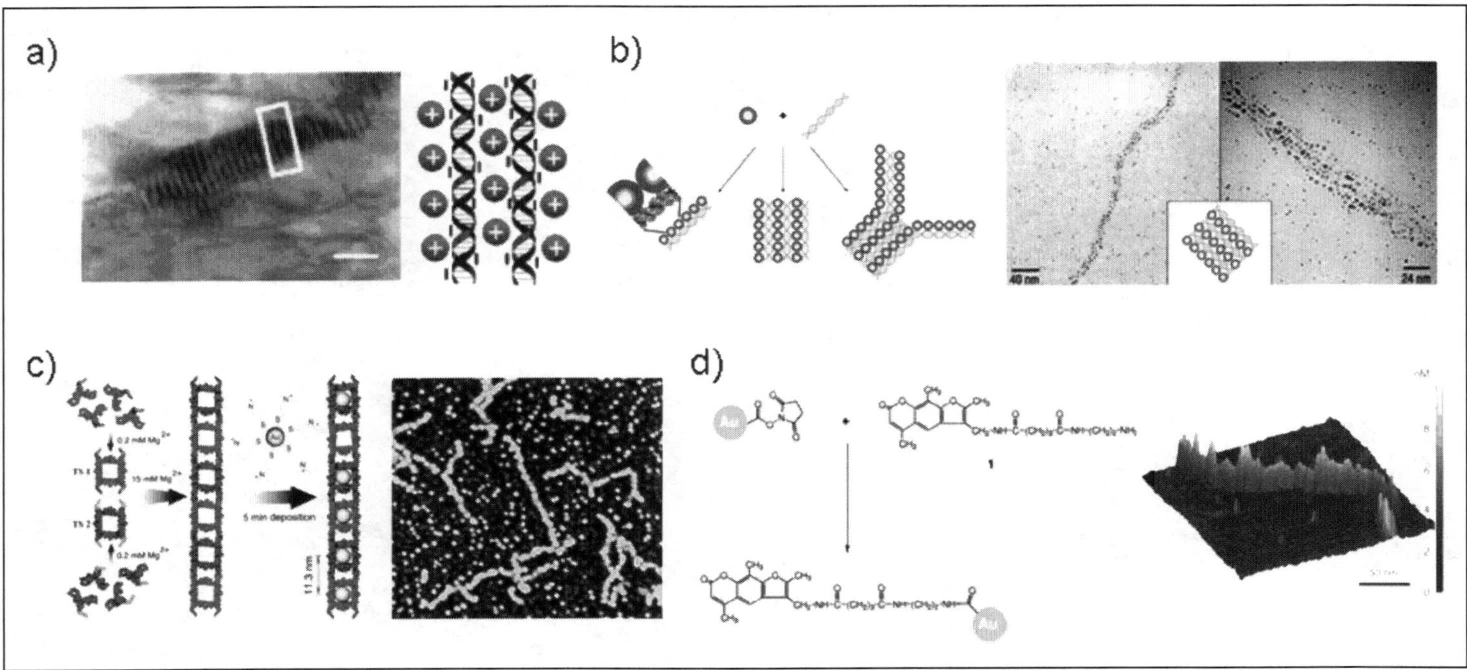

Figure 3. AuNP assemblies driven by electrostatic interactions. a) TEM image and schematic showing DNA-gold nanoparticle complex film showing linear assemblies (reprinted with permission from: Kumar A et al. Advanced Materials 2001; 13:341-44[9]). b) Electrostatic assembly of gold nanoparticles in different architectures and TEM images of linear assemblies (reprinted from: Warner MG, Hutchison JE. Nature Materials 2:272-77, ©2003 with permission from Nature Publishing Group[10]). c) Supramolecular assembly of cationic gold nanoparticles templated by RNA TS ladder (reprinted in part from: Koyfman AY et al. J Am Chem Soc 127:11886-87, ©2005 with permission from the American Chemical Society[12]). d) Strategy of covalent linking and AFM image showing linear arrays of gold nanoparticles (reprinted in part with permission from: Patolsky F et al. Angew Chem Int Ed 2002; 41:2323-2327[13]).

ligation reacting thiol modified DNA oligonucleotide with monomaleimido-modified Au particles to get the AuNP-DNA conjugates. AuNP-DNA conjugates are used extensively nowadays to pattern nanoparticles of different sizes onto scaffolds of varying shapes and geometries.

Metal nanoparticles possess surface plasmons. A solution of AuNPs appears red and shows a UV absorption peak at ~520 nm. When the AuNPs are brought sufficiently close, they exhibit plasmon coupling and the UV absorption red shift to ~565 nm and the solution becomes purple. Mirkin and coworkers[14] were first to exploit optical properties of AuNPs and reported controlled aggregation of AuNPs using AuNP-DNA conjugates. The reversibility of the aggregation process was visualized by thermal denaturation of the assembly.

Covalent attachment of the AuNPs with DNA opened innumerable strategies to assemble metal nanoparticles in various one and two dimensional patterns with tunable periodicities.

One Dimensional AuNP Ensembles Templated by DNA

Alivisatos and coworkers[17] came up with a protocol that AuNPs can be conjugated to discrete and known number of DNA oligonucleotides by adjusting DNA:AuNP ratio. Separation of the AuNP-DNA conjugates with discrete number of DNA was achieved by agarose gel electrophoresis. AuNP-DNA conjugates were then extracted out of the gel. AuNPs of different

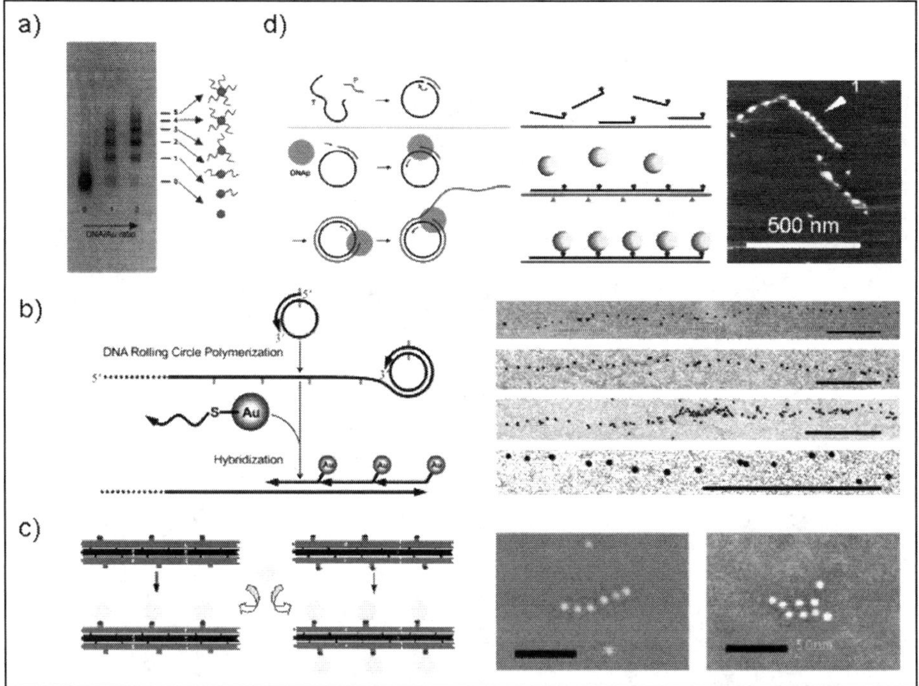

Figure 4. One dimensional assemblies of AuNPs patterns. a) Electrophoretic separation of AuNP with discrete number of DNA strands (reprinted in part from: Zanchet D et al. Nano Letters 1:32-35, ©2001 with permission from the American Chemical Society[17]). b) 1D array of AuNPs on RCA template (reprinted with permission from: Deng Z et al. Angew Chem Int Ed 2005; 44:3582-3585[18]). c) Arrays of AuNPs mediated by biotin-streptavidin interactions (reprinted in part from: Li H et al. J Am Chem Soc 126:418-19, ©2004 with permission from the American Chemical Society[19]). d) Large 1D arrays on RCA template mediated by biotin-streptavidin interactions (reprinted in part from: Beyer S et al. Nano Letters 5:719-722, ©2005, with permission from the American Chemical Society[20]).

sizes and conjugation with DNA of varying lengths was evaluated using gel electrophoresis methods. DNA of length more than 50 bases is required for accurate resolution of the conjugate of AuNP with a single copy of DNA from the bare AuNPs (Fig. 4a).

Alivisatos and coworkers[16] created nanoparticle assemblies containing discrete number of AuNPs in a variety of homo and hetero, dimers and trimers, depicting spatial control on the nanoparticles.

Mao and coworkers[18] came up with an innovative approach to synthesize long DNA templates by rolling circle amplification (RCA). In RCA, under isothermal conditions, a circular ssDNA acts as a template for DNA polymerase, and a DNA polymerase progresses around the circular template continuously and generates a long, linear DNA with repetitive sequences complementary to the circular template. AuNPs were each conjugated to a single copy of the DNA sequence complementary to the repeating unit of the RCA product: When they were hybridized to the long DNA, a linear arrays of metal nanoparticles (up to 4 μm long) were generated (Fig. 4b).

Yan and coworkers[19] used biomolecular interactions mediated by protein ligands to assemble AuNPs in one dimensional arrays. Biotin-streptavidin is a well known pair of ligands with binding affinity $k_d = 10^{-15}$ M. A linear array that is self-assembled from a TX molecule was used as a scaffold. Each TX molecule has two stem loops that each contains two biotin groups. The biotin modified TX linear array was incubated with streptavidin conjugated AuNPs, such that the AuNPs were organized along the TX array at precisely controlled positions. Depending on the selective modification of the TX loop with biotin, single or double layer arrays of AuNPs were generated (Fig. 4c).

Simmel and coworkers[20] have combined the strong affinity of biotin-streptavidin with RCA method to produce longer AuNP arrays. A linear DNA template with periodic repeats was synthesized using RCA. Short strand of DNA with sequences complementary to the template was modified with a biotin group at one end. Following hybridization of the long template DNA with its complementary partner, a duplex DNA with biotin molecules arranged at periodic positions was produced. Incubation of streptavidin conjugated AuNPs yielded large arrays of metal nanoparticles organized in a linear fashion (Fig. 4d).

Two Dimensional Arrays of AuNPs Using DNA as a Scaffold

Future nanoelectronic circuitry and a revolution in circuit fabrication technology using biomoleculer nanolithography led to the success of patterning metal nanoparticles in two and three dimensional architectures.

Kiehl and coworkers reported a few examples of 2D assemblies of AuNPs using DNA scaffold of predesigned periodicities. An earlier effort involved the use of 2D DNA crystal composed of double-crossover molecules (ABCD system) containing DNA hairpins.[22] Thiol modified DNA oligonucleotide was covalently linked to 1.4 nm monomaleimido derivatized AuNPs. The AuNP modified DNA was annealed with other strands to form 2D crystal, thereby aligning AuNPs in a row along the DX 'B' tile.

Another report used the same scaffold, ABCD system, wherein 6 nm AuNPs were organized along the 'B' tiles carrying a single stranded A_{15}.[23] DNA scaffold was assembled first and deposited on substrate. Multiple copies of DNA oligo with the sequence (T_{15}) were attached to AuNPs through the gold-thiol chemistry. The AuNPs covered with DNA oligos were then hybridized to the A_{15} probes, which resulted in parallel lines of AuNPs with inter-line distance of ~ 63 nm and an interparticle distance within the line of ~10-30 nm (Fig. 5a). Conjugation with multiple copies of DNA increased the probability of AuNPs assembly to the scaffold, but this may also cause undesired folding or overlapping of the scaffold. Therefore the hybridization of the AuNPs to the DNA array was done only after depositing the DNA array to the surface.

In a further step towards building nanoelectronic circuitry, the team employed the same strategy to align multi-component arrays of different sizes of nanoparticles on the scaffold.[24]

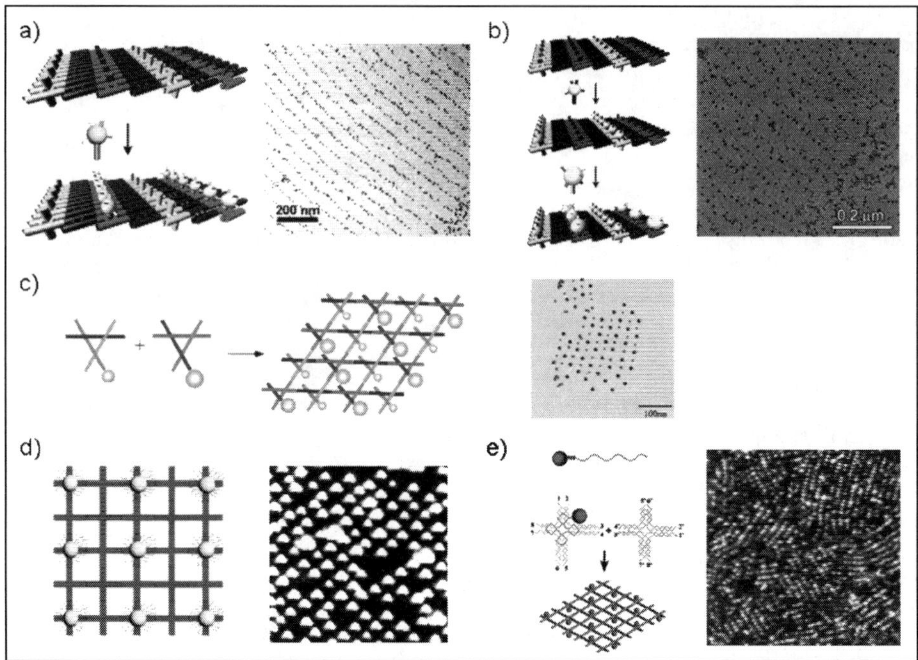

Figure 5. Two dimensional arrays of AuNPs on DNA scaffold. a) 2D arrays of AuNPs (reprinted in part from: Pinto YY et al. Nano Letters 4:2343-47, ©2004 with permission from the American Chemical Society[23]). b) Arrays showing assembly of multicomponent architecture (reprinted in part from: Pinto YY et al. Nano Letters 5:2399-02 ©2005 with permission from the American Chemical Society[24]). c) AuNP arrays templated by robust DNA scaffold (reprinted in part from: Zheng J et al. Nano Letters 6:1502-04, ©2006 with permission from the American Chemical Society[25]). d) Assembly of AuNPs mediated by DNA tiles in 2D arrays (reprinted in part from: Zhang J et al. Nano Letters 6:248-51, ©2006 with permission from the American Chemical Society[26]). e) Larger 2D arrays of AuNPs organized by DNA tiles where one DNA tile is modified with one gold nanoparticle prior to self-assembly process as shown in schematic above (reprinted with permission from: Sharma J et al. Angew Chem Int Ed 2006; 45:730-35[28]).

AuNPs, 5 nm and 10 nm, were modified with multiple copies of oligos of different sequences. The 2D DNA scaffold, bearing hairpin loops on 'B' and 'D' tiles with probe strands complementary to the strands conjugated on the AuNPs. Following deposition of the scaffold and the addition of DNA conjugated AuNPs, large alternate arrays of AuNPs with striped pattern of 5 and 10 nm particle sizes were formed with no signs of cross-contamination (Fig. 5b).

A more robust 2D array of triangular shaped DNA tiles with diamond cavities was also used as a scaffold.[25] Two ends of a triangle tile were involved in self-assembly process by mutual sticky ends cohesion and the third end was modified with a AuNP. In this study, AuNPs were modified with a single copy of DNA oligonucleotide by a method described by Alivisatos.[17] Ensembles with mono and multi-component system have been generated (Fig. 5c).

Yan and coworkers[26] have used another scaffold which have periodic square like cavities with tailorable dimensions (Fig. 5d). The design consisted of two cross-shaped tiles (tiles A and B) that self-assemble into a 2D crystal by joining alternatively with each other.[27] Interparticle distances between the constituent tiles can be controlled by adding a few DNA turns to each arm of the B tile, thus DNA 2D crystals with various geometries can be assembled using this system. Tile A was modified with a single stranded A_{15} base sequence which served as a hybridization site for T_{15} modified AuNPs. Self-assembled 2D nanogrid was deposited on mica and AuNPs modified with T_{15} were added onto the surface to allow the modified AuNPs to bind the hybridization sites. AuNPs were arranged in a periodic fashion but show larger inter-particle distance between the neighboring AuNPs than the original design. This was explained based on the balance between the repulsion between the highly negatively charged AuNPs (modified with multiple copies of DNA oligos) and the attraction force of the base-paring between the single stranded probe on the DNA tile and the DNA oligos on the AuNPs.

As described earlier, to achieve more precise control of the positions of the nanoparticles on the scaffold, nanoparticle modified with a single copy of DNA was desirable. Yan and coworkers[28] synthesized and separated AuNPs with a single copy of DNA, which were then passivated with a layer of thiol modified T_5 oligonucleotide to render these AuNPs stable against high salt concentration. DNA-Au conjugates mixed with other DNA strands were allowed to self-assemble into 2D scaffolds with different geometries, periodicities and interparticle distances. AFM images show arrays of AuNPs organized in predesigned manner (Fig. 5e).

Accomplishing complex architectures with biomolecular nanolithography would be more feasible if one can conjugate more than one strand of DNA to AuNPs, proposed by Niemeyer et al.[29] Hybrids with two to seven different DNA oligonucleotides attached to AuNPs were generated. These hybrids were considered versatile as they can hybridize to more than one complementary target with similar hybridization capabilities. The group proved their hypothesis by hybridizing AuNPs with two DNA oligonucleotides to assemble monolayer of AuNPs on the solid substrate.

Patterning a discrete number of nanoparticles in a deliberately designed, finite-sized architecture is critical in nanocircuit fabrication. An endeavor undertaken by Sleiman et al[30] resulted in the hexagon pattern of AuNPs templated by DNA template designed by sequential self-assembly manner. To construct a template with well-defined architecture, an organic unit (vertex) was chemically synthesized that exhibited a dihedral angle of 120°. Two DNA arms were attached to the two ends of the vertex molecule. The sequences of the arms were designed in such a way that one arm on each building block had complementary sequence to an arm on the adjacent building block. AuNPs monofunctionalized with succinimidyl ester moiety were conjugated to one arm of the DNA modified with an amine group. Six AuNPs conjugated DNA building blocks were synthesized and assembled in a sequential manner to accomplish hexagon pattern of AuNPs.

Conclusions

AuNPs can be arranged in linear or complex fashions templated by DNA scaffold. These assemblies are mediated either by electrostatic interactions or by covalently modifying AuNPs with discrete or multiple copies of DNA oligonucleotide. Using DNA scaffold with tunable dimensions, an ensemble with predefined pattern and precisely defined periodicities and interparticle distances can be accomplished. For an efficient chip circuitry, it is desirable that nanoparticles of various sizes, shapes and properties are aligned in regular arrays with well-defined interparticle distances. Self-assembly of Au-DNA conjugates opens up a new era of research where nanoparticles of different sizes can be assembled in precise periodicities and complexities for applications in the field of miniature nanoelectronics, photonics and to study the optical and optoelectronic phenomenon observed in proximal distances.

Acknowledgements

The authors thank the financial supports by National Science Foundation and Arizona State University to H. Y.

References

1. Seeman NC. DNA in a material world. Nature 2003; 421:427-31.
2. Yan H. Nucleic acid nanotechnology. Science 2004; 306:2048-49.

3. Seeman NC. Nucleic acid junctions and lattices. J Theor Biol 1982; 99:237-47.
4. Lin C, Liu Y, Rinker S et al. DNA tile based self-assembly: Building complex nanoarchitectures. Chem Phys Chem 2006; 7:1641-47.
5. Seeman NC. Nucleic acid nanostructures and topology. Angew Chem Int Ed 1998; 37:3220-38.
6. Feldkamp U, Niemeyer CM. Rational design of DNA nanoarchitectures. Angew Chem Int Ed 2006; 45:1856-76.
7. Simmel FC, Dittmer WU. DNA nanodevices. Small 2005; 1:284-99.
8. Kiehl RA. Nanoparticle electronic architectures assembled by DNA. J Nanoparticle Res 2000; 2:331-32.
9. Kumar A, Pattarkine M, Bhadbhade M et al. Linear superclusters of colloidal gold particles by electrostatic assembly on DNA templates. Advanced Materials 2001; 13:341-44.
10. Warner MG, Hutchison JE. Linear assemblies of nanoparticles electrostatically organized on DNA scaffolds. Nature Materials 2003; 2:272-77.
11. Nakao H, Shiigi H, Yamamoto Y et al. Highly ordered assemblies of Au nanoparticles organized on DNA. Nano Letters 2003; 3:1391-94.
12. Koyfman AY, Braun G, Magonov S et al. Controlled spacing of cationic gold nanoparticles by nanocrown RNA. J Am Chem Soc 2005; 127:11886-87.
13. Patolsky F, Weizmann Y, Lioubashevski et al. Au-nanoparticle nanowires based on DNA and polylysine templates. Angew Chem Int Ed 2002; 41:2323-2327.
14. Mirkin CA, Letsinger RL, Mucic RC et al. A DNA-based method for rationally assembling nanoparticles into macroscopic materials. Nature 1996; 382:607-09.
15. Alivisatos AP, Johnsson KP, Peng X et al. Organisation of 'nanocrystal molecules using DNA. Nature 1996; 382:609-611.
16. Loweth CJ, Caldwell WB, Peng X et al. DNA-based assembly of gold nanocrystals. Angew Chem Int Ed 1999; 38:1808-12.
17. Zanchet D, Micheel CM, Parak WJ et al. Electrophoretic isolation of discrete Au nanocrystal/DNA conjugates. Nano Letters 2001; 1:32-35.
18. Deng Z, Tian Y, Lee SH et al. DNA-encoded self-assembly of gold nanoparticles into one-dimensional arrays. Angew Chem Int Ed 2005; 44:3582-3585.
19. Li H, Park SH, Reif JH et al. DNA-templated self-assembly of protein and nanoparticle linear arrays. J Am Chem Soc 2004; 126:418-19.
20. Beyer S, Nickels P, Simmel FC. Periodic DNA nanotemplates synthesized by rolling circle amplification. Nano Letters 2005; 5:719-722.
21. Niemeyer CM, Bürger W, Peplies J. Covalent DNA-streptavidin conjugates as building blocks for novel biometallic nanostructures. Angew Chem Int Ed 1998; 37:2265-68.
22. Xiao S, Liu F, Rosen AE et al. Self assembly of metallic nanoparticle arrays by DNA scaffolding. J Nanoparticle Res 2002; 4:313-17.
23. Pinto YY, Le JD, Seeman NC et al. DNA-templated self-assembly of metallic nanocomponents arrays on a surface. Nano Letters 2004; 4:2343-47.
24. Pinto YY, Le JD, Seeman NC et al. Sequence-encoded self-assembly of multiple-nanocomponent arrays by 2D DNA scaffolding. Nano Letters 2005; 5:2399-02.
25. Zheng J, Constantinou PE, Micheel C et al. Two-dimensional nanoparticle arrays show the organizational power of robust DNA motifs. Nano Letters 2006; 6:1502-04.
26. Zhang J, Liu Y, Ke Y et al. Periodic square-like gold nanoparticle arrays templated by self-assembled 2D DNA nanogrids on a surface. Nano Letters 2006; 6:248-51.
27. Yan H, Park SH, Finkelstein G et al. DNA-templated self-assembly of protein arrays and highly conductive nanowires. Science 2003; 301:1882-84.
28. Sharma J, Chhabra R, Liu Y et al. DNA-templated self-assembly of two-dimensional and periodical gold nanoparticle arrays. Angew Chem Int Ed 2006; 45:730-35.
29. Niemeyer CM, Ceyhan B, Hazarika P. Oligofunctional DNA-gold nanoparticle conjugates. Angew Chem Int Ed 2003; 42:5766-70.
30. Aldaye FA, Sleiman HF. Sequential self-assembly of a DNA hexagon as a template for the organization of gold nanoparticles. Angew Chem Int Ed 2006; 45:2204-09.

CHAPTER 9

Liposomes in Biology and Medicine

Reto A. Schwendener*

Abstract

Drug delivery systems (DDS) have become important tools for the specific delivery of a large number of drug molecules. Since their discovery in the 1960s liposomes were recognized as models to study biological membranes and as versatile DDS of both hydrophilic and lipophilic molecules. Liposomes—nanosized unilamellar phospholipid bilayer vesicles—undoubtedly represent the most extensively studied and advanced drug delivery vehicles. After a long period of research and development efforts, liposome-formulated drugs have now entered the clinics to treat cancer and systemic or local fungal infections, mainly because they are biologically inert and biocompatible and practically do not cause unwanted toxic or antigenic reactions. A novel, up-coming and promising therapy approach for the treatment of solid tumors is the depletion of macrophages, particularly tumor associated macrophages with bisphosphonate-containing liposomes. In the advent of the use of genetic material as therapeutic molecules the development of delivery systems to target such novel drug molecules to cells or to target organs becomes increasingly important. Liposomes, in particular lipid-DNA complexes termed lipoplexes, compete successfully with viral gene transfection systems in this field of application. Future DDS will mostly be based on protein, peptide and DNA therapeutics and their next generation analogs and derivatives. Due to their versatility and vast body of known properties liposome-based formulations will continue to occupy a leading role among the large selection of emerging DDS.

State of the Art of Nanosized Drug Delivery Systems

The first microencapsulated drugs were introduced in the 1950s and polymer based slow release systems appeared shortly thereafter. Soon after their discovery in the 1960s by A.D. Bangham and colleagues, liposomes—phospholipid bilayer nanocontainers with spherical shape properties—were recognized as potential drug delivery systems (DDS).[1,2] Since then a tremendous amount of work on applications of liposomes has been accomplished. Due to their versatility nanosized small unilamellar liposomes are used as models to study biological and biophysical membrane properties and as carriers of drugs for therapeutic applications. Liposomes undoubtedly represent today the most extensively and advanced drug delivery vehicles. Liposome-formulated drugs have entered the clinics to treat cancer and systemic or local fungal infections, mainly because they are biologically inert, biocompatible and practically do not cause unwanted toxic or antigenic reactions and, most importantly, industrial large-scale production of liposome formulated drugs has allowed their advance in the pharmaceutical industry.[3-5]

In the advent of the use of genetic material (DNA, ribozymes, DNAzymes, aptamers, (antisense-) oligonucleotides, small interfering RNAs) as therapeutic molecules the development of delivery systems to target these molecules to cells or to target organs becomes increasingly

*Reto A. Schwendener—Institute of Molecular Cancer Research, University of Zurich, Winterthurerstrasse 190, CH-8057 Zurich, Switzerland. Email: rschwendener@imcr.unizh.ch

Bio-Applications of Nanoparticles, edited by Warren C.W. Chan. ©2007 Landes Bioscience and Springer Science+Business Media.

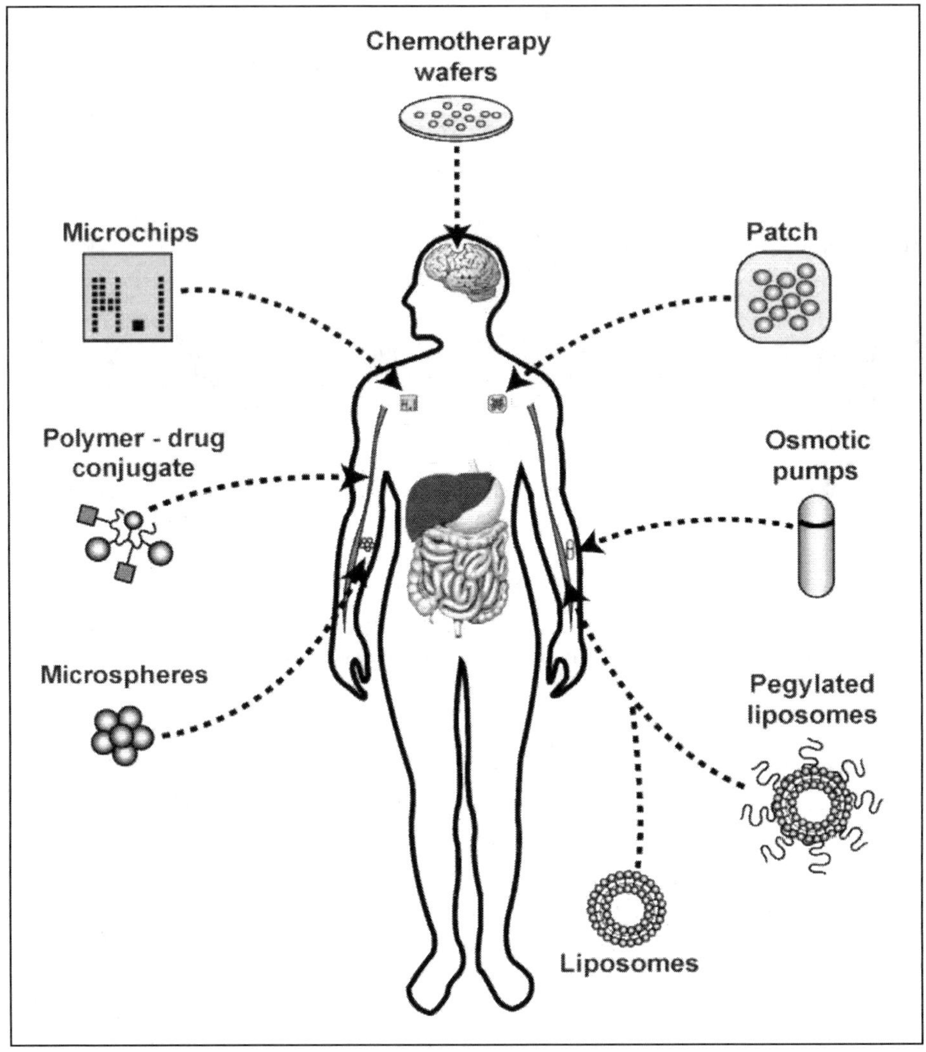

Figure 1. Modern drug delivery systems. There are a variety of different delivery strategies that are either currently being used or are in the testing stage to treat human cancers and other diseases. Examples of these include polymer microspheres, polymer wafers, osmotic pumps, liposomal systems, polymer/drug targeting moiety conjugates, and controlled release microchips. Reprinted from: Moses MA et al. Cancer Cell 4:337-341; ©2003, with permission from Elsevier.[12]

important. In this field the liposomes, in particular lipid-DNA complexes termed lipoplexes (see below), compete with viral gene transfection systems. Nanoparticles, nanospheres, polymersomes, nanogels, micelles, dendrimers, and virosomes are other main types of nanocarrier systems used for drug delivery.[6-11] As schematically shown in Figure 1, modern DDS including polymer-drug conjugates, liposomes, osmotic pumps, microchips, wafers, transdermal patches and other systems vary in their concepts, compositions, shapes, sizes, drug loading capacity as well as in their pharmacokinetic and organ distribution properties.[12] All DDS, however, pursue the aim of improving drug delivery for the benefit of the patient.

The major applications of DDS comprise drugs that possess nonideal physico-chemical and pharmacological properties such as (1) Poor solubility; (2) Tissue damage caused by unintentional extravasation of drugs; (3) Loss of drug activity following administration; (4) Unfavorable pharmacokinetic properties and poor biodistribution and (5) Lack of selectivity for target organs or tissues. Systemic drug distribution may cause toxic side effects and low drug concentrations at target tissues; this could lead to suboptimal therapeutic effects.

The formulation of pharmacologically active drug molecules in DDS can improve or abolish these unfavorable properties. However, there are also disadvantages in the development of particulate drug carriers, such as system complexity, unwanted biologic and immunologic effects, stability, costs of development and scale-up, as well as intellectual property issues. In the limited format of this chapter it is not possible to cover all methods and references from the vast field of liposome technology. Hence, we concentrate on summarizing use and properties of liposomes as DDS for the delivery of cytotoxic molecules for cancer therapy.

Evolution of Liposomes in Cancer Therapy

Liposomes have become known as one of the most versatile tools for the delivery of pharmacologically active molecules. Since their discovery in the 1970s their potential for the delivery of cytotoxic drugs in cancer therapy has been recognized.[4,5,11,13,14]

Liposomes are spherical vesicles that consist of an aqueous compartment enclosed by a phospholipid bilayer. If multiple bilayers of lipids are formed around a primary core, the structures that are generated are termed multilamellar vesicles (MLVs). MLVs are formed spontaneously upon reconstitution of dry lipid films in aqueous media. Small (nanosized) unilamellar vesicles (SUVs) of scaleable mean diameters of 20 to 500 nanometers are produced by high pressure extrusion of MLVs through polycarbonate membranes. SUVs are also obtained by ultrasonication, by detergent dialysis and by many other, less important methods. Hydrophilic and hydrophobic drugs can both be entrapped in liposomes. Since the composition of the liposome bilayers can be varied with a huge selection of different phospholipids and additional intercalating molecules, liposomal delivery systems are of high versatility and customized formulations can easily be engineered to obtain desired sizes, surface charge, membrane composition and morphology providing them with high versatility such as long circulation half-life, sustained and targeted drug delivery or diagnostic imaging properties.[11,15-19]

As schematically shown in Figure 2, the liposomes evolved from rather simple compositions (Fig. 2A,B) to highly sophisticated multi-component systems. The state-of-the-art liposomes used for parenteral drug delivery are the long circulating ("stealth") liposomes (Fig. 2C-E). Stealth liposomes are sterically stabilized formulations that include polyethylene glycol (PEG)-conjugated lipids or other hydrophilic coating molecules. The surface grafted polymers create an impermeable, highly hydrophilic layer on the outer liposome surface. The prominent properties of long-circulating liposomes are dose-independent, nonsaturable, log-linear pharmacokinetics and increased bioavailability. Pegylation prevents or retards opsonization and recognition of the liposomal vesicles by the monocytic phagocyte system (MPS).[11] Due to their long circulation time in blood and the enhanced drug permeability and retention effect in tumor tissues, PEG-liposomes accumulate at high concentrations (up to 10% of the injected dose per organ) in tumors.[20-24]

Immunoliposomes (Fig. 2B,D) are complex drug or gene delivery systems that are developed for specific cell targeting by attachment of functionalized antibodies or antibody fragments to the outer surface of the liposomes. The modern immunoliposomes are PEG-liposomes to which receptor specific molecules are attached, preferably at the distal tips of the PEG chains (Fig. 2E,J). Immunoliposomes target cell specific receptors and facilitate receptor-mediated endocytosis for cell uptake.[25]

A variety of tumor-specific antibodies have been used for targeting of liposomes to tumor cells or molecules located in the tumor stroma. In earlier studies whole IgG antibodies were linked to the liposome surface by various coupling methods.[26] Today the most advanced

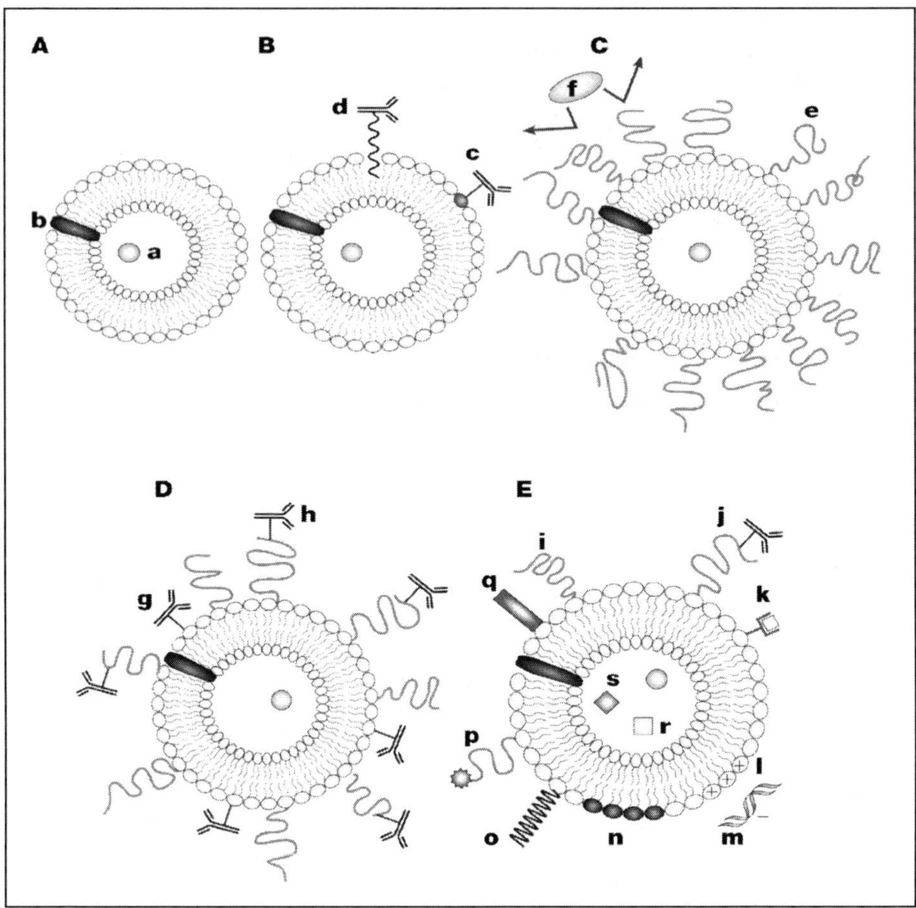

Figure 2. Evolution of liposomes. A) Early traditional 'plain' liposomes with water soluble drug (a) entrapped into the aqueous liposome interior, and lipophilic drug (b) incorporated into the liposomal membrane. B) Antibody-targeted immunoliposome with antibody covalently coupled (c) to the reactive phospholipids in the membrane, or hydrophobically anchored (d) into the liposomal membrane after preliminary modification with a hydrophobic moiety. C) Long-circulating liposome grafted with a protective polymer (e) such as PEG, which shields the liposome surface from the interaction with opsonizing proteins (f). D) Long-circulating immunoliposome simultaneously bearing both protective polymer and antibody, which can be attached to the liposome surface (g) or, preferably, to the distal end of the grafted polymeric chain (h). E) New-generation liposome, the surface of which can be modified (separately or simultaneously) by different ways. Among these modifications are: the attachment of protective polymer (i) or protective polymer and targeting ligand, such as antibody (j); the attachment/incorporation of a diagnostic label (k); the incorporation of positively charged lipids (l) allowing for the complexation with DNA yielding lipoplex structures (m); the incorporation of stimuli-sensitive lipids (n); the attachment of a stimuli-sensitive polymer (o); the attachment of a cell-penetrating peptide (p); the incorporation of viral components (q). In addition to a drug, liposomes can be loaded with magnetic particles (r) for magnetic targeting and/or with colloidal gold, silver particles or fluorescent molecules (s) for microscopic analysis. Adapted from: Torchilin VP. Nat Rev Drug Discov 4:145-160; ©2005, with permission from Nature Publishing Group.[11]

immunoliposomes are the anti-p185/HER2 liposomes that target the herceptin receptor which is over-expressed in various cancers, especially breast cancer. Long-circulating immunoliposomes targeted to HER2 (ErbB2, Neu) have been prepared by conjugation of anti-HER2 MAb fragments (Fab' or single chain Fv, scFv) to liposome-grafted polyethylene glycol chains. MAb fragment conjugation did not affect the biodistribution or long-circulating properties of i.v.-administered liposomes.[27-29] The epidermal growth factor receptor (EGFR) is another target for immunoliposomes that bind to and internalize in tumor cells that over-express EGFR. Anti-EGFR immunoliposomes have been constructed modularly with Fab' fragments of the antibody cetuximab.[30-32] A large number of antibodies directed against other target molecules expressed on colon,[33] B-cell lymphoma,[34] and neuroblastoma[35] tumors have been used for the preparation of immunoliposomes. In order to target the ED-B isoform of fibronectin, which is exclusively expressed in the extracellular matrix of solid tumors, we constructed immunoliposomes decorated with scFv antibody fragments directed against ED-B fibronectin and successfully used these DDS for targeted delivery of cytotoxic drugs into tumors in vivo.[36] We also developed specific antibodies and immunoliposomes for specific targeting of tumor endothelial marker (TEM1) and the vascular endothelial growth factor receptor-2 (VEGFR-2).[37,38] Tissue-specific gene delivery using immunoliposomes has also been achieved with folate[39,40] and transferrin[41] receptor specific immunoliposomes. Additionally, tumor vasculature targeted immunoliposome therapy was shown to be effective with liposomal doxorubicin.[42-44]

Various cell uptake mechanisms for liposomes have been described.[8,45] Due to their particulate properties, phagocytic uptake mechanisms (phagocytose, endocytose, pinocytose) are predominant, however cell membrane adhesion and fusion can also occur. In the phagocytic uptake pathway liposomes are captured at the cell surface followed by endosomal and lysosomal uptake. Drug liberation into the cytoplasm depends on the lipid composition of the liposomes. To release encapsulated material into the cytoplasm of a cell, pH-sensitive liposomes can be generated by addition of dioleylphosphatidyl-ethanolamine (DOPE) to liposomes composed of acidic lipids such as cholesterylhemisuccinate (CHEMS) or oleic acid and other lipids. At a pH of 7, these lipids possess the typical bilayer structure; however, upon endosomal compartmentalization (pH becomes more acidic) they undergo protonation and collapse into nonbilayer structures. This leads to the disruption and destabilization of the endosomal membrane, which in turn promotes rapid release of encapsulated molecules into the cytoplasm.[8]

Cell penetrating peptides (CPPs) have proven to be efficient intracellular delivery systems overcoming the lipophilic barrier of cell membranes. CPPs can deliver a wide range of large cargo molecules such as proteins, peptides, oligonucleotides and even small nanoparticles as liposomes to a variety of cell types and to different cellular compartments. The CPPs are basic, lysine- or arginine- rich amphipathic peptides originating from different sources. CPPs can either form complexes with many different types of molecules (peptides, proteins, plasmids, oligonucleotides, siRNA, dyes etc.) or they can be covalently linked to these cargo molecules.[46,47] Liposomes have also been decorated with the TAT[48] or pAntp CPPs,[49] demonstrating higher cell uptake rates in vitro. Regrettably, their usefulness as drug delivery systems is hampered by their ability to penetrate virtually any cell type both in vitro and in vivo in a nonspecific mode. This feature complicates CPP applications as target specific drug delivery systems; therefore therapeutic applications seem unlikely, unless their target cell specificity can be significantly improved.

Liposomes as Carriers of Lipophilic and Amphiphilic Nucleoside Analogs

The majority of applications of liposomes as therapeutic DDS are based on the encapsulation of water soluble cytotoxic molecules within the trapped aqueous volume of the liposomes. Liposomes loaded with cytotoxic anti-tumor drugs doxorubicine, mitoxantrone, topotecan, irinotecan and cytarabine are examples of clinically applied chemotherapeutic liposome formulations.[5,11,15,16,22,50-55] For a current summary of clinically used liposomal anti-cancer formulations (see ref. 22). In contrast to the extensive exploitation of the trapped aqueous volume of the liposomes that serves as nanocontainer for water soluble molecules, the phospholipid bilayer has not been given the same attention for its use as carrier matrix for lipophilic drugs. Hence, the development of liposomal drug formulations with lipophilic drugs is less popular. This difference may have several reasons with the main reason being that the chemistry required to transform water soluble molecules into lipophilic compounds allowing incorporation into the lipid bilayer core is difficult. The most favorable chemical modifications consist in the attachment of long chain fatty acyl or alkyl residues, for example saturated or unsaturated fatty acids, preferably palmitic or stearic acid and alkylamines, preferably hexadecyl- or octadecylamine to a suitable functional group of the hydrophilic part of the drug molecule. Some recent examples of lipophilic modifications of antitumor drugs and their formulation in liposomes are gemcitabine, 5-iodo-2'-deoxyuridine, methotrexate, paclitaxel and a lipophilic topoisomerase inhibitor.[56-62]

Drugs that are highly lipophilic by their own nature, e.g., taxanes and epothilones can only be used therapeutically by addition of possibly toxic solubilizing agents (e.g., Cremophor EL) in complex pharmaceutical formulations.[63,64] One of several feasible means of obtaining nontoxic parenterally applicable formulations of such drugs is their incorporation into the bilayer matrix of phospholipid liposomes.[65]

Nucleoside analogs are a major class of chemotherapeutic agents for the treatment of cancer and viral diseases. Natural endogenous nucleosides must be phosphorylated to corresponding 5'-triphosphates in order to be incorporated into the DNA or RNA synthesised within the cell. Nucleoside analogs are in essence prodrugs since they must undergo the same transformations in the cytoplasm similarly to the natural nucleosides before becoming active. We chose the approach of chemical transformation of water-soluble nucleosides of known cytotoxic and antiviral properties into lipophilic drugs or prodrugs, thus reversing the paradigm of transforming lipophilic molecules into hydrophilic derivatives. The first cytotoxic nucleoside we chose is 1-β-D-arabinofuranosyl cytosine (ara-C) because its major clinical disadvantages are a very short plasma half-life and rapid inactivation. To reduce these limitations, a large number of 5'- and N^4-substituted ara-C derivatives have been synthesized and characterized in the past (reviewed in ref. 66). Of a series of N^4-alkyl-ara-C derivatives with alkyl chain lengths ranging between 6 and 22 C-atoms, N^4-octadecyl-ara-C (NOAC) exerted the strongest anti-tumor activity after oral and parenteral therapy in several mouse tumor models and showed to have distinct pharmacological properties compared to ara-C.[67,68] Of note, (ref. 69) provides an excellent review of the topic.

Consequently, we further modified NOAC by the synthesis of a new generation of lipophilic/amphiphilic heterodinucleoside phosphate derivatives, termed "duplex drugs" that combine the clinically used antitumor drugs ara-C and 5-fluorodeoxyuridine (5-FdU) with NOAC to the heterodinucleoside phosphates arabinocytidylyl-N^4-octadecyl-1-β-D-arabinofuranosyl cytosine (ara-C-NOAC) and 2'-deoxy-5-fluorouridylyl-N^4-octadecyl-1-β-D-arabinofuranosyl cytosine (5-FdU-NOAC).[70,71] Ethynylcytidine (1-(3-C-ethynyl-β-D-ribo-pentafuranosyl)-cytosine, ETC) is a novel nucleoside that was found to be highly cytotoxic.[72] Its combination with NOAC yields the lipophilic duplex drug ETC-NOAC (3'-C-ethynylcytidylyl-(5'→5')-N^4-octadecyl-1-β-D-arabinofuranosyl cytosine). Due to the combination of the effects of both active molecules that can be released into the cytoplasm as monomers or as the corresponding monophosphates (MP), the cytotoxic activity of the duplex drugs is expected to be more pronounced as compared to the monomeric drugs. Further, it can be anticipated that the monophosphorylated nucleosides ara-CMP, 5-FdU-MP and ETC-MP, respectively, are directly released in the cell after enzymatic cleavage of the parent drugs. Thus, monophosphorylated molecules do not need to pass the first phosphorylation step, which is known to be rate limiting.[73,74] The lipophilic side chains warrant a stable incorporation of these duplex drugs into liposomes, allowing the exploitation of the liposome formulations advantages.

We conclude that the chemical modification of water-soluble molecules by attachment of long lipophilic chains and their stable incorporation into bilayer membranes of small unilamellar liposomes represent very promising examples of taking advantage of the high loading capacity lipid bilayers offer for lipophilic drugs. The combination of chemical modifications of water soluble drugs with their pharmaceutical formulation in liposomes is a valuable method for the development of novel pharmaceutical preparations not only for the treatment of tumors or infectious diseases, but also for many other disorders.

Liposome-Mediated Depletion of Tumor Associated Macrophages

The physical depletion of macrophages located in organs of the monocytic phagocyte system (MPS; spleen, liver, lymph nodes, bone marrow) by liposome encapsulated clodronate (clodrolip) has become an important and reliable method to study the roles of macrophages in the immune system and in inflammatory processes.[75-78] Even though the infiltration of macrophages into solid tumors and their pro-tumorigenic function has been described three decades ago, their use as potential therapeutical targets is only now being discussed.[79-82] Tumor cells shed chemokines that attract macrophages from the peripheral circulation. These macrophages infiltrate the stroma of solid tumors and accumulate in hypoxic tumor tissue. Tumor associated macrophages (TAMs) play a pivotal role in tumor growth and metastasis by promoting tumor angiogenesis. Recently, we have investigated whether the depletion of TAMs would inhibit tumor angiogenesis and consequently tumor growth. We show that TAM depletion mediated by clodrolip inhibits tumor growth, presumably through blocking tumor angiogenesis and promoting tumor cell starving. Clodrolip are liposomes containing the drug dichloromethylenebisphosphonic acid (also known as clodronate). In our experiments, tumor bearing mice were treated with clodrolip as single therapy in comparison to free clodronate and in combination with anti-VEGF single chain fragment antibodies,

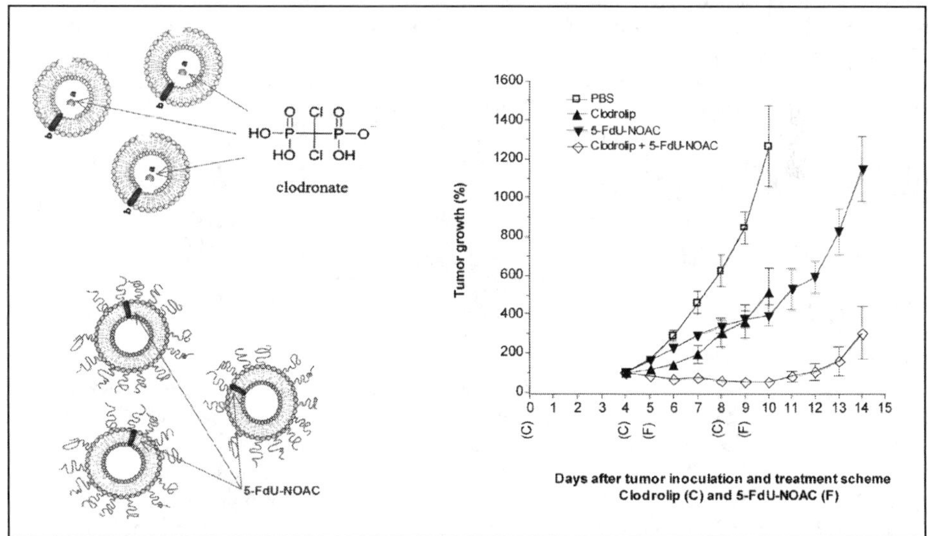

Figure 3. Depletion of tumor associated macrophages in combination with liposomal chemo-therapy. Treatment of F9 teratocarcinoma tumors in syngeneic Sv129 mice, either with clodronate in plain liposomes (clodrolip) and 5-FdU-NOAC in pegylated long circulating liposomes alone (black triangles) or in combination (open diamonds). Phosphate buffer (PBS) treated controls are shown with open squares. The experiment was performed as described in reference 36.

resulting in drastic tumor growth inhibition and exhaustion of TAM cell populations.[83] In a representative experiment shown in Figure 3 we treated mice bearing syngeneic F9 teratocarcinoma tumors with clodronate and the lipophilic heterodinucleotide duplex drug 5-FdU-NOAC, both applied in liposome formulations. Macrophage depletion combined with a cytotoxic therapy was highly effective in this tumor model. Based on our results we conclude that clodrolip mediated depletion of TAMs in concert with cytotoxic or anti-angiogenic treatment regimens represents a new and highly effective therapeutic modality for the treatment of solid tumors and prevention of metastasis. Further, this is an interesting tool for the study of macrophage function in solid tumors.

Cationic Liposomes and Lipoplexes as DNA Delivery Systems

Liposomes can be used as DNA drug delivery systems either by entrapping DNA-based therapeutics inside the aqueous liposome core or by complexing them to positively charged lipids (lipoplexes, see below). Liposomes offer significant advantages over viral delivery systems. They are generally nonimmunogenic because of the absence of protein components. Liposome encapsulated DNA molecules are protected from nuclease activity for enhanced biological stability. Cationic polymers have an enormous potential for DNA complexation and have shown to be useful as nonviral vectors for gene therapy applications. In past years, liposomes composed of cationic lipids, termed lipoplexes, have routinely been utilized for the delivery of nucleic acids such as plasmids, oligodeoxynucleotides and siRNA to cells in culture and in vivo. A large number of these reagents are commercially available or can be formulated in the laboratory.[7-9,84-91] The majority of cationic lipid-DNA complexes form a multilayered structure with DNA molecules intercalated between the cationic lipids. An inverted hexagonal structure with single DNA strands encapsulated in lipid tubules is observed rarely.[92] Together with other advantages the lipoplexes have the ability to transfer very large genes into cells. However, the understanding of their mechanisms of action is still incomplete and their cell transfection efficiencies remain low compared to those of viruses. Despite the appreciable success of cationic lipids in gene transfer, toxicity is a main issue for both in vitro and in vivo applications. Inflammatory toxicity represents a typical effect associated with systemic administration of lipoplexes. Recent results indicate that lipoplex gene delivery systems mediate uptake of plasmid DNA by the liver, mainly by the phagocytic Kupffer cells, in which a large amount of cytokines is produced.[93] In addition, these complexes are immunostimulatory, a property that may either be harmful or beneficial. Another disadvantageous property of lipoplex mediated gene transfer is the low transfection efficiency; this has been attributed to the heterogeneity and instability of the lipoplex formulations. Lipoplex size heterogeneity also adversely affects their quality control, scale-up, and long-term shelf stability, which are important issues for pharmaceutical development. Another unwanted property of cationic lipids is the rapid inactivation of their cargo in the presence of serum proteins.[94] Development of optimized cationic lipids that are safe to use for in vivo applications is an ongoing process. A cautionary note to the potential dangers of all viral gene products, transgenes, viral proteins and peptides and CpG DNA sequences in siRNA or plasmids formulated in liposomes or other DDS has to be given. Immune responses induced by these molecules may lead to problems such as transient gene expression, nonefficient readministration of the same vectors and to severe side-effects in clinical trials.[95] Due to their particulate nature, the DDS are recognized as foreign and thus elicit immune reactions of the host organism. However, the immunomodulating activities of the DDS depend largely on their composition, size and homogeneity. Synthetic polymers can exhibit significant immunomodulatory activity, whereas liposomes prepared with natural phospholipids and cholesterol are known to be less immunogenic.

Outlook and Future Directions

The development of DDS is an ongoing challenging venture that combines multidisciplinary research efforts in various areas including bioengineering, nanotechnology, biomaterials, pharmaceutics, biochemistry, and cell and molecular biology. Specific characteristics of pathological processes and cell or tissue types that are subject of therapeutic interventions govern the path from target selection to the development of specific DDS formulations. The identification of novel cellular targets, for example easily accessible vascular endothelial cells, in contrast to tumor cells or other less targetable tissues, will lead to optimized pharmaceutical drug delivery formulations and preparation technologies. Refinement of DDS in order to overcome unwanted properties such as toxicity, nonspecific tissue distribution and uncontrolled release of entrapped active molecules will be the major challenges in the field. Future DDS will mostly be based on protein, peptide and DNA therapeutics and their next generation analogs and derivatives. Liposome-based formulations will continue to occupy a leading role among the large selection of emerging DDS due to their versatility and vast body of known properties.

Acknowledgements

The author wishes to acknowledge the contributions of Herbert Schott, Cornelia Marty and Steffen Zeisberger.

References

1. Bangham AD, Standish MM, Watkins JC. Diffusion of univalent ions across the lamellae of swollen phospholipids. J Mol Biol 1965; 13:238-252.
2. Sessa G, Weissmann G. Phospholipid spherules (liposomes) as a model for biological membranes. J Lipid Res 1968; 9:310-318.
3. Bangham A. Liposomes: Realizing their promise. Hosp Pract 1992; 27:51-62.
4. Kim S. Liposomes as carriers of cancer chemotherapy. Drugs 1993; 46:618-638.
5. Drummond DC, Meyer O. Hong K et al. Optimizing liposomes for delivery of chemotherapeutic agents to solid tumors. Pharmacol Rev 1999; 51:691-743.
6. Ding BS, Dziubla T, Shuvaev VV et al. Advanced drug delivery systems that target the vascular endothelium. Mol Intervent 2006; 6:98-112.
7. Patil SD, Rhodes DG, Burgess DJ. DNA-based therapeutics and DNA delivery systems: A comprehensive review. AAPS J 2006; 7:E61-E77.
8. Torchilin VP. Recent approaches to intracellular delivery of drugs and DNA and organelle targeting. Ann Rev Biomed Eng 2006; 8:1.1-1.31.
9. Tiera MJ, Winnik FO, Fernandes JC. Synthetic and natural polycations for gene therapy: State of the art and new perspectives. Curr Gene Ther 2006; 6:59-71.
10. Sahoo SK, Labhasetwar V. Nanotech approaches to drug delivery and imaging. Drug Discov Today 2003; 8:1112-1120.
11. Torchilin VP. Recent advances with liposomes as pharmaceutical carriers. Nat Rev Drug Discov 2005; 4:145-160.
12. Moses MA, Brem H, Langer R. Advancing the field of drug delivery: Taking aim at cancer. Cancer Cell 2003; 4:337-341.
13. Gregoriadis G, Wills EJ, Swain CP et al. Drug-carrier potential of liposomes in cancer chemotherapy. Lancet 1974; 1(7870):1313-1316.
14. Rahman YE, Cerny EA, Tollaksen SL et al. Liposome-encapsulated actinomycin D: Potential in cancer chemotherapy. Proc Soc Exp Biol Med 1974; 146:1173-1176.
15. Allen TM. Ligand-targeted therapeutics in anticancer therapy. Nat Rev Canc 2002; 2:750-763.
16. Fenske DB, Cullis PR. Entrapment of small molecules and nucleic acid-based drugs in liposomes. Meth Enzymol 2005; 391:7-40.
17. Barratt G. Colloidal drug carriers: Achievements and perspectives. CMLS Cell Mol Life Sci 2003; 60:21-37.
18. Ramsay EC, Dos Santos N, Dragowska WH et al. The formulation of lipid-based nanotechnologies for the delivery of fixed dose anticancer drug combinations. Curr Drug Deliv 2:341-351.
19. Allen TM. Liposomal drug formulations: Rationale for development and whatA color version of this figure is available online at www.eurekah.com.A color version of this figure is available online at www.eurekah.com.5; we can expect for the future. Drugs 1998; 56:747-756.
20. Gabizon A, Martin F. P:olyethylene glycol-coated (pegylated) liposomal doxorubicin. Rationale for use in solid tumors. Drugs 1997; 54(S4):15-21.
21. Coukell AJ, Spencer CM. Polyethylene glycol-liposomal doxorubicin. Drugs 1997; 53:520-538.
22. Hofheinz RD, Gnad-Vogt SU, Beyer U et al. Liposomal encapsulated anti-cancer drugs. Anti-Cancer Drugs 2005; 16:691-707.

23. Moghimi SM, Szebeni J. Stealth liposomes and long circulating nanoparticles: Critical issues in pharmacokinetics, opsonization and protein-binding properties. Prog Lipid Res 2003; 42:463-478.
24. Sapra P, Allen TM. Ligand-targeted liposomal anticancer drugs. Prog Lipid Res 2003; 42:439-462.
25. Harasym TO, Bally MB, Tardi P. Clearance properties of liposomes involving conjugated proteins -for targeting. Adv Drug Deliv Rev 1998; 32:99-118.
26. Schwendener RA, Trub T, Schott H et al. Comparative studies of the preparation of immunoliposomes with the use of two bifunctional coupling agents and investigation of in vitro immunoliposome-target cell binding by cytofluorometry and electron microscopy. Biochim Biophys Acta 1990; 1026:69-79.
27. Park JW, Hong K, Carter P et al. Development of anti-p185HER2 immunoliposomes for cancer therapy. Proc Natl Acad Sci USA 1995; 92:1327-1331.
28. Kirpotin D, Park JW, Hong K et al. Sterically stabilized anti-HER2 immunoliposomes: Design and targeting to human breast cancer cells in vitro. Biochem 1997; 36:66-75.
29. Park JW, Benz CC, Martin FJ. Future directions of liposome- and immunoliposome-based cancer therapeutics. Semin Oncol 2004; 6(S13):196-205.
30. Mamot C, Drummond DC, Greiser U et al. Epidermal growth factor receptor (EGFR)-targeted immunoliposomes mediate specific and efficient drug delivery to EGFR- and EGFRvIII-overexpressing tumor cells. Cancer Res 2003; 63:3154-3161.
31. Mamot C, Drummond DC, Noble CO et al. Epidermal growth factor receptor-targeted immunoliposomes significantly enhance the efficacy of multiple anticancer drugs in vivo. Cancer Res 2005; 65:11631-11638.
32. Mamot C, Ritschard R, Kung W et al. EGFR-targeted immunoliposomes derived from the monoclonal antibody EMD72000 mediate specific and efficient drug delivery to a variety of colorectal cancer cells. J Drug Target 2006; 14:215-223.
33. Hosokawa S, Tagawa T, Niki H et al. Efficacy of immunoliposomes on cancer models in a cell-surface-antigen-density-dependent manner. Br J Canc 2003; 89:1545-1551.
34. Sapra P, Moase EH, Ma J et al. Improved therapeutic responses in a xenograft model of human B lymphoma (Namalwa) for liposomal vincristine versus liposomal doxorubicin targeted via anti-CD19 IgG2a or Fab' fragments. Clin Cancer Res 2004; 10:1100-1111.
35. Pastorino F, Brignole C, Marimpietri D et al. Doxorubicin-loaded Fab' fragments of anti-disaloganglioside immunoliposomes selectively inhibit the growth and dissemination of human neuroblastoma in nude mice. Cancer Res 2003; 63:86-92.
36. Marty C, Odermatt B, Schott H et al. Cytotoxic targeting of F9 teratocarcinoma tumours with anti-ED-B fibronectin scFv antibody modified liposomes. Br J Canc 2002; 87:106-112.
37. Marty C, Langer-Machova Z, Sigrist S et al. Isolation and characterization of a scFv antibody specific for tumor endothelial marker 1 (TEM1), a new reagent for targeted tumor therapy. Cancer Lett 2006; 235:298-308.
38. Rubio Demirovic A, Marty C, Console S et al. Targeting human cancer cells with VEGF receptor-2-directed liposomes. Oncol Rep 2005; 13:319-24.
39. Gabizon A, Horowitz AT, Goren D et al. In vivo fate of folate-targeted polyethylene glycol liposomes in tumor-bearing mice. Clin Canc Res 2003; 9:6551-6559.
40. Leamon CP, Cooper SR, Hardee GE. Folate-liposome-mediated antisense oligodeoxynucleotide targeting to cancer cells: Evaluation in vitro and in vivo. Bioconjug Chem 2003; 14:738-747.
41. Xu L, Huang CC, Huang W et al. Systemic tumor-targeted gene delivery by anti-transferrin receptor scFv-immunoliposomes. Mol Canc Ther 2002; 1:337-346.
42. Pastorino F, Brignole C, Marimpietri D et al. Vascular damage and anti-angiogenic effects of tumor vessel-targeted liposomal chemotherapy. Cancer Res 2003; 63:7400-7409.
43. Schiffelers RM, Fens MH, Janssen AP et al. Liposomal targeting of angiogenic vasculature. Curr Drug Deliv 2005; 2:363-368.
44. Schiffelers RM, Koning GA, Ten Hagen TL et al. Anti-tumor efficacy of tumor vasculature-targeted liposomal doxorubicin. J Control Release 2003; 91:115-122.
45. Huth US, Schubert R, Peschka-Suss R. Investigating the uptake and intracellular fate of pH-sensitive liposomes by flow cytometry and spectral bio-imaging. J Control Release 2006; 110:490-504.
46. Järver P, Langel U. Cell-penerating peptides - A brief introduction. Biochim Biophys Acta 2006; 1758:260-263.
47. Wagstaff KM, Jans DA. Protein transduction: Cell penetrating peptides and their therapeutic applications. Curr Med Chem 2006; 13:1371-1387.
48. Torchilin VP, Rammohan R, Weissig V et al. TAT peptide on the surface of liposomes affords their efficient intracellular delivery even at low temperature and in the presence of metabolic inhibitors. Proc Nat Acad Sci USA 2001; 98:8786-8791.

49. Marty C, Meylan C, Schott H et al. Enhanced heparan sulfate proteoglycan-mediated uptake of cell-penetrating peptide-modified liposomes. CMLS Cell Mol Life Sci 2004; 61:1785-1794.
50. Rose PG. Pegylated liposomal doxorubicin: Optimizing the dosing schedule in ovarian cancer. Oncologist 2005; 10:205-214.
51. Schwendener RA, Fiebig HH, Berger MR et al. Evaluation of incorporation characteristics of mitoxantrone into unilamellar liposomes and analysis of their pharmacokinetic properties, acute toxicity, and antitumor efficacy. Canc Chemother Pharmacol 1991; 27:429-439.
52. Pestalozzi B, Schwendener R, Sauter C. Phase I/II study of liposome-complexed mitoxantrone in patients with advanced breast cancer. Ann Oncol 1992; 3:419-421.
53. Liu JJ, Hong RL, Cheng WF et al. Simple and efficient liposomal encapsulation of topotecan by ammonium sulfate gradient: Stability, pharmacokinetic and therapeutic evaluation. Anti-Cancer Drugs 2002; 13:709-717.
54. Drummond DC, Noble CO, Guo Z et al. Development of a highly active nanoliposomal irinotecan using a novel intraliposomal stabilization strategy. Cancer Res 2006; 66:3271-3277.
55. Rueda Dominguez A, Olmos Hidalgo D, Viciana Garrido R et al. Liposomal cytarabine (DepoCyte) for the treatment of neoplastic meningitis. Clin Transl Oncol 2005; 7:232-238.
56. Immordino ML, Brusa P, Rocco F et al. Preparation, characterization, cytotoxicity and pharmacokinetics of liposomes containing lipophilic gemcitabine prodrugs. J Control Release 2004; 100:331-346.
57. Bergman AM, Kuiper, CM, Noordhuis P et al. Antiproliferative activity and mechanism of action of fatty acid derivatives of gemcitabine in leukemia and solid tumor cell lines and in human xenografts. Nucleos Nucleot Nucl Acids 2004; 23:1329-1333.
58. Harrington KJ, Syrigos KN, Uster PS et al. Targeted radiosensitisation by pegylated liposome-encapsulated 3', 5'-O-dipalmitoyl 5-iodo-2'-deoxyuridine in a head and neck cancer xenograft model. Br J Canc 2004; 91:366-373.
59. Pignatello R, Puleo A, Puglisi G et al. Effect of liposomal delivery on in vitro antitumor activity of lipophilic conjugates of methotrexate with lipoamino acids. Drug Deliv 2003; 10:95-100.
60. Stevens PJ, Sekido M, Lee RJ. A folate receptor-targeted lipid nanoparticle formulation for a lipophilic paclitaxel prodrug. Pharm Res 2004; 21:2153-2157.
61. Lundberg BB, Risovic V, Ramaswamy M et al. A lipophilic paclitaxel derivative incorporated in a lipid emulsion for parenteral administration. J Control Release 2003; 86:93-100.
62. Lopez-Barcons LA, Zhang J, Siriwitayawan G et al. The novel highly lipophilic topoisomerase I inhibitor DB67 is effective in the treatment of liver metastases of murine CT-26 colon carcinoma. Neoplasia 2004; 6:457-467.
63. Fahr A, van Hoogevest P, May S et al. Transfer of lipophilic drugs between liposomal membranes and biological interfaces: Consequences for drug delivery. Eur J Pharm Sci 2005; 26:251-265.
64. Ten Tije AJ, Verweij J, Loos WJ et al. Pharmacological effects of formulation vehicles: Implications for cancer chemotherapy. Clin Pharmacokinet 2003; 42:665-685.
65. Strickley RG. Solubilizing excipients in oral and injectable formulations. Pharm Res 2004; 21:201-230.
66. Hamada A, Kawaguchi T, Nakano M. Clinical pharmacokinetics of cytarabine formulations. Clin Pharmacokinet 2002; 41:705-718.
67. Schwendener RA, Schott H. Lipophilic 1-β-D-arabinofuranosyl cytosine derivatives in liposomal formulations for oral and parenteral antileukemic therapy in the murine L1210 leukemia model. J Canc Res Clin Oncol 1996; 122:723-726.
68. Schwendener RA, Friedl K, Depenbrock H et al. In vitro activity of liposomal N^{4}octadecyl-1-β-D-arabinofuranosyl cytosine (NOAC), a new lipophilic derivative of 1-β-D-arabino-furanosyl cytosine on biopsized clonogenic human tumor cells and hematopoietic precursor cells. Invest New Drugs 2001; 19:203-210.
69. Schwendener RA, Schott H. Lipophilic arabinofuranosyl cytosine derivatives in liposomes. Meth Enzymol 2005; 391:58-70.
70. Cattaneo-Pangrazzi RMC, Schott H, Wunderli-Allenspach H et al. The novel heterodinucleoside dimer 5-FdU-NOAC is a potent cytotoxic drug and a p53-independent inducer of apoptosis in the androgen-independent human prostate cancer cell lines PC-3 and DU-145. Prostate 2000; 45:8-18.
71. Cattaneo-Pangrazzi RMC, Schott H, Wunderli-Allenspach H et al. New amphiphilic heterodinucleoside phosphate dimers of 5-fluorodeoxyuridine (5FdUrd): Cell cycle dependent cytotoxicity and induction of apoptosis in PC-3 prostate tumor cells. Biochem Pharmacol 2000; 60:1887-1896.
72. Takatori S, Kanda H, Takenaka K et al. Antitumor mechanisms and metabolism of the novel antitumor nucleoside analogues, 1-(3-C-ethynyl-β-D-ribo-pentofuranosyl)cytosine and 1-(3-C-ethynyl-β-D-ribopento-furanosyl)-uracil. Canc Chemother Pharmacol 1999; 44:97-104.

73. Krise JP, Stella VJ. Prodrugs of phosphates, phosphonates, and phosphinates. Adv Drug Deliv Rev 1996; 19:287-310.
74. Wagner CR, Iyer VV, McIntee EJ. Pronucleotides: Toward the in vivo delivery of antiviral and anticancer nucleotides. Med Res Rev 2000; 20:417-451.
75. Seiler P, Aichele P, Odermatt B et al. Crucial role of marginal zone macrophages and marginal zone metallophils in the clearance of lymphocytic choriomeningitis virus infection. Eur J Immunol 1997; 27:2626-2633.
76. Roscic-Mrkic B, Schwendener RA, Odermatt B et al. Roles of macrophages in measles virus infection of genetically modified mice. J Virol 2001; 75:3343-3351.
77. Tyner JW, Uchida O, Kajiwara N et al. CCL5/CCR5 interaction provides anti-apoptotic signals for macrophage survival during viral infection. Nat Med 2005; 11:1180-1187.
78. Van Rooijen N, Kors N, Kraal G. Macrophage subset repopulation in the spleen: Differential kinetics after liposome-mediated elimination. J Leukocyte Biol 1989; 45:97-104.
79. Mantovani A, Allavena P, Sica A. Tumour-associated macrophages as a prototypic type II polarised phagocytic population: Role in tumour progression. Eur J Canc 2004; 40:1660-1667.
80. Joyce JA. Therapeutic targeting of the tumor microenvironment. Canc Cell 2005; 7:513-520.
81. Balkwill F. Cancer and the chemokine network. Nat Rev Canc 2004; 4:540-550.
82. Pollard JW. Tumour-educated macrophages promote tumour progression and metastasis. Nat Rev Canc 2004; 4:71-78.
83. Zeisberger SM, Odermatt B, Marty C et al. Clodronate-liposome-mediated depletion of tumour-associated macrophages: A new and highly effective antiangiogenic therapy approach. Br J Canc 2006; 95:272-281.
84. Ewert K, Evans HM, Ahmad A et al. Lipoplex structures and their distinct cellular pathways. Adv Genet 2005; 53:119-155.
85. Spagnou S, Miller AD, Keller M. Lipidic carriers of siRNA: Differences in the formulation, cellular uptake, and delivery with plasmid DNA. Biochemistry 2004; 43:13348-13356.
86. Shuey DJ, McCallus DE, Giordano T. RNAi: Gene-silencing in therapeutic intervention. Drug Discov Today 2002; 7:1040-1046.
87. May S, Ben-Shaul A. Modeling of cationic lipid-DNA complexes. Curr Med Chem 2004; 11:151-167.
88. Tranchant I, Thompson B, Nicolazzi C et al. Physicochemical optimisation of plasmid delivery by cationic lipids. J Gene Med 2004; 6:S24-S35.
89. Zhdanov RI, Podobed OV, Vlassov VV. Cationic lipid-DNA complexes-lipoplexes-for gene transfer and therapy. Bioelectrochem 2002; 58:53-64.
90. Matsuura M, Yamazaki Y, Sugiyama M et al. Polycation liposome-mediated gene transfer in vivo. Biochim Biophys Acta 2003; 1612:136-143.
91. Yu W, Pirollo KF, Rait A et al. A sterically stabilized immunolipoplex for systemic administration of a therapeutic gene. Gene Ther 2004; 11:1434-1440.
92. Safinya CR. Structures of lipid-DNA complexes: Supramolecular assembly and gene delivery. Curr Opin Struct Biol 2001; 11:440-448.
93. Zhang JS, Liu F, Huang L. Implications of pharmacokinetic behavior of lipoplex for its inflammatory toxicity. Adv Drug Deliv Rev 2005; 57:689-698.
94. Audouy SA, de Leij LF, Hoekstra D et al. In vivo characteristics of cationic liposomes as delivery vectors for gene therapy. Pharm Res 2002; 9:1599-1605.
95. Zhou H, Liu D, Liang C. Challenges and strategies: The immune responses in gene therapy. Med Res Rev 2004; 24:748-761.

CHAPTER 10

Fluorescent Nanoparticle for Bacteria and DNA Detection

Wenjun Zhao, Lin Wang and Weihong Tan*

Abstract

Using bioconjugated dye-doped silica nanoparticles (NPs), we have developed a bioassay for the accurate determination of a single bacterial cell within 20 minutes without any signal amplification or sample enrichment. The antibody-conjugated NPs can specifically and quantitatively detect bacteria, such as *Escherichia coli* O157:H7 from beef through antibody-antigen recognition. Dye-doped silica NPs have also been successfully used for DNA detection at sub-fentomolar concentrations. Our results demonstrate the potential of dye-doped silica NPs for broad applications in practical biotechnological and medical applications in various biodetection systems. The ultimate goal of integrating bionanotechnology into complex biological systems will emerge as a revolutionary tool for ultrasensitive detection of disease markers and infectious agents.

Introduction

Highly sensitive fluorescence-based detection techniques have been extensively used in both biological research and clinical diagnosis; among these, organic fluorophores are the most commonly used signal transduction tool. Although organic fluorophores are versatile and easy to use, they suffer from problems inherent to their molecular nature that limits their sensitivity, photostability and bioconjugation possibilities.[1] For instance, in most cases only one or several fluorophores can attach to a biomolecule without interfering with the target's binding specificity or causing it to precipitate. Thus a biomolecule recognition event is signaled by only one or a few flourophores bound to the target. If a sample has only trace amounts of the biological analyte of interest, additional steps required for signal amplification can be time-consuming and impede analyte quantification.[2] Furthermore when exposed to a continuous light source or the complex environment inside living cells, organic fluorophores degrade and photobleach rapidly. These two factors can result in false-positive or false-negative signals and can affect prolonged cell monitoring and 3-D optical sectioning imaging. Moreover, although most organic fluorophores can be conjugated with biomolecules such as DNA and proteins, different conjugation chemistry must be used to attach the fluorophore to a given biomolecule of interest. This chemistry can be too difficult, time-consuming, and/or expensive for routine applications. All of these limitations have hindered the use of fluorophores for certain biomedical research and disease diagnosis.

*Corresponding Author: Weihong Tan—Center for Research at The Bio/Nano Interface, Department of Chemistry, Shands Cancer Center and UF Genetics Institute, University of Florida, Gainesville, Florida 32611-7200, U.S.A. Email: tan@chem.ufl.edu

Bio-Applications of Nanoparticles, edited by Warren C.W. Chan. ©2007 Landes Bioscience and Springer Science+Business Media.

NPs have shown great promise in bioanalysis and diagnostic applications because of their unique optical properties, high surface-to-volume ratio, and other size-dependent properties. When designed with surface modifications, these special properties provide probes for ultrasensitive and selective bioassays.[2] Dye-doped silica NPs are just such a class of materials. Varying in size between 2 and 200 nm in diameter, these NPs contain large quantities of dye molecules housed inside a polymer or silica matrix. This structure gives them a fluorescence signal that is up to 10^4 times that of an organic fluorophore.[7] When dye-doped NP probes are used, a biomolecule recognition event signaled by one NP will link thousands of dye molecules to one target molecule, thus greatly enhance the fluorescence signal. This signal enhancement facilitates ultrasensitive analyte determination and the monitoring of rare biological events that are otherwise undetectable with existing fluorescence labeling techniques. The polymer and silica matrix serves as a protective shell, or dye isolator, limiting the photobleaching effect of the outside environment and light exposure on the flourophores.

Compared with polymer NPs, silica NPs possess several advantages. Silica NPs are easy to separate via centrifugation during particle preparation, surface modification, and other solution treatment processes because silica is denser than polystyrene (e.g., 1.96 g/cm^3 for silica vs. 1.05 g/cm^3 for polystyrene). Furthermore silica NPs are more hydrophilic, biocompatible, not subject to microbial attack, and no swelling or porosity change occurs with changes in pH.[8] This article will examine the current diverse applications of silica NPs in molecular recognition as well as offer an outlook on the potential uses of dye-doped silica NPs for bioanalysis and biotechnology.

NP Preparation and Bioconjugation

Dye-doped silica NPs can be prepared by two general synthetic routes: the Stöber process or the microemulsion process. In 1968, Stöber et al. introduced a method for synthesizing fairly monodisperse silica NPs with diameters ranging between 50 nm and 2 μm. The details of the mechanism of Stöber-based NP formation have been extensively investigated,[10-12] and the method has been optimized to synthesize dye-doped silica NPs by covalently attaching fluorescent organic dye molecules to the silica matrix.[13-15] Dye-doped silica NPs can also be synthesized by hydrolyzing TEOS (tetraethyl orthosilicate) in a reverse-micelle or water-in-oil (w/o) microemulsion system—a homogeneous mixture of water, oil, and surfactant molecules.[7,17-21]

For bioanalysis and biotechnological applications, dye-doped silica NPs must be linked to target biorecognition elements such as antibodies and DNA molecules. Many of these molecules can be physically adsorbed onto the silica NP surface. However, covalent attachment of biorecognition elements to the particle surface is preferred, not only to avoid desorption from the particle surface but also to control the number and orientation of the immobilized biorecognition elements. For covalent attachment, the particle surface first needs to be modified with suitable functional groups (e.g., thiol, amine, and carboxyl groups). This is typically done by applying a stable additional silica coating (post-coating) that contains the functional group(s) of interest. For the Stöber NPs, surface modification is usually done after NP synthesis to avoid potential secondary nucleation. Surface modification of microemulsion NPs can be achieved in the same manner or via direct hydrolysis and cocondensation of TEOS and other organosilanes in the microemulsion solution.[22]

After the NPs are modified with different functional groups, they can act as a scaffold for the grafting of biological moieties (DNA oligonucleotides, aptamers, antibodies, peptides, etc.) by means of standard covalent bioconjugation schemes. For example, carboxyl-modified NPs have pendent carboxylic acids, making them suitable for covalent coupling of proteins and other amine-containing biomolecules via water-soluble carbodiimide reagents.[23] Disulfide-modified oligonucleotides can be immobilized onto thiol-functionalized NPs by disulfide-coupling chemistry.[24] Amine-modified NPs can be coupled to a wide variety of haptens and drugs via succinimidyl esters and iso(thio)cyanates. Other approaches use electrostatic interactions between NPs and charged adapter molecules[25,26] or between NPs and proteins modified to incorporate charged domains. The bioconjugation or labeling strategy is rationally designed on the basis of the biomolecular function of the surface-attached entities. For example, protein recognition sites are oriented away from the NP surface to ensure a sufficient binding efficiency.[27] After the bioconjugation step, the NPs can be separated from unbound biomolecules by centrifugation, dialysis, filtration, or other separation techniques.

Bioconjugated NPs for Bacteria Detection

We have developed a novel fluorescent NP-based immunoassay for the precise and rapid detection of specific bacterial species.[28] The analysis of bacteria is vital for food safety, clinical diagnosis, and anti-bioterrorism. However, traditional methods for the detection of trace amounts of bacteria require laborious amplification or enrichment of the target bacteria in the sample.[29-31] Recently, many attempts have been made to improve the sensitivity of bacteria detection without the need for target amplification and enrichment,[32-34] but rapid bacterial detection down to the single-cell level has been a challenging task. The two major challenges for the rapid detection of a single bacterium are the achievement of (i) short to real-time detection and (ii) ultrasensitivity in bioanalysis. To reduce the time required for target detection, a minimal amount of sample manipulation is essential. The sensitivity of the detection method has to be high enough to eliminate the need for signal amplification and sample enrichment steps and also allows for the accurate identification of a single bacterium in a short period.

By using bioconjugated NPs, we have developed a fluorescence-based immunoassay for the accurate determination of a single bacterial cell within 20 minutes without any amplification or enrichment. *Escherichia coli* O157:H7 is chosen as the target bacterium as it is one of the most dangerous agents of food-borne diseases. Several of the reported outbreaks of *E. coli* O157:H7 infections have led to death especially among children and the elderly. Given the low infectious dose of *E. coli* O157:H7 (about 10-100 cells), the presence of even a single bacterium in food may pose a serious health risk. Therefore, easy, rapid, and sensitive detection of trace amounts of *E. coli* O157:H7 is critical to minimize or eliminate potential infections.

NPs approximately 60 nm in size were first conjugated to a monoclonal antibody (mAb) specific against *E. coli* O157:H7. These NP-antibody conjugates bound only to the antigen on the *E. coli* O157:H7 bacteria surface (Fig. 1). As each NP encapsulates thousands of fluorescent dye molecules in a protective silica matrix, each time a NP binds with an antigen it provides a much higher signal than conventional immunoassays. Because the mAb was highly selective for the antigen on the bacteria surface the antibody-conjugated NPs specifically associated with

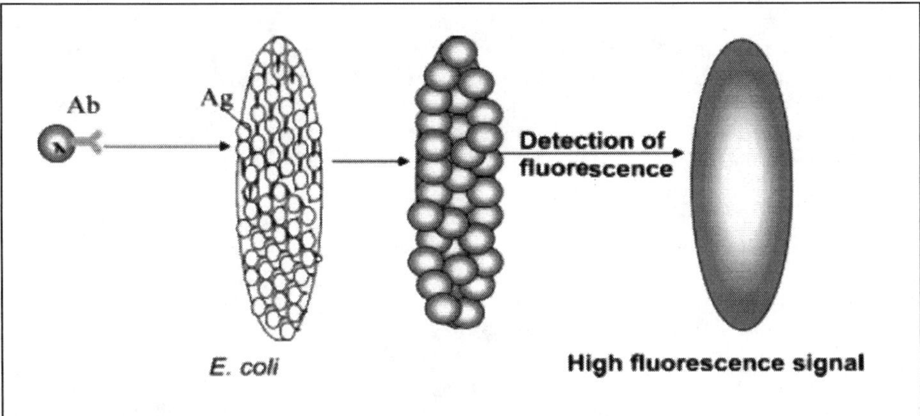

Figure 1. Single bacterium detection scheme. One single bacterium will bind with many NPs for signaling.

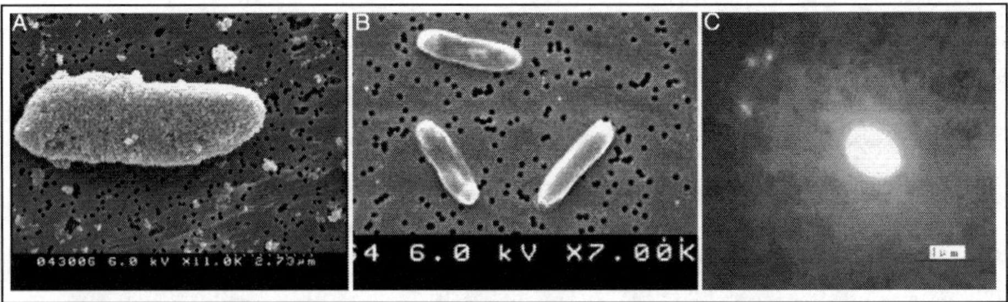

Figure 2. Images of bacterial cells. A) Scanning electron microscope image of *E. coli* O157:H7 cell incubated with antibody-conjugated NPs. B) Scanning electron microscope image of *E. coli* DH5α cell (negative control) incubated with NPs conjugated with antibody for *E. coli* O157:H7. (C) Fluorescence image of *E. coli* O157:H7 after incubation with antibody-conjugated NPs. The fluorescence intensity is strong, enabling single-bacterium cell identification in an aqueous solution.

E. coli O157:H7 cell surfaces (Fig. 2A) but not with other closely related species such as *E. coli* DH5α, which lacks the surface O157:H7 antigen (Fig. 2B). The scanning electron microscope image of the *E. coli* O157:H7 cell after incubation with the NPs showed that there were thousands of antibody-conjugated NPs bound to a single bacterium, providing significant fluorescent signal amplification as compared with a single dye molecule. The NP-based signal amplification can be easily seen with a fluorescence microscope, as shown in Figure 2C. In solution-based experiments for bacteria detection, we have demonstrated that signal amplification by the antibody-conjugated NPs is >1,000 times greater than that produced with dye molecule-labeled antibody. In the comparison experiment, a conventional organic fluorophore—tetramethylrhodamine was chosen to label the mAb to *E. coli* O157:H7. The signal from the bacteria of this conventionally labeled fluorophore was much weaker. The high fluorescence signal enhancement by the NP-based antibody can provide a foundation for the rapid detection of a single bacterium in solution samples. In addition to the high intensity of the fluorescence signal, the NPs were highly photostable because of the protective function of the silica matrix. After 20 minutes of continuous excitation, the fluorescence intensity remained constant.

High-Throughput and Quantitative Detection of Bacteria

In order to adapt this NP-based assay for routine use in toxicology screenings, bioterrorism detection, or medical diagnosis, high-throughput determination of multiple bacterial samples is critical. The single-bacterium assay described above can be easily adapted for multiple-sample determination by testing many aliquots of samples simultaneously. With the plate reader fluorometer, we were able to detect >300 samples at one time with a single-bacterium detection limit. The identification of a bacterium was based on the fluorescence intensities measured in each well of a 384-well plate. Using this method, we were able to determine the existence of a single bacterium with 99.99% accuracy when compared with the widely used colony forming units/ml count (CFU) on agar plates. Our results were further confirmed using a laboratory-made flow-cytometer which precisely detected single bacterial cells with antibody-conjugated NPs bound to them and recording the fluorescent spike when the cells flew through a detection channel. Laser excitation and an optical design for the collection of the fluorescence emission in the orthogonal direction of the forward scattered light beam made the cytometric analysis more efficient and accurate. In the current scheme, a micrometer-sized capillary flow cell and the narrow focusing of the excitation light beam reduces the probing volume of the sample to a few picoliters. Moreover, this design decreases the chance of detecting two or multiple events simultaneously. The total time for the sample detection and analysis with the present system was 60 seconds, which minimized the duration of the bacterial assay even further.

Detection of Bacteria from Beef Samples

To test the usefulness of our bioassay method for bacteria detection in real samples, we determined the number of *E. coli* O157:H7 in several spiked ground beef samples. The recovery rate of the spiked bacteria from the ground beef increased from 50% to 90% as the number of spiked bacteria increased from 2 to 400. The recovered samples were equally divided into two portions as described above. One portion was analyzed with a CFU count on LB agar plates and the other portion was subjected to fluorescence detection with the antibody-conjugated NPs. It should be noted that the colony morphology of the *E. coli* O157:H7 on LB agar was easily distinguishable from other bacteria derived from the ground beef. The number of bacterial cells determined by the two methods was highly correlated, with a correlation factor of 0.99. This result clearly demonstrates that the bacterium assay based on bioconjugated NPs can be used to effectively detect a single bacterium in solution recovered from a ground beef sample within 20 minutes. We conducted both positive and negative control experiments to confirm that the effects of potential interference, such as fat in the ground beef, were negligible.

Using NPs, we have developed an assay tool which enables the detection of one bacterium cell per given sample within 20 minutes using only a spectrofluorometer. In addition, we have designed a simple flow cytometry device to detect antibody-conjugated NPs bound to single bacterial cells. The two detection methods correlated extremely well. Furthermore, we were able to detect multiple samples with high throughput by using a 384-well microplate format. To show the usefulness of this assay, we have accurately detected 1-400 *E. coli* O157 bacterial cells in spiked ground beef samples.

In summary, dye-doped silica nanoparticles can be used as a tool for fast and ultrasensitive bacteria detection. A single bacterium can be detected quickly and accurately without any signal amplification or sample enrichment. Antibody-bioconjugated NPs provide significant signal amplification for the bioanalysis and enumeration of bacteria. This bioassay is rapid (<20 minutes from bacteria binding to detection and analysis), convenient, and highly selective. Furthermore, because multiple samples can be analyzed simultaneously, this assay is adaptable to high-throughput bioanalysis for multiple pathogens. In addition, the accurate and reliable detection of trace amounts of *E. coli* O157:H7 in spiked ground beef samples demonstrates the practical usefulness of this assay system.

Using Bioconjugated NPs to Detect DNA

In recent years high-throughput technologies such as microarrays have changed the ways researchers address biological problems. DNA microarrays have been widely used in genomics, drug discovery, and clinical diagnosis. However, accurate detection of low-abundance targets remains a challenge due to the sensitivity limitation of current techniques. NPs, with their intense fluorescence signal, can be applied to DNA hybridization assays offering distinct advantages over traditional fluorescence-based techniques. The scheme is based on a sandwich assay:[1,2] capture DNA is immobilized onto a glass surface and hybridized with an unlabeled target DNA complementary to both the capture and the probe DNA sequences. Probe DNA attached to dye-doped silica NPs is then added. The sandwich assay is employed because it eliminates the need to label the target. The scheme is shown in Figure 3. One probe DNA hybridizes one target DNA and thus immobilizes one dye-doped NP to the surface, bringing many dye molecules for signaling. Via fluorescence intensity monitoring or counting the number of fluorescent spots from the surface-bound dye-doped NPs, DNA target molecules can be detected down to sub-picomolar levels; single nucleotide polymorphism analysis is also feasible.[35]

This strategy could be extended to the use of NPs as fluorescent labels for DNA and protein microarray technology to meet the critical demand for enhanced sensitivity. Rubpy-doped silica NPs have been used in a one-color microarray system showing good correlation with the traditional phycoerythrin labeling method, but with better photostability and a 20 times lower detection limit. Cy3- and Cy5-doped silica NPs have also been used for two-color microarray detection in a sandwich assay format and exhibited 10 times higher sensitivity than conventional

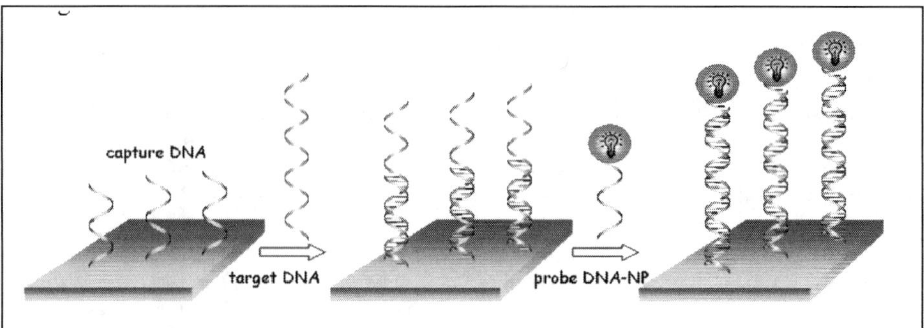

Figure 3. Schematic of a sandwich DNA assay based on NP.

cyanine dyes.[36] Although the technology of NP-based microarrays is still in its infancy, these advances show the great potential of NPs for microarray analysis in genetic screening, proteomics, and medical diagnostics.

Challenges and Trends

The demand for highly sensitive nonradiation based systems for signaling in diverse fields has driven the design of nanomaterials away from electronics and computers and more toward biomedical fields. When nanomaterials are combined with biomolecules for bioanalysis and biotechnology applications, they have demonstrated unique advantages. Each of the NPs can emit an extremely strong fluorescence signal, enabling us to achieve enormous signal amplification for ultrasensitive target detection and for monitoring rare events that are otherwise undetectable with existing labeling techniques. Furthermore, the nanoscale size of the particles minimizes physical interference with the biological recognition events, whereas the nature of silica particles enables us to easily modify the surface for conjugation with various biomolecules for a wide range of applications in the bioassay systems. Moreover, the potential to prepare the NPs with any existing fluorophores provides the diversity of NPs for various applications. By integrating nanotechnology into complex biological systems, we can achieve the detection and prevention of disease at the earliest stage of its development.

Dye-doped silica NPs have been successfully used in bioanalysis but are far from reaching their full potential. Their applications are bound to evolve rapidly as researchers improve their ability to manipulate and apply these NPs. Improvements in the NP dispersion should prevent agglomeration, decrease background noise, and reduce nonspecific adhesion to surfaces. When these photostable and highly fluorescent dye-doped silica NPs are better integrated into the complex biological world, they will have far-reaching impacts in the areas of bioanalysis, molecular imaging, and biotechnology.

References

1. Lin W, Keming W, Swadeshmukul S et al. Watching silica nanoparticles glow in the biological world. Anal Chem 2006; 1:647-654.
2. Tan W, Wang K, He X et al. Bionanotechnology based on silica nanoparticles. Med Res Rev 2004; 5:621-638.
3. Dabbousi BO, Rodriguez Viejo J, Mikulec FV et al. (CdSe)ZnS core-shell quantum dots: Synthesis and characterization of a size series of highly luminescent nanocrystallites. J Phys Chem B 1997; 101:9463-9475.
4. Godovsky DY. Device applications of polymer-nanocomposites. Adv Polym Sci 2000; 153:163-205.
5. Bruchez MJ, Moronne M, Alivisatos AP et al. Semiconductor nanocrystals as fluorescent biological labels. Science 1998; 281:2013-2016.
6. Dahan M, Laurence T, Pinaud F et al. Time-gated biological imaging by use of colloidal quantum dots. Optics Letters 2001; 26:825-827.
7. Zhao X, Bagwe RP, Tan W. Development of organic-dye-doped silica nanoparticles in a reverse microemulsion. Adv Mater 2004; 16:173-176.

8. Jain TK, Roy I, De TK et al. Nanometer silica particles encapsulating active compounds: A novel ceramic drug carrier. J Am Chem Soc 1998; 120:11, (092-11,095).

9. Tan W, Wang K, He X et al. Bionanotechnology based on silica nanoparticles. Med Res Rev 2004; 24:621-638.

10. Tan CG, Bowen BD, Epstein N. Production of monodisperse colloidal silica spheres: Effect of temperature. J Colloid Interface Sci 1987; 118:290-293.

11. Coenen S, De KC. Synthesis and growth of colloidal silica particles. J Colloid Interface Sci 1988; 124:104-110.

12. Van BA, Kentgens AP. Particle morphology and chemical microstructure of colloidal silica spheres made from alkoxysilanes. J Noncryst Solids 1992; 149:161-178.

13. Van BA, Imhof A, Hage W et al. Three-dimensional imaging of submicrometer colloidal particles in concentrated suspensions using confocal scanning laser microscopy. Langmuir 1992; 8:1514-1517.

14. Van BA, Vrij A. Synthesis and characterization of colloidal dispersions of fluorescent, monodisperse silica spheres. Langmuir 1992; 8:2921-31.

15. Verhaegh NA, Blaaderen AV. Dispersions of rhodamine-labeled silica spheres: Synthesis, characterization, and fluorescence confocal scanning laser microscopy. Langmuir 1994; 10:1427-38.

16. Nyffenegger R, Quellet C, Ricka J. Synthesis of fluorescent, monodisperse, colloidal silica particles. J Colloid Interface Sci 1993; 1:150-157.

17. Santra S, Zhang P, Wang K et al. Conjugation of biomolecules with luminophore-doped silica nanoparticles for photostable biomarkers. Anal Chem 2001; 20:4988-93.

18. Santra S, Wang K, Tapec R et al. Development of novel dye-doped silica nanoparticles for biomarker application. J Biomed Opt 2001; 6:160-166.

19. Santra S, Tapec R, Theodoropoulou N et al. Synthesis and characterization of silica-coated iron oxide nanoparticles in microemulsion: The effect of nonionic surfactants. Langmuir 2001; 17:2900-2906.

20. He X, Wang K, Tan W et al. Bioconjugated nanoparticles for DNA protection from cleavage. J Am Chem Soc 2003; 24:7168-7169.

21. Tapec R, Zhao XJ, Tan W. Development of organic dye-doped silica nanoparticles for bioanalysis and biosensors. J Am Chem Soc 2002; 2:405-409.

22. Deng G, Markowitz Michael, Kust PR et al. Control of surface expression of functional groups on silica particles. Mater Sci Eng C 2000; 11:165-172.

23. Gerion D, Pinaud F, Williams SC et al. Synthesis and properties of biocompatible water-soluble silica-coated CdSe/ZnS semiconductor quantum dots. J Phys Chem B 2001; 105:8861-8871.

24. Hilliard LR, Zhao X, Tan W. Immobilization of oligonucleotides onto silica nanoparticles for DNA hybridization studies. Anal Chim Acta 2002; 470:51-56.

25. Roy I, Ohulchanskyy TY, Bharali DJ et al. Optical tracking of organically modified silica nanoparticles as DNA carriers: A nonviral, nanomedicine approach for gene delivery. Proc Natl Acad Sci USA 2005; 102:279-284.

26. Zhu S, Xiang, J, Li X et al. Poly(l-lysine)-modified silica nanoparticles for the delivery of antisense oligonucleotides. Biotechnol Appl Biochem 2004; 39:179-187.

27. Ye Z, Tan M, Wang G et al. Preparation, characterization, and time-resolved fluorometric application of silica-coated terbium(III) fluorescent nanoparticles. Anal Chem 2004; 76:513-518.

28. Zhao X, Hilliard LR, Mechery SJ et al. A rapid bioassay for single bacterial cell quantitation using bioconjugated nanoparticles. Proc Natl Acad Sci USA 2004; 101:15, (027-15,032).

29. Phillips CA. The epidemiology, detection and control of Escherichia coli O157. J Sci Food Agric 1999; 79:1367-1381.

30. Deisingh AK, Thompson M. Strategies for the detection of Escherichia coli O157:H7 in foods. J Appl Microbiol 2004; 96:419-429.

31. Iqbal SS, Mayo MW, Bruno JG et al. A review of molecular recognition technologies for detection of biological threat agents. Biosens Bioelectron 2000; 15:549-578.

32. Song JM, Vo DT. Miniature biochip system for detection of Escherichia coli O157:H7 based on antibody-immobilized capillary reactors and enzyme-linked immunosorbent assay. Anal Chim Acta 2004; 1:115-121.

33. Muhammad T, Zarini A, Evangelyn C. Fabrication of a disposable biosensor for Escherichia coli O157:H7 detection. IEEE Sens J 2003; 3:345-351.

34. Weimer BC, Walsh MK, Beer C et al. Solid-phase capture of proteins, spores, and bacteria. Appl Environ Microbiol 2001; 3:1300-1307.

35. Zhao X, Tapec DR, Tan W. Ultrasensitive DNA detection using highly fluorescent bioconjugated nanoparticles. J Am Chem Soc 2003; 38:11474-11475.

36. Zhou X, Zhou J. Improving the signal Sensitivity and photostability of DNA hybridizations on microarrays by using dye-doped core-shell silica nanoparticles. Anal Chem 2004; 18:5302-5312.

CHAPTER 11

Dendrimer 101

Lajos P. Balogh*

Abstract

In this chapter dendrimer basics are reviewed. It is impossible to describe, refer to or even list the related literature (our "dendrimer" database consists of over 7,000 papers and patents), thus selection of references has been made based on personal preference, often choosing clarity over details and overarching principles instead of detailed chemical structures. A large number of excellent reviews are available to those who are interested in more detail in particular areas.[1-26] References will take the reader to original scientific papers that provide more detail about a particular topic, describe experiments, and draw conclusions reflecting each author's personal views. Even listing of books and reviews must be partial, as they number in the hundreds.

Brief History and Definitions

Historically it was Vögtle's principal work on cascade molecules[27] that opened the door for dendrimers. The name, dendrimer, originates from the Greek words dendron (meaning tree) and meros (meaning part). Originally, the name was coined by Vogel[28] and it literally means: "tree-like".

According to Tomalia,[29] dendritic polymers, also called dendrimers, have three distinguishing architectural features: an initiator core, interior "layers" (subsequent generations) composed of repeating units that are radially attached to the initiator core, and an exterior "layer" or surface of terminal functionality attached to the outermost generation. De Brabander and Meijer[30] defined a "dendrimer" as a highly branched (i.e., degree of branching, DB >1), highly functionalized macromolecule with a well-defined structure. According to Majoral,[31] dendrimers are highly branched regular three-dimensional monodisperse macromolecules with a branch occurring at each monomer unit. Newkome[32] have even proposed a systematic nomenclature based on existing IUPAC rules.

In fact, the term "dendrimer" is interpreted by other scientists with a much broader meaning than the ones described earlier. The term "dendrimer" may refer to a structure, a molecule, or to a certain material (just like a graph of a tree, an actual tree and a forest).

A "dendritic structure" is a geometric representation of regular, repeated and ideal symmetrical branches (dendrons, cascade molecules) connected to a core. Ideal dendrimer structures can be described with exact mathematical expressions, using the terminology of polymer chemistry.[33-35]

*Lajos P. Balogh—Department of Radiation Medicine, Roswell Park Cancer Institute, Department of Oncology, University at Buffalo SUNY, Elm and Carlton Streets, Buffalo, NY 14263, U.S.A. Email: lajos.balogh@roswellpark.org

Bio-Applications of Nanoparticles, edited by Warren C.W. Chan. ©2007 Landes Bioscience and Springer Science+Business Media.

$$DP = N_{RU} = N_c \left[\frac{N_b^{G+1} - 1}{N_b - 1} \right]$$

$$MW = M_c + N_c \left[M_{RU} \left(\frac{N_b^{G} - 1}{N_b - 1} \right) + M_E \left(\frac{N_b^{G}}{N_b - 1} \right) \right]$$

$$N_{EG} = N_c N_b^{G}$$

$N_{EG} = N_c N_b^{G}$

M = molecular weight at generation G,

M = molecular weight at generation G,

M_c = molecular weight of the core,

N_c = the core functionality,

N_b = the degree of branching

M_{RU} is the molecular weight of the repeat unit

ME is molecular weight of the terminal groups

N_{EG} is the number of repeating units. If the core functionality equals to one, dendrons (cascade molecules) result.

Dendrons have a simple cascade structure culminating in a focal point. Attachment of these focal points to any core (atom, molecule, chain, or another dendrimer) results in a dendrimer (structure, molecule, or material). Dendrons of a dendrimer do not have to be identical. The higher the core functionality and the branching functionality, and the shorter the connectors are, the quicker high density ensues in the dendrimer.

Dendritic structures are composed from branching sites and connectors forming symmetric patterns. Repeated motifs that are in equal distance from the core (moving through the connections) are called generations. (The name refers to the stepwise pair of reaction sequences they can be generated by.) Every dendrimer molecule can individually be described by its own chemical composition and chemical structure.

Chemical structure of a dendrimer molecule refers to the spatial arrangement of atoms and their connecting chemical bonds. In geometric and chemical representations of large dendrimer structures in 2D space, symmetry and connectivity are often overemphasized at the expense of angles (that are often distorted) and actual bond lengths (that may be altered). Images of molecular models are computer-generated, thus atomic sizes, angles and distances are kept at proper values. These images are more informative of the general shape and surface structure, but the interior usually cannot be viewed.

Dendrimer families are synthesized using a variety of different synthetic routes, typically applying repeated sequences of reaction (and purification) steps. Some of the significant families are the poly(amidoamine) (PAMAM - Tomalia),[36] arborols (Newkome),[20,37] poly(propyleneimine) (PPI, (De Brabander and Meijer),[30] Frechet's polyether dendrimer,[15] and Majoral's phosphorus-based dendrimers.[31,38]

Constructing a dendritic network from atoms connected by bonds necessarily invokes chemical properties. Therefore dendrimer molecules of various families always have certain properties that are characteristic to their family. Consecutive repeated sequences result in generations. Members (generations) of a dendrimer family contain the same molecular motifs, but their size and mass are different and therefore may have very diverse physical properties (also depending on their termini). For the above reasons, none of the observed properties of any family member can directly be generalized to "dendrimers."

A formal dendrimer nomenclature was suggested later by Wilks for the DuPont SCION database,[39] but these nomenclatures are rarely used in common practice or in the scientific literature. Meijer's naming system for poly(propylene imine) dendrimers follows this scheme:

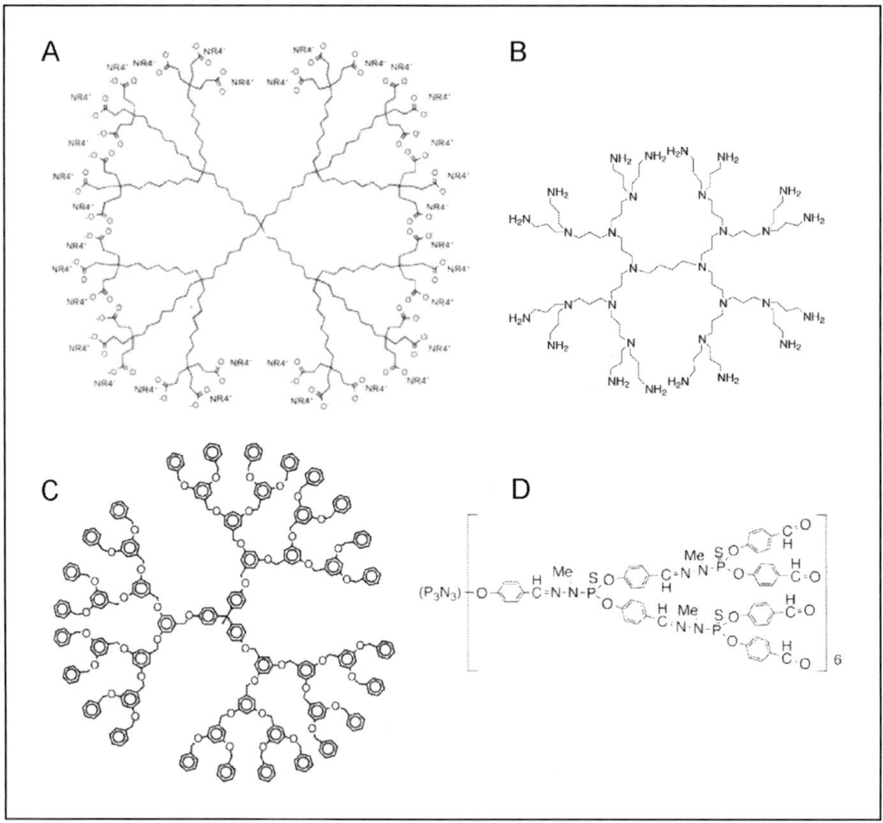

Figure 1. Chemical structures for ideal molecules of four additional significant dendrimer families. A) arborol[5] B) Polypropyleneimine (PPI),[8] Frechet's polyether dendrimer,[15] and structure of a dendron of Majoral's phosphorus-based dendrimers.[16] A) Reprinted with permission from Boas U et al, Chem Soc Rev 2004; 33:43-63, © Royal Society of Chemistry. B) Reprinted with permission from Cloninger MJ, Curr Opinion in Chem Biol 2002; 6:742-748 © Elsevier. C) Reprinted with permission from Grayson SM et al, Chem Rev 2001; 101:3819-3867, Scheme 2 © 2001 American Chemical Society. D) Reprinted with permission from Majoral JP et al, Chem Rev 1999; 99:845-880, Scheme 50 © 1999 American Chemical Society.

Core-*dendr*-termini$_{\#}$ (e.g., DAB-*dendr*-$(NH_2)_{64}$ is a diaminobutane core dendritic poly(propyleneimine) having 64 primary amine termini, also named as DAB-*dendr*-$(PA)_{64}$, where PA stands for polyamine. PAMAM manufacturer Dendritech, MI, uses lot numbers to really identify their products made in separate batches, in which the generations are numbered: e.g., PAMAM 0602-03-G5.0-LD is the third batch of generation five amine terminated PAMAM dendrimer manufactured in June 2002 in technical quality. Sigma-Aldrich material names start with a symbol and added comments on core or termini, e.g., Family-(core)-Termini, followed by linear or "empirical" formulas and formula weights of an ideal dendrimer (not the real average molecular mass). The linear (ideal) formula (or a fraction of it) starts with the word "monomer," then the number of generation followed by the word *"dendri"* referring to the structure, finished with the ideal/theoretical termini (Fig. 1).

Synthetic Strategies

Dendritic molecules can be constructed in virtually any desired composition.[14] To date, hundreds of dendrimer families and dendritic structures have been reported in the literature. Sometimes only a part of the macromolecule has dendritic architecture, and some authors have less stringent ideas regarding what constitutes a dendrimer. The major synthesis strategies remain the divergent and the convergent routes. Both employ either nonreactive or protecting groups either on A or B components of AB_n building blocks. Branching is created during reaction steps by bi- or multifunctional reactants.

The divergent route[34] leads to higher yields with lower dendritic purity.[41] Convergent strategies[15] usually result in much more uniform molecules, although in lower overall yield (a great number of other strategies combining the above strategies have also been reported) (Fig. 2).

Commercial manufacturers use the divergent approach. In divergent synthesis, the mass of produced dendrimers increase in every generation (it nearly doubles minus the loss due to purification steps). Convergent strategies usually produce dendrimer materials with higher dendritic purity (better purification is possible because the dendrons and the product dendrimers are more different) but during the synthesis the absolute mass of the products decreases. A comprehensive description for convergent strategies is available (see ref. 15).

A dendrimer material is a particular substance that is used for practical purposes. Dendrimer materials always contain molecules with both ideal and nonideal structures. Dendrimer materials are made in multistep processes, where synthesis of every generation includes at least two reactions and one or two purification steps. As only perfect reactions (complete conversion with a 100% yield) would result in completely ideal molecules, or perfect purification (total elimination of molecules with less than perfect structure) can lead to perfect materials, any dendrimer material is a mixture of very similar dendrimer molecules.

To describe better the purity of dendrimer materials, the term "dendritic purity" has also been introduced.[41] *Dendritic purity* is defined as the number of dendrimer molecules of perfect structure divided by the total number of dendrimers multiplied by 100 %. As an example, Meijer et al have calculated the polydispersity and dendritic purity for a poly(propyleneimine) (DAB-*dendr*-$(NH_2)_{64}$) dendrimer. In this case, polydispersity of PD=1.0018 and dendritic purity of 15.3 was found from the deconvoluted mass spectra. The overall branching efficiency, as typically used for hyperbranched polymers[42] was calculated to be a = 0.987, i.e., 98.7%-a of all the branching was ideal (Fig. 3).

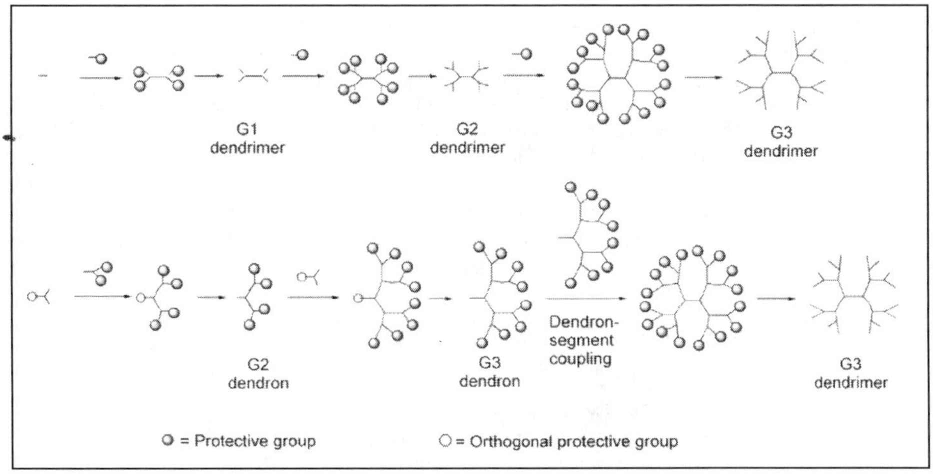

Figure 2. Divergent and convergent strategies (reprinted with permission from Boas U et al, Chem Soc Rev 2004; 33:43-63, © Royal Society of Chemistry).

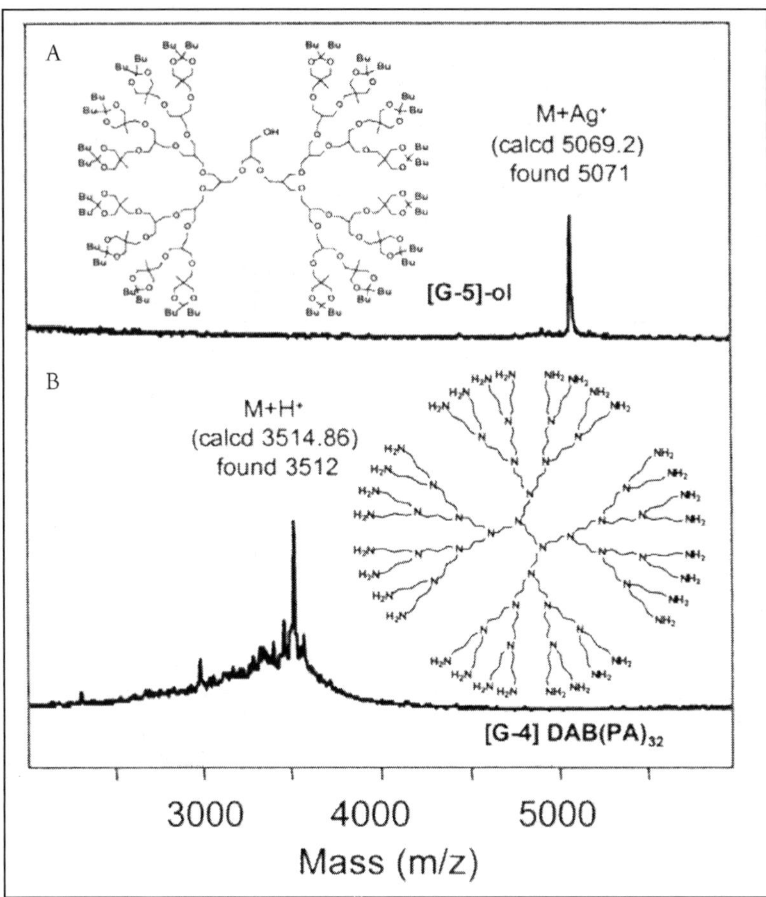

Figure 3. A) shows the mass spectrum of a convergently synthesized fifth generation aliphatic polyether containing two dendrons. In the mass spectrum, only one peak is observed, corresponding to the mass of the ideal molecule plus a silver cation. B) displays the mass spectrum of a commercial grade fourth generation poly(propylene imine) dendrimer indicating the presence of a measurable amount of molecules with nonideal structures.[15] (Reprinted with permission from Grayson SM et al, Convergent Dendrons and Dendrimers: from Synthesis to Applications, Chem Rev 2001; 101:3819-3867, F4 © 2001 American Chemical Society.)

Synthesis of Commercial Dendrimers

Commercial dendrimers are sold as given generations of a particular family. Of the many divergent syntheses studied to date, a few appear to be particularly noteworthy. These include Dow's poly(amidoamine) PAMAM[28] and DSM's poly(propylene imine) PPI dendrimers,[30] Perstorp™ aliphatic polyesters[43] are also made in industrial amounts. As every batch of manufactured dendrimers has its unique synthesis history, all commercial dendrimer materials are somewhat different. It is up to the end user whether these differences are relevant or irrelevant from the point of view of the intended application (Fig. 4).

Although a number of additional very effective synthesis strategies using larger building blocks such as "hypercore",[45] "hypermonomer"[46] and the "double exponential dendrimer growth"[47] have been introduced to create dendritic structures, commercial manufacturing still favors the divergent strategy.

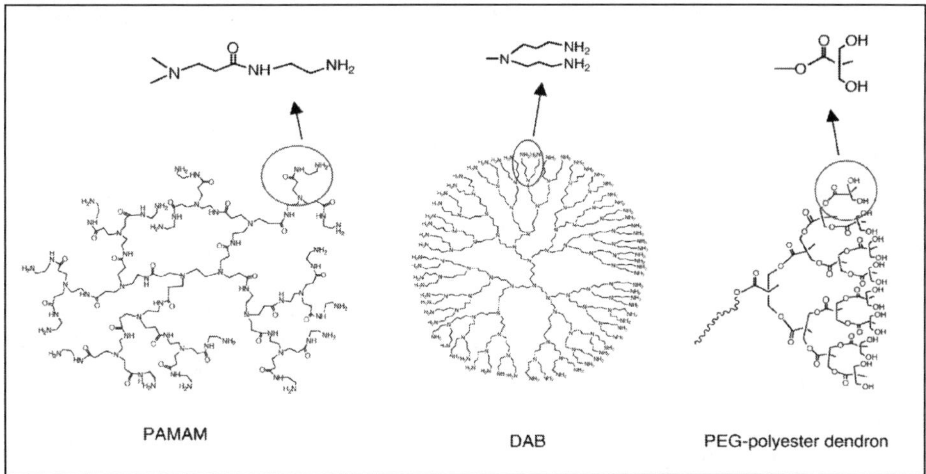

Figure 4. Nominal (ideal) chemical structures of Starburst™, Astramol™, and Perstorp™ materials.[44] (Reprinted with permission from Duncan R et al, Advanced Drug Delivery Reviews 2005; 57:2215-2237 F1c © Elsevier).

Ideal dendritic growth and its possibility have excessively been debated in the literature. Theoretically, ideal dendritic growth may occur only until the steric hindrance between existing terminal groups prevents the attachment of incoming monomers to all available termini (for divergent synthesis) or all the dendrons (for convergent strategies). De Gennes[48] has predicted that after the molecules reached the "dense-packed" state, ideal growth cannot be continued for dendrimer molecules because of the steric hindrance between the "closely packed" termini. A common misunderstanding that dendritic growth cannot continue—it actually can, but not from all termini.

Ideal dendritic growth can only occur when the symmetry is not broken by side-reactions or incomplete conversion. However, in practice, only a certain number of reactions are successful. For convergent syntheses incomplete coupling results in lower molecular weight side-products, that usually can be separated, thus purity can be maintained in convergent synthesis (only the yield will be lower). For divergent approaches, such as PPI and PAMAM dendrimers, side-reactions result in monosubstituted branching sites ("missing arms") and loops[49] in addition to the ideal disubstitution (Fig. 5).

Using the divergent approach, the molecular mass keeps doubling, but the dendritic purity gradually declines. For example, synthesis of PPI dendrimers utilizes both solution chemistry and hydrogenation in gas phase. While successive reactions result in products of high dendritic purity PPI materials, vaporization of molecules is increasingly harder for the consecutive generations, and completion of the -CN -> -CH$_2$NH$_2$ catalytic reduction step is practically impossible after generation five. Low generation Astramol materials are very pure, but high generation materials may also contain a small amount of -CN end groups.

Structural Diversities of Molecules in Dendrimer Materials

The reason why structural diversities are present in dendrimer materials is simple: there are no perfect chemical reactions (Figs. 6,7).

Every dendrimer synthesis is based on iterative, or stepwise, procedures involving a reaction and a subsequent purification step. Let's assume that we achieve the same degree of perfection in every single step. Purification of reaction products may result in the removal of unwanted molecules, but any purification method has to be based on differences in molecular property. The more expressed these properties (e.g., size, chromatographic behavior, volatility, etc.) are,

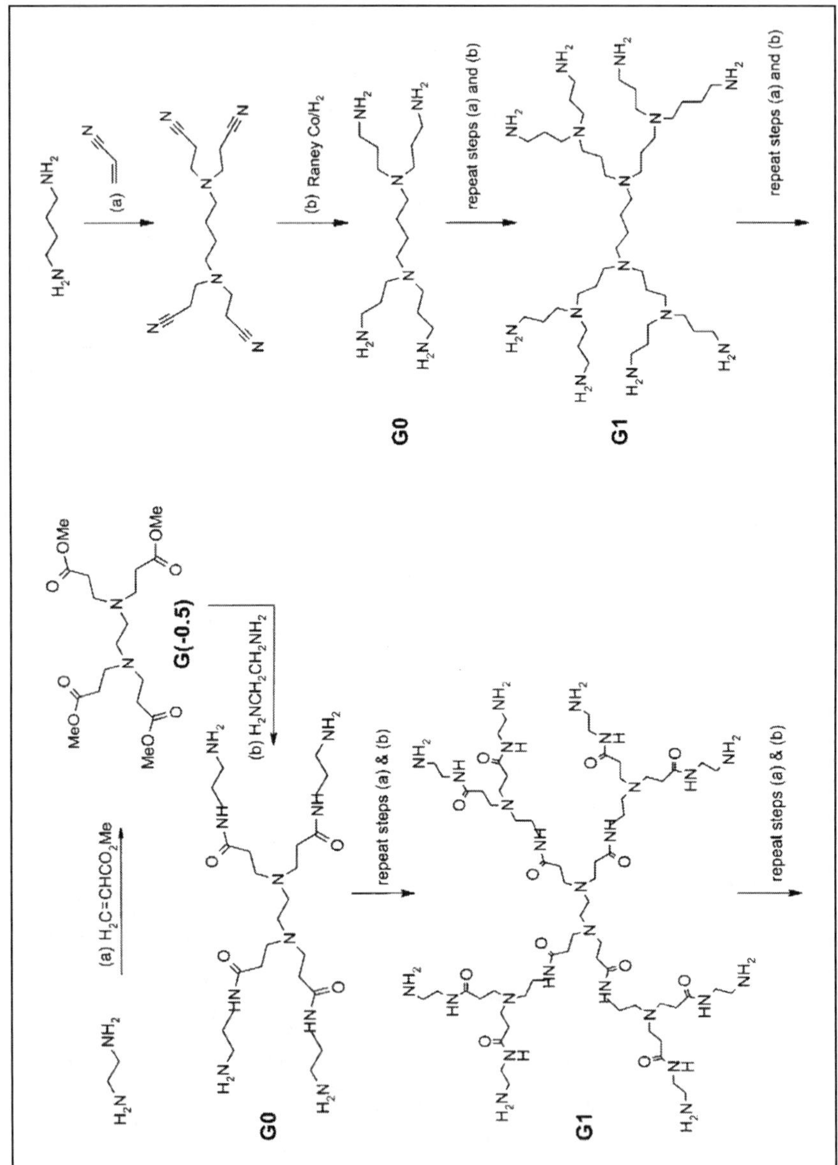

Figure 5. Synthetic schemes of poly(amidoamine) and poly(propyleneimine) dendrimers.[50] (Reprinted with permission from Crooks RM et al, Topics in Current Chemistry 2001; 212:81-135, © Elsevier.)

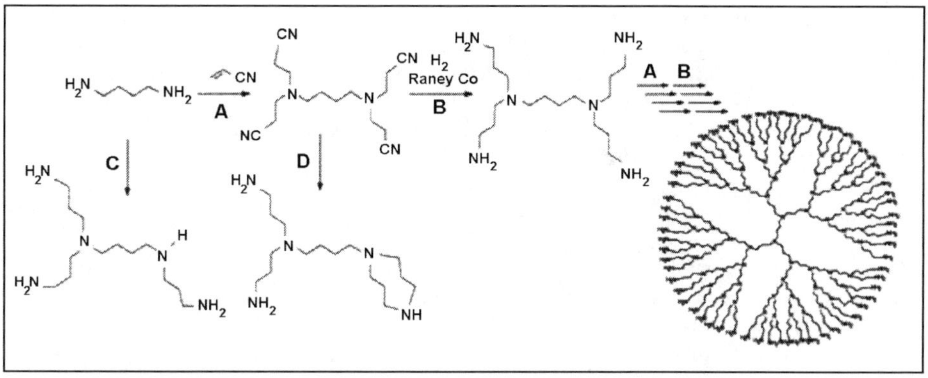

Figure 6. Synthesis and possible side-reactions in PPI manufacturing.[41] (Reprinted with permission from Hummelen JC et al, Chem Eur J 1997; 3(9):1489-1493, Scheme I © WILEY-VCH Verlag GmbH.)

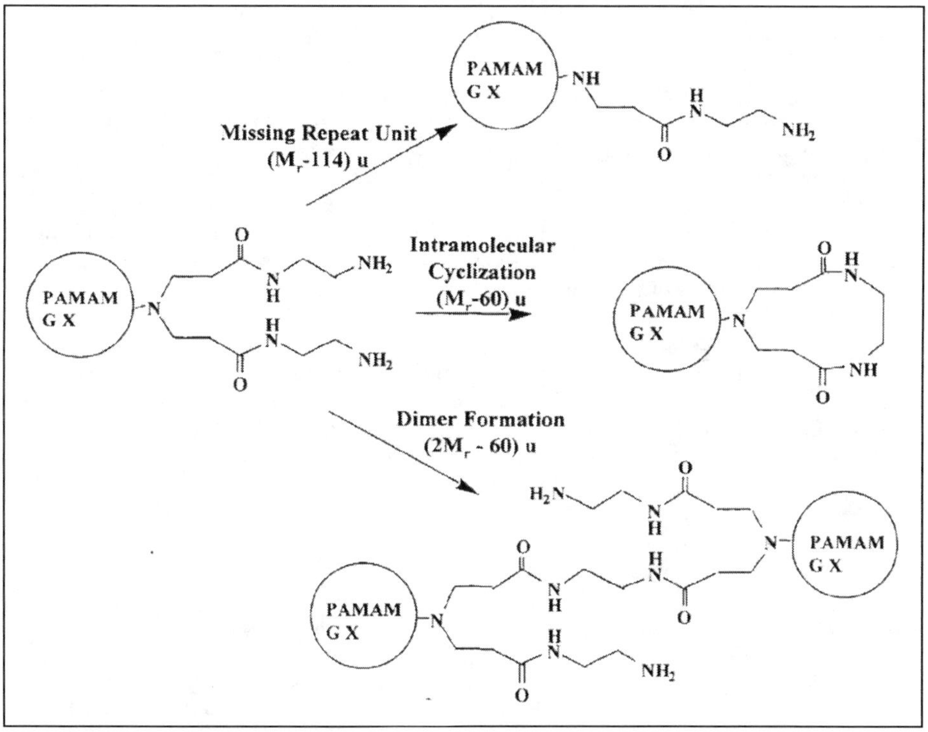

Figure 7. Origin of skeletal diversity in PAMAM materials.[51] (Reprinted with permission from Tolic LP et al, Internat J Mass Spectr Ion Proc 1997; 165/166:405-418 © Elsevier).

the more efficient the separation method. Repeating the same synthetic procedures for a divergent synthesis, the same small structural differences in every generation gradually give rise to relatively smaller and smaller differences in respective properties among the molecules of the material. As purifications turn out to be less and less efficient, more and more molecules with nonideal skeletons become part of the material. Reacting molecules with nonideal skeletons leads to molecules with increasing degree of nonideality, etc.

In any multistep synthesis, the final yield can be calculated as the product of individual conversions. For a generation five dendrimer (assuming divergent strategy and 90% yield of perfect structures in every step, the final yield of perfect structures is only $0.9^{10} = 0.3487$ (only 34% of the molecules would be ideal).

There are three possible types of structural diversities present in dendrimer materials: generational, skeletal, and substitutional diversity[52] (Fig. 8).

In any given dendrimer material there may be different generations present, either due to incomplete conversion of intermediates or due to "trailing" generations (PAMAM) and/or imperfect purification steps. One of the reactants in PAMAM dendrimer synthesis is ethylenediamine. This bifunctional molecule always produces a small amount of dimers whose mass is very close to molecules in the next generation (e.g., a $(PAMAM\_E5.NH_2)_2$ molecule is almost identical with $PAMAM\_E6.NH_2$).

Skeletal diversity arises during dendrimer synthesis and they cannot be corrected or removed by those, who buy the materials—these nonideal molecules will always be present in the subsequent products.

Substitutional diversity arises during post-modification reactions, when less than 100% of the available terminal groups are substituted. Incomplete substitution of all Z-groups may be due to steric hindrance, when not all the terminal groups can react, or intentional, when less then equivalent reactant was added to the dendrimer. Exact locations of the new substituents may differ from molecule to molecule, and as a consequence, the dendrimer material is composed of slightly different molecules. Substitutional diversity always exists in a dendrimer material after partial substitution. Composition of these dendrimers should be described by an average composition and a distribution function.

Dendritic Properties

The presence of a large number of terminal groups and the limitations or a complete lack of interpenetration results in solubility, viscosity, and thermal behaviors different from those of more classical polymers.[7] These properties are referred to as "dendritic" properties.

"Dendritic" properties are the consequence of:
- the structure (cascade symmetry of dendrons),
- the uniformity of molecules (nearly perfect dendritic molecules and narrow polydispersity of the materials),
- the collective and cooperative actions of internal and external functional groups (interactions and reactions);
- the large number of termini (-Z groups) that can provide multiple functionalities to the molecules,
- the high local concentrations of these internal/external functional groups;
- the ability of solvated MGD and HGD molecules to behave as dendrimer "pseudophase";
- the soft to hard (organic) nanoparticle character of MGD and HGD subclasses.

If chemical composition, architecture and size determine properties, than family, structure, generation, and surface substitution must determine dendrimer properties. A number of characteristic general properties of dendrimers can be concluded from the geometric structure itself, being caused by the symmetry and high degree of branching. These properties are the function of following parameters: the core structure, the connector length, the degree of branching, etc., and they appear in all the families (for example, the tendency of dendrons to act in a concerted fashion.[19,54]

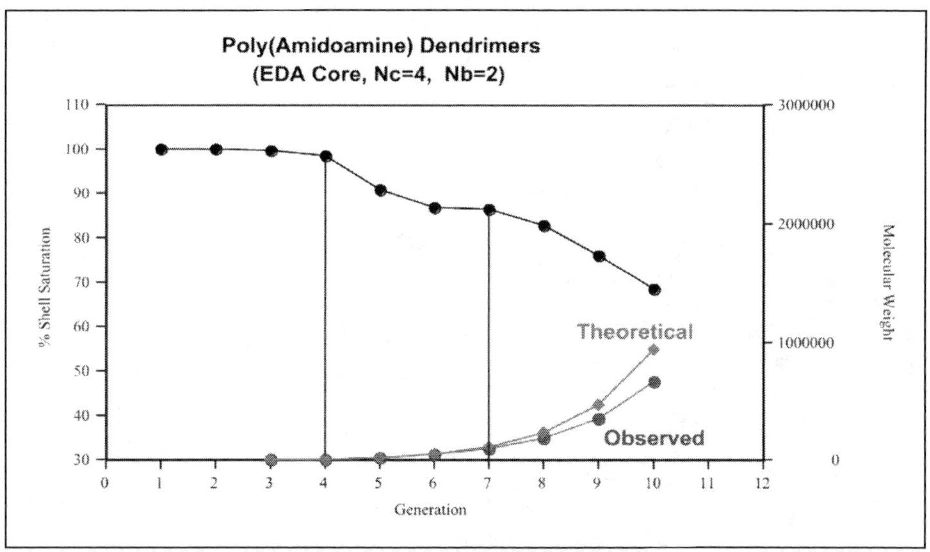

Figure 8. Comparison of theoretical molecular mass, observed molecular mass, and % "shell filling" (i.e., observed efficiency of reacting the termini) for "gold standards" of ethylenediamine core PAMAM dendrimers as a function of generation.[53] (Reprinted with permission from Tomalia DA et al, PNAS, 2002; 99(8):5081-5087 © The National Academy of Sciences of the United States of America.)

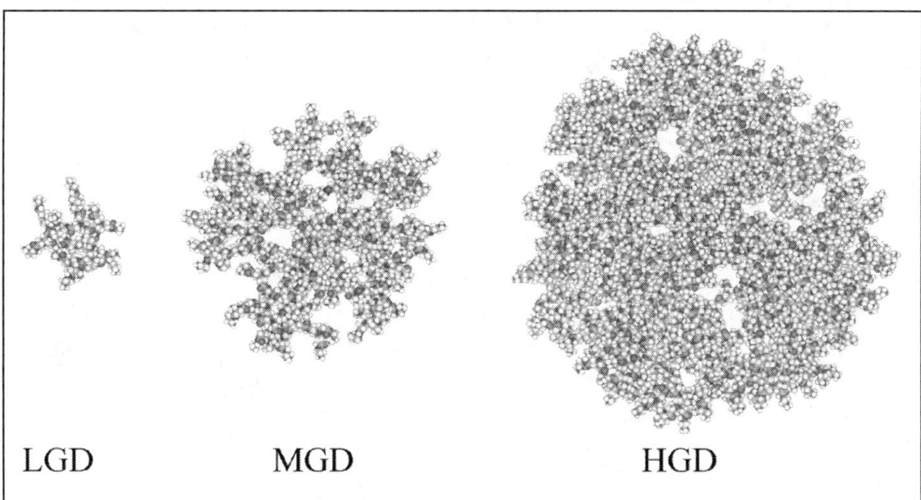

Figure 9. Illustration of LGD, MGD, and HGD dendrimer molecules. (Space filling molecular models of ethylenediamine core PAMAM generation 2, generation 5, and generation 7 primary amine terminated ideal dendrimer molecules.)

To classify properties, it is practical to distinguish classes of 'low", "high", and "middle" generation dendrimer molecules as they carry similar properties. These classes can be found in all families, but at different generational level because of the different structural elements and connector lenghts (Fig. 9).

Low Generation Dendrimer Molecules (LGD)

Low generation dendrimer molecules (LGD) are relatively small molecules or low molecular weight polymers (oligomers) with fully penetrable networks and a few reactive sites that are all equally accessible. Historically, these structures were made first, but they were not appreciated at that time, because "dendritic properties" cannot be and (consequently were not) observed.

High Generation Dendrimer Molecules, (HGD)

High generation dendrimer molecules, (HGD) are spheroidal organic nanoparticles with high structural symmetry and high relative density. In these dense organic nanoparticles the kinetic accessibility of groups on the surface is much greater than those are in the interior space.

The terms "surface groups" and "terminal groups" have different meaning: "Surface" refers to a spatial position, (which is usually temporary or statistical, due to molecular dynamics) and "terminal" refers to a permanent position in the chemical structure.

Medium Generation Dendrimer Molecules (MGD)

Medium generation dendrimer molecules (MGD), are polymers in between LGD and HGD, and may display characteristics of both groups. These dendrimers display behavior of "soft colloidal" particles.[55,56]

"Low", "middle", and "high" in this classification does not refer to the numerical value of a generation in a specific family. Any family will have members in LGD, MGD, and HGD classes, with gradual transitions of properties between them. The difference rather comes from the decreasing accessibility of interior and terminal functions, of the increasing persistency of the globular shape and of the increasing rigidity of the molecules with increasing generation number.

In LGD molecules the accessibility of the internal and external functions is equal. Solvent molecules between the branches are present in high excess compared to atoms belonging to the dendrimer molecule. Solvent molecules "within" the dendrimer molecule are minimally different from the free solvent pool around the molecules, as most of them does not interact with dendrimer atoms. In HGD molecules the accessibility of the surface is much greater than that of the interior—reactants added to the solution will react first on the surface-exposed functions and they need time to diffuse into the HGD. These molecules can adequately described as organic nanoparticles.

Properties of MGD molecules are transitional in between LGD and HGD molecules, and MGD molecules often can be approximated as soft, deformable particles or emulsion droplets.

Exempt from this rule are dendrimer molecules built from rigid sub-units and connectors. The cascade (tree-like) structural elements of dendrimer molecules, which is made up of repeating rigid units arranged in a hierarchical, self-similar fashion around a core raises the possibility of their application as artificial antennae systems.[57-59] The two basic properties here are (1) the (potentially) large number of absorbing units, which number grows exponentially with generation, (2) the high probability for the capture of light, and (C) the relatively short distance of the periphery from the center, where a fluorescent trap, a reaction center or a chemical sensor can be located.[60]

When the structure is built from electrically conducting units[61] like highly branched phenylacetylene dendrimers, the symmetric and ordered exciton funnel, with a well-directed energy gradient towards the core. Spectroscopic evidence indicated transfer efficiency of 98% from the photoabsorbing dendrimer backbone to the core (Fig. 10).

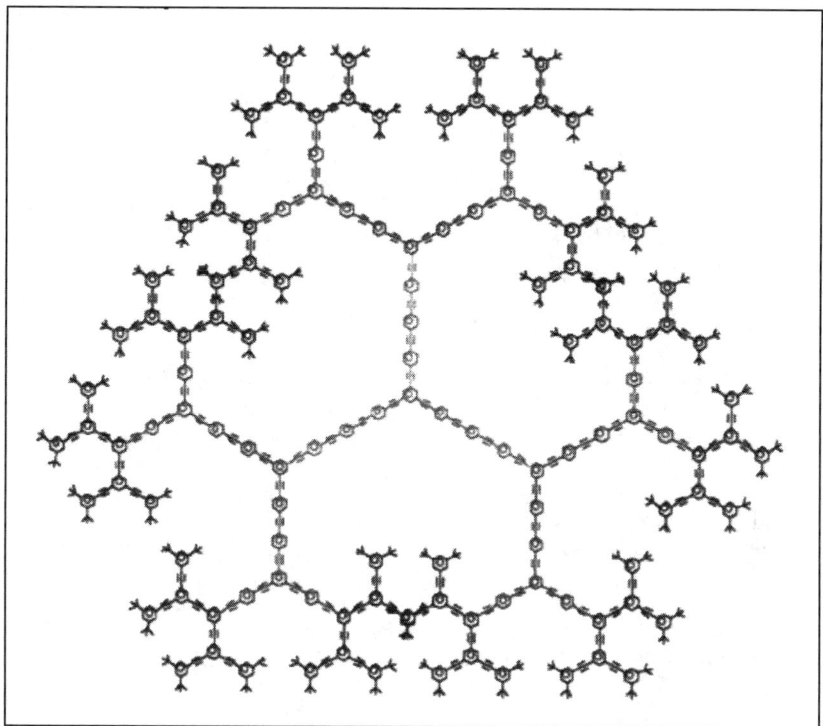

Figure 10. Chemical structure of an aromatic conjugated dendrimer, referred to as an extended Bethe dendrimer. It contains 127 benzene rings and a total of 2676 atoms. The peripheral benzene rings are terminated with *tert*-butyl groups.[61] (Reprinted from Shortreed MR et al, J Phys Chem B 1997; 101:6318-6322 © American Chemical Society.)

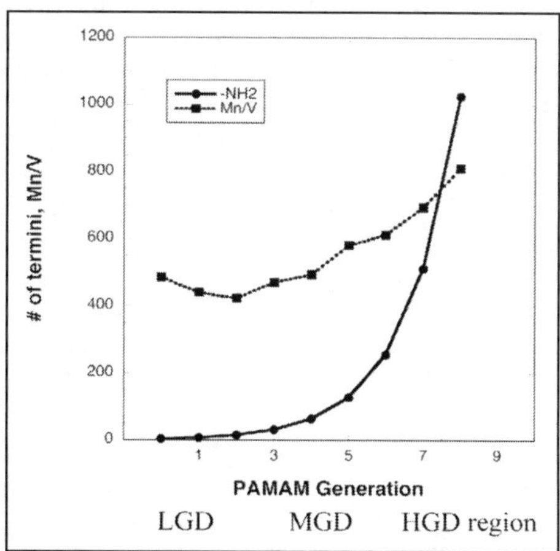

Figure 11. Change of number of terminal groups and mass per unit volume (i.e., calculated average density) for ideal PAMAM molecules as a function of generation. Volume has been calculated for spheres, i.e., assuming fully protonated molecules in water.[35]

Physical properties of dendrimer materials are different from linear molecules[62] and from conventional polymers with low degree of branching.[63] These properties also are different for different families and different generations (subclasses) as the consequence of increasingly expanding structure and evolving spherical geometry. In dendrimer molecules, the number of terminal functionalities increases exponentially with increasing generations, and simultaneously the overall density of the dendrimer molecules increases (Fig. 11).

The number of terminal groups exponentially increases as a function of generation, but the flexibility of the molecules and the accessibility of all functional groups decreases. The high level of symmetry, increasing overall density and increasing rigidity leads to decreasing interpenetration between molecules, as a function of generation. Although formally the density of terminal groups (number of -Z groups per surface area) increases, the number of interaction termini between molecules decreases as HGD molecules (organic nanoparticles) cannot penetrate each other, and can interact between at their surface.

Properties of Molecules; Good Solvent—Bad Solvent

End groups of solvated and flexible dendrimers in nonionic state are always "backfolded", i.e., terminal groups can also be found in the interior. In good solvents, the molecules are expanded, but not disentangled. In a bad solvent (where due to the low solubility the concentration is low), the molecules are smaller, due to this "backfolding" (Fig. 12).

In subsequent generations of dendrimers the mass increases exponentially while the volume increases cubically (after adoption of a globular shape). Thus, a maximum is observed in the plot of log intrinsic viscosity vs. molecular weight. The reciprocal of intrinsic viscosity, corresponding to the solution density, would therefore be expected to exhibit a minimum. This leads also to the prediction of a minimum in the refractive index and in the refractive index increment, dn/dc, for dendritic molecules as the generation number increases.[65]

Hydrodynamic radius of PAMAM dendrimers in different solvents were measured by holographic relaxation spectroscopy (HRS)[66] (Fig. 13).

Protonation of Polyionic Dendrimers Results in Nanophase-Separation as a Function of pH

Typical examples of polyionic dendrimers are those families, in which the branching atom is nitrogen, or phosphorous. Interior nitrogen branching sites can be protonated in solution, when they subsequently acquire positive charge (and attract anions), or can be deprotonated, (when they offer free electron pairs for coordinative binding of metal ions). Carboxyl terminal units can be deprotonated to offer large number of negative charges (Fig. 14).

Because identical charges repel each other and these electrostatic interactions are strong, protonation of cationic sites and deprotonation of anionic sites may induce large structural changes in the molecules to reach energy minimum, making them expand or collapse.

This phenomenon may result in complete nanophase-separation. For example, PPI dendrimers become completely and easily soluble in water when protonated, PAMAM dendrimers may shrink or expand depending on pH.[67-69] This scaling law holds for various solvent conditions as has been observed in HRS studies on PAMAM dendrimer in water, methanol, and butanol, and is supported by Brownian dynamics simulations on model dendrimers under varying solvent conditions.[64]

Physical Properties of Dendrimer Materials

In neat state—unlike in conventional (linear or branched) polymers—there is no interpenetration for HGD dendrimers. Dendrimers in solid phase have a very different structure than in solution because removal of solvent molecules results in the collapse of the spheroidal molecules. These materials display polymer-like glass transition temperatures (T_G) during the first run of DSC (Differential Scanning Calorimetry), but not on the consecutive runs, as it takes an extremely long annealing time until the materials solidify and recover (Fig. 15).

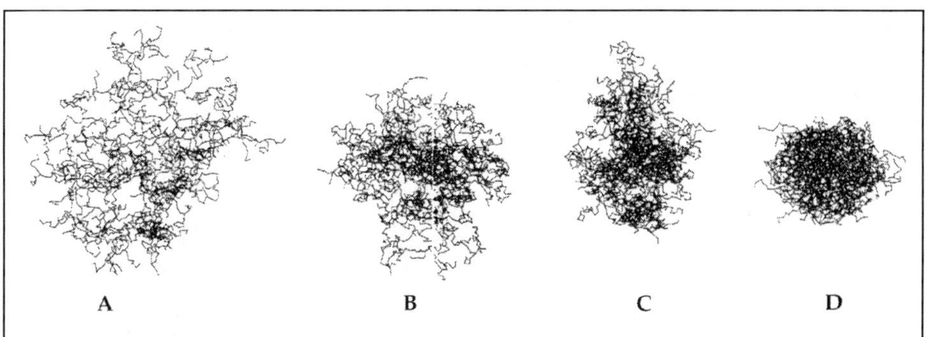

Figure 12. Simulated configurations of generation six dendrimers in varying solvent quality. Projections on the z = 0 plane of the typical configurations of generation six dendrimers under various solvent conditions: A) athermal (very good solvent); B) T = 4.0 (good solvent); C) T = 3.0 (Theta-solvent); D) T = 2.0 (poor solvent-large monomer attraction).[64] In very good solvents (the athermal case), there is no significance to the value of the temperature.[64] (Reproduced with permission from Murat M et al, Macromolecules 1996; 29:1278-1285, Figure 3, © American Chemical Society).

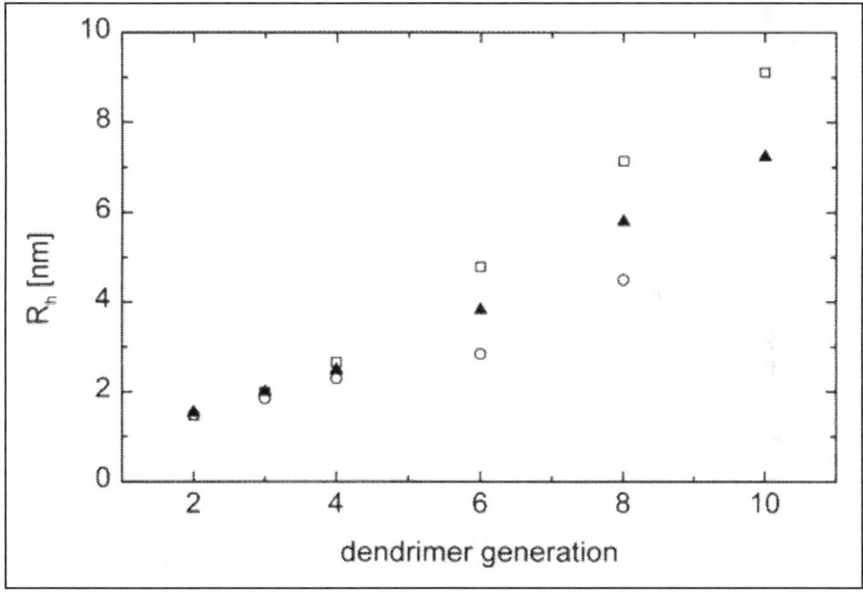

Figure 13. Hydrodynamic radius of PAMAM dendrimers of different generations in water (□), methanol (▲), and butanol (O), as measured by holographic relaxation spectroscopy.[66] (Reprinted with permission from Stechemesser S et al, Macromolecules 1997; 30:2204-2206, Fig. 3a © American Chemical Society).

Due to their symmetry and spherical shape, solid dendrimers generally are amorphous and do not form crystals. Their glass transition temperature (Tg) depends on their family, and usually is not high. When in solid state, they do not have outstanding mechanical properties. However, as part of crosslinked polymeric systems, they may improve mechanical properties dramatically (e.g., aliphatic polyester PERSTORP[70] improves epoxies). When melted, they

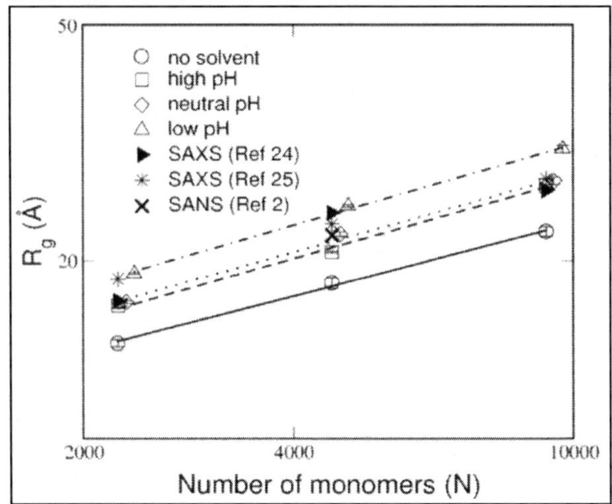

Figure 14. The radius of gyration R_g as a function of number of monomers in the dendrimer using a log-log scale. The lines are the best fit to a power law dependence with exponents ~0.33. This scaling exponent indicates a compact space-filling object under various solvent conditions.[102] (Reprinted with permission from Maiti PK et al, Macromolecules 2005; 38:979-991 © American Chemical Society.)

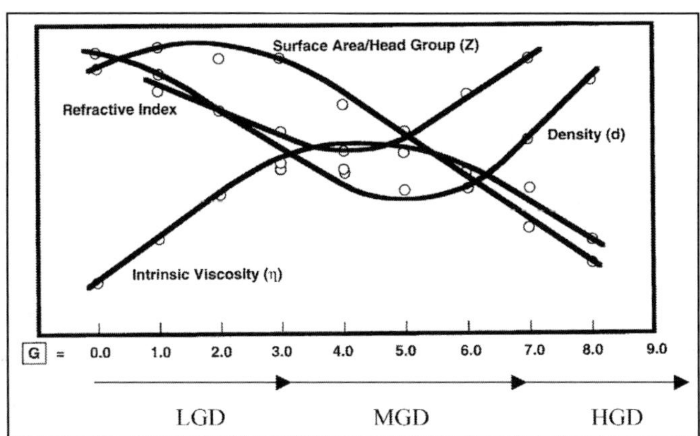

Figure 15. Calculated and measured properties of PAMAM dendrimers in solution.[35] The arrows illustrate the LGD-MGD-HGD regimes for PAMAM dendrimers. Dendrimer properties on the left are very different from those on the right, while in between are transitional. (Reprinted with permission from Tomalia DA et al, Agnew Chem Int Ed Engl 1990; 29:138-175, © WILEY-VCH.)

behave as ideal Newtonian liquids.[71,72] Dendrimer materials can play a key role in medicine and drug delivery, due to the variety of surface functional groups that can facilitate and modulate the delivery process.

Dendrimers as manufactured cannot be used directly for biomedical applications and have to be transformed into nanodevices using post-synthetic modifications. This may involve the modification of the exterior, the interior or both. Major methods are covalent attachment of multiple functional molecules to the dendrimer termini, or physically solubilizing drugs by ionic or hydrophobic interactions.

Many different molecules and building blocks can be covalently bound to the termini of dendrimers to form nanodevices. A nanodevice is a polyfunctional macromolecule/nanoparticle that is able to perform different functions. To reactive terminal groups various substituents can be attached to, such as simple molecules that regulate charge (e.g., hemisuccinates or acetylation), polymers (e.g., PEG) that modify solubility and biologic compatibility, dyes that collect light, and complexing ligands that make imaging possible, targeting moieties (peptides, sugars, antibodies) that modify biokinetics, etc.

A few possible transformations are illustrated in Figure 16.

Listing of all these applications is out of the scope of this chapter and the reader is referred to other reviews (refs. 2,11,23,73).

Dendrimers interact with biologic entities primarily through their solvated surfaces, and the interior only contributes to the resulting system (e.g., by changing the endosomal pH as a secondary effect). Biocompatibility and toxicity are primary issues.[44] The two primarily important factor in determining the biodistribution of dendrimers are size and charge.[74,75] Too much positive surface charge may generate holes in the cellular bilayer, which makes positively charged dendrimers toxic.[76]

Year by year, increasing number of biomedical applications are reported for dendrimers.[77] There are two basic ways dendrimers can work for drug delivery; (1) either by multiple covalent modifications of the termini, (which then may be exposed on the surface), or (2) initiate a physical binding in the dendrimer phase ("encapsulation", "encrustment", "entrapment", etc).[78,79]

The basis of entrapment is that MGD and HGD dendrimers can form a pseudo-phase of different polarity ("dendrimer phase") in solutions, in which a large number of solvent molecules reside.[80] These nanoscopic domains can interact with guest molecules either by very strong electrostatic interactions, hydrogen bonding or hydrophopic interactions.[81] This ("dendrimer phase" of solvated nonionic dendrimers)—depending on its polar or apolar character—

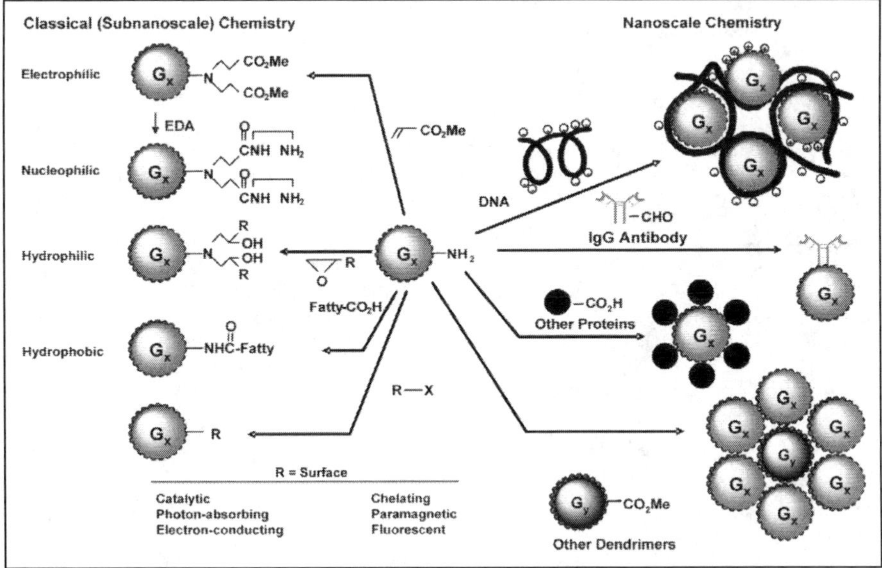

Figure 16. Options for modifying amine-terminated dendrimers by utilizing classical sub-nanoscale and nanoscale reagents.[63] (Reprinted with permission from Tomalia DA, Prog Polym Sci 2005; 30:294-324 (Fig. 21) © Elsevier.)

is able to dissolve/solubilize smaller molecules (e.g., apolar dyes or drugs) by hydrophobic interactions.

Solubilization of apolar drugs is especially important for drug delivery, as an immense number of anticancer drugs have low solubility. Dendrimers with large internal dipole moments are able to solvate polar molecules.[17] The inclusion complex and the receptor specific approach may have very different delivery profiles.[82] Some of these examples of encapsulated drugs tried are doxorubicin,[83] 5-fluoroacil,[84] niclosamide,[85] cis-platine,[86] nicotinic acid,[87] ketoprofen,[88] etc.

Receptor specific targeting of anticancer drugs may improve therapeutic response considerably. These include synthetically available molecules such as folate,[89-91] antibodies,[92] and various peptides.[93] Examples of covalently bound drugs are ibuprofen,[94] methotrexate,[91,95] sialic acid,[96] and taxol.[97]

Ionic dendrimers (e.g., those with -N< branching sites) can bind acids or metal ions (cations by donor-acceptor interactions or anions when protonated) and can form nanocomposites.[79,80,97-99] The high local concentrations of interior nitrogens may preorient inorganic ions in the interior, which then can be immobilized in the form of nanocomposites. Silver/dendrimer nanocomposites have been used for cell labeling (Lesniak, 2005) and for radiation therapy.[101]

Conclusions

Summarizing, "dendritic properties" are the manifestation of nanoscale droplet/particle properties of medium and high generation dendrimers, due to symmetry and macromolecular character. Regardless of semantics, the importance of creating complex nanosized devices by controlling properties of dendrimers and dendrimer derivatives through synthetic manipulation of structure and composition is becoming increasingly important for everyday life, especially in medicine. Dendrimers are fascinating structures, interesting molecules, challenging materials and promising tools for the present and future that provide a unique opportunity to improve the quality of life.

References

1. Dendrimer database at the Center for Molecular Design and Recognition web site (http://www.dendrimers.com).
2. David K Smith, ed. Recent Developments in Dendrimer Chemistry: Tetrahedron. 2005; 59(22).
3. Tomalia DA, Frechet JMJ, eds. Dendrimers and Dendritic Polymers. Prog Polym Sci 2005; 30.
4. Aulenta F, Hayes W, Rannard S. Dendrimers: A new class of nanoscopic containers and delivery devices. Eur Polym J 2003; 39:1741-1771.
5. Boas U, Peter M, Heegaard H. Dendrimers in drug research. Chem Soc Rev 2004; 33:43-63.
6. Bosman AW, Janssen HM, Meijer EW. About dendrimers: Structure, physical properties, and applications. Chem Rev 1999; 99(7):1665-1688.
7. Caminade AM, Laurent R, Majoral JP. Characterization of dendrimers. Adv Drug Deliv Reviews 2005; 57:2130-2146.
8. Cloninger MJ. Biological applications of dendrimers. Curr Opin in Chem Biol 2002; 6:742-748.
9. Crespo L, Sanclimens G, Pons M et al. Peptide and amide bond-containing dendrimers. Chem Rev 2005; 105:1663-1681.
10. Niederhafner P, Sebestik J, Jezek J. Peptide dendrimers. J Peptide Sci 2005; 11:757-788.
11. Florence T, ed. Dendrimers: A versatile targeting platform. Adv Drug Deliv Rev Vol 2005; 57(15).
12. Dvornic PR. PAMAMOS: The first commercial silicon-containing dendrimers and their applications. J Polym Sci Part A: Polym Chem 2006; 44(9):2755-2773.
13. Dykes GM. Dendrimers: A review of their appeal and applications. J Chem Technol Biotechnol 2001; 76:903-918.
14. Frechet JMJ, Tomalia DA, eds. Dendrimers and Other Dendritic Polymers. New York: John Wiley and Sons Ltd, 2002.
15. Grayson SM, Frechet JMJ. Convergent dendrons and dendrimers: From synthesis to applications. Chem Rev 2001; 101:3819-3867.
16. Gorman C. Metallodendrimers: Structural diversity and functional behavior. Adv Mater 1998; 10(4):295-309.

17. Majoral JP, Caminade AM. Dendrimers containing heteroatoms (Si, P, B, Ge, or Bi). Chem Rev 1999; 99:845-880.
18. Maurice WPL, Baars A, Meijer EW. Dendrimers II - Architecture, nanostructure and supramolecular chemistry. Topics in Curr Chem 2000; 210:131-182.
19. Newkome GR, Moorefield CN, Vögtle F. Dendrimers and dendrons: Concepts, syntheses, applications. Wiley VCH Weinheim 2001.
20. Newkome GR, Moorefield CN, Vögtle F. Dendritic macromolecules: Concepts, syntheses, perspectives. New York: Wiley-VCH, 1996.
21. Newkome GR, ed. Advances in Dendritic Macromolecules. Amsterdam, London, New York, Oxford, Shannon, Tokyo: JAI Press, Inc., (An Imprint of Elsevier Science), 2002:5.
22. Tomalia DA, Dvornic P. Dendritic polymers, divergent synthesis (Starburst polyamidoamine dendrimers). In: Salamone JC, ed. Polymeric Materials Encyclopedia, Vol. 3 (D-E). New York; CRC Press, 1996:1814-1830.
23. Tomalia DA, Frechet JMJ. Introduction to "Dendrimers and Dendritic Polymers". Prog Polym Sci 2005; 30:217-219.
24. Vögtle F, ed. Dendrimers: Top Curr Chem. Berlin: Springer and dendrimer related volumes thereof, 1998:197.
25. Vögtle F, Gestermann S, Hesse R et al. Functional dendrimers. Prog Polym Sci 2000; 25:987-1041.
26. Zeng F, Zimmerman SC. Dendrimers in supramolecular chemistry: From molecular recognition to self-assembly. Chem Rev 1997; 97:1681-1712.
27. Buhleier E, Wehner W, Vögtle F. "Cascade"- and "Nonskid-Chain-like" synthesis of molecular cavity topologies. Synthesis 1978; (3):155-158.
28. Tomalia DA, Baker H, Dewald J et al. A new class of polymers: Starburst®-dendritic macromolecules. P Polym J (Tokyo) 1985; 17:117-132.
29. Tomalia DA, Baker H, Dewald J et al. Dendritic macromolecules: Synthesis of starburst dendrimers. Macromol 1986; 19:2466-2468.
30. de Brabander-van den Berg EMM, Meijer EW. Poly(propylene imine) dendrimers: Large-scale synthesis by hetereogeneously catalyzed hydrogenations. Angew Chem Int Ed 1993; 32:1308-1310.
31. Majoral JP, Caminade AM. Divergent approaches to phosphorus-containing dendrimers and their functionalization. In: Vögtle F, ed. Dendrimers: Top Curr Chem. Berlin: Springer, 1998:197:79-124.
32. Newkome GR. A Systematic nomenclature for cascade polymers. J Polym Sci Part A: Polym Chem 1993; 31(3):641-651.
33. Tomalia DA, Dewald JR. Dense Star Polymers Having Core: Core Branches. Terminal Groups USP, 1983, (4,507,466).
34. Tomalia DA, Dewald JR, Hall MJ et al. First SPSJ Int Polym: Conference. Kyoto, Japan: 1984:65.
35. Tomalia DA, Naylor AM, Goddard IIIrd WA. Starburst dendrimers: Molecular-level control of size, shape, surface chemistry, topology, and flexibility from atoms to macroscopic matter. Angew Chem Int Ed Engl 1990; 29:138-175.
36. Tomalia ct al. US 4,558,120 A, 4,568,737A, US 4,587,329A, 4,631,337A etc.
37. Newkome GR, Yao Z, Baker GR et al. Cascade molecules: A new approach to micelles, A [27]-Arborol. J Org Chem 1985; 50:2003-2004.
38. Launay N, Caminade AM, Lahana R et al. A general synthetic strategy for neutral phosphorus-containing dendrimers. Angew Chem Int Ed Eng 1994; 33:1589-1592.
39. Schultz JL, Wilks ES. Dendritic and star polymers: Classification, nomenclature, structure representation, and registration in the DuPont SCION database. J Chem Inf Comput Sci 1998; 38:85-99.
40. Newkome GR, Young JK, Baker GR et al. Cascade polymers.' pH dependence of hydrodynamic radii of acid terminated dendrimers. Macromol 1993; 26:2394-2396.
41. Hummelen JC, van Dongen JLJ, Meijer EW. Electrospray mass spectrometry of poly(propylene imine) dendrimers - The issue of dendritic purity or polydispersity. Chem Eur J 1997; 3(9):1489-1493.
42. Frey H. Degree of branching in hyperbranched polymers. 2. Enhancement of the DB: Scope and limitations. Acta Polymerica 1997; 48(8):298-309.
43. Boogh L, Pettersson B, Månson JAE. Dendritic hyperbranched polymers as tougheners for epoxy resins. Polymer 1999; 40:2249.
44. Duncan R, Izzo L. Dendrimer biocompatibility and toxicity. Adv Drug Deliv Rev 2005; 57:2215-2237.
45. Wooley KL, Hawker CJ, Frechet JMJ. Hyperbranched macromolecules via a novel double-stage convergent growth approach. J Am Chem Soc 1991; 113:4252-4261.
46. Wooley KL, Hawker CJ, Frechet JMJ. A "Branched-Monomer Approach" for the rapid synthesis of dendrimers. Angew Chem Int Ed 1994; 33:82-85.

47. Kawaguchi T, Walker KL, Charles L et al. Double exponential dendrimer growth. J Am Chem Soc 1995; 117:2159-2165.
48. deGennes PG, Hervet HJ. Statistics of starburst polymers. J Phys Lett (Paris) 1983; 44:L351-L360.
49. Peterson JV, Allikmaa JS, Pehk T et al. Structural deviations in poly(amidoamine) dendrimers: A MALDI-TOF MS analysis. Eur Polym J 2003; 39:33-42.
50. Crooks RM, Buford I, Lemon IIIrd et al. Dendrimer-encapsulated metals and semiconductors: Synthesis. Characterization, and Applications Topics in Current Chemistry 2001; (212):81-135.
51. Tolic LP, Anderson GA, Smith RD et al. Electrospray ionization fourier transform ion cyclotron resonance mass spectrometric characterization of high molecular mass starburst dendrimers. Internat J Mass Spectr Ion Proc 1997; 165/166:405-418.
52. Balogh LP, Shi X, Bányai I et al. Generational, skeletal and substitutional diversities in generation one poly(amidoamine) dendrimers. Polymer 2005; 46(9):3022-3034.
53. Tomalia DA, Brothers IInd HM, Piehler LT et al. Partial shell-filled core-shell tecto(dendrimers) - A strategy to surface differentiated nano-clefts and cusps. PNAS 2002; 99(8):5081-5087.
54. Mansfield ML. Dendron segregation in model dendrimers. Polymer 1994; 35:1827-1830.
55. Ballauff M, Likos CN. Dendrimers in solution: Insight from theory and simulation. Angew Chem Int Ed Eng 2004; 43:2998-3020.
56. Harreis HM, Likos CN, Ballauff M. Can dendrimers be viewed as compact colloids? A simulation study of the fluctuations in a dendrimer of fourth generation. J Chem Phys 2003; 118:1979-88.
57. Balzani S, Campagna G, Denti A et al. Harvesting sunlight by artificial supramolecular antennae. Sol Energy Mater Sol cells 1995; 38:159-173.
58. Stewart GM, Fox MA. Chromophore-labeled dendrons as light harvesting antennae. J Am Chem Soc 1996; 118:4354-4360.
59. Kopelman R, Shortreed M, Shi ZY et al. Spectroscopic evidence for excitonic localization in fractal antenna supermolecules. Phys Rev Letters 1997; 78(7):1239-1242.
60. Bar-Haim A, Klafter J. Dendrimers as light harvesting antennae. J Luminescence 1998; 76-77:197-200.
61. Shortreed MR, Swallen SF, Shi ZY et al. Directed energy transfer funnels in dendrimeric antenna supermolecules. J Phys Chem B 1997; 101:6318-6322.
62. Hawker CJ, Malmström EE, Frank CW et al. Exact linear analogs of dendritic polyether macromolecules: Design, synthesis, and unique properties. J Am Chem Soc 1997; 119:9903-9904.
63. Tomalia DA. Birth of a new macromolecular architecture: Dendrimers as quantized building blocks for nanoscale synthetic polymer chemistry. Prog Polym Sci 2005; 30:294-324.
64. Murat M, Grest GS. Molecular dynamics study of dendrimer molecules in solvents of varying quality. Macromolecules 1996; 29:1278-1285.
65. Wooley KL, Hawker CJ, Frechet JMJ. Unsymmetrical three-dimensional macromolecules: Preparation and characterization of strongly dipolar dendritic macromolecules. J Am Chem Soc 1993; 115:11496-11505.
66. Stechemesser S, Eimer W. Solvent-dependent swelling of poly(amidoamine) Starburst dendrimers. Macromol 1997; 30:2204-2206.
67. Borkovec M, Koper GJM. Proton binding characteristics of branched polyelectrolytes. Macromol 1997; 30:2151-2158.
68. van Duijvenbode RC, Rajanayagam A, Koper GJM et al. Synthesis and protonation behavior of carboxylate-functionalized poly(propyleneimine) dendrimers. Macromol 2000; 33:46-52.
69. Cakara D, Kleimann J, Borkovec M. Microscopic protonation equilibria of poly(amidoamine) dendrimers from macroscopic titrations. Macromol 2003; 36:4201-4207.
70. Jannerfeldt G, Boogh L, Månson JAE. Influence of hyperbranched polymers on the interfacial tension of polypropylene/polyamide-6 blends. J Polym Sci Part B: Polym Phys 1999; 37:2069-2077.
71. Uppuluri S, Morrison FA, Dvornic PR. Rheology of dendrimers. 2. Bulk polyamidoamine dendrimers under steady shear, creep, and dynamic oscillatory shear. Macromolecules 2000; 33:2551-2560.
72. Uppuluri S, Keinath SE, Tomalia DA et al. Rheology of dendrimers. I. Newtonian flow behavior of medium and highly concentrated solutions of polyamidoamine (pamam) dendrimers in ethylenediamine (eda) solvent. Macromolecules 1998; 31:4498-4510.
73. Amiji M, ed. Nanotechnology in Cancer Therapy. Boca Raton, London, New York: CRC Press (Taylor and Francis Group), 2006.
74. Nigavekar SS, Balogh L, Khan MK. 3H dendrimer nanoparticle organ/tumor distribution. Pharm Res 2004; 21(3):476-83.
75. Khan MK, Nigavekar SS, Balogh LP. In vivo biodistribution of dendrimers and dendrimer nanocomposites - Implications for cancer imaging and therapy. Technol Canc Res and Treatment 2005; 4(6):603-613.

76. Hong S, Bielinska AU, Mecke A et al. Interaction of poly(amidoamine) dendrimers with supported lipid bilayers and cells: Hole formation and the relation to transport. Bioconj Chem 2004; 15:774-782.
77. Stiriba SE, Frey H, Haag R. Dendritic polymers in biomedical applications: From potential to clinical use in diagnostics and therapy. Angew Chem Int Ed 2002; 41(8):1329-1334.
78. Balogh LP, Swanson DR, Spindler R et al. Formation and characterization of dendrimer-based water soluble inorganic nanocomposites. Proc ACS PMSE 1997; 77:118-9.
79. Beck-Tan N, Balogh L, Trevino S. Structure of metallo-organic nanocomposites produced from dendrimer complexes. Proc ACS PMSE 1997; 77:120.
80. Balogh L, Tomalia DA. Poly(amidoamine) dendrimer-templated nanocomposites. 1. Synthesis of zerovalent copper nanoclusters. J Am Chem Soc 1998; 120:7355-7356.
81. Gupta U, Agashe HB, Jain NK et al. Dendrimers: Novel polymeric nanoarchitectures for solubility enhancement. Biomacromol 2006; 7(3):649-658.
82. Patri AK, Kukowska-Latallo JF, Baker Jr JR. Targeted drug delivery with dendrimers: Comparison of the release kinetics of covalently conjugated drug and noncovalent drug inclusion complex. Adv Drug Deliv Rev 2005; 57:2203-2214.
83. Muggia FM. Doxorubicin-polymer conjugates: Further demonstration of the concept of enhanced permeability and retention. Clin Canc Res 1999; 5(1):7-8.
84. Bhadra D, Bhadra S, Jain S et al. A PEGylated dendritic nanoparticulate carrier of fluorouracil. Internat J of Pharmaceutics 2003; 257(1-2):111-124.
85. Devarakonda B, Hill RA, Liebenberg W et al. Comparison of the aqueous solubilization of practically insoluble niclosamide by polyamidoamine (PAMAM) dendrimers and cyclodextrins. Internat J Pharmaceutics 2005; 304:193-209.
86. Malik N, Evagorou EG, Duncan R. Dendrimer-platinate: A novel approach to cancer chemotherapy. Anti-cancer Drugs 1999; 10(8):767-775.
87. Yiyun C, Tongwen X. Solubility of nicotinic acid in polyamidoamine dendrimer solutions. Eur J Med Chem 2005; 40:1384-1389.
88. Yiyun C, Tongwen X, Rongqiang F. Polyamidoamine dendrimers used as solubility enhancers of ketoprofen. Eur J Med Chem 2005; 40:1390-1393.
89. Quintana A, Raczka E, Piehler L et al. Design and function of a dendrimer-based therapeutic nanodevice targeted to tumor cells through the folate receptor. Pharm Res 2002; 19(9):1310-1316.
90. Wiener EC, Konda S, Shadron A et al. Targeting dendrimer-chelates to tumors and tumor cells expressing the high-affinity folate receptor. Invest Radiol 1997; 32(12):748-54.
91. Kukowska-Latallo JF, Candido KA, Cao Z et al. Nanoparticle targeting of anticancer drug improves therapeutic response in animal model of human epithelial cancer. Cancer Res 2005; 65(12):5317-5324.
92. Patri AK, Myc A, Beals J et al. Synthesis and in vitro testing of J591 antibody-dendrimer conjugates for targeted prostate cancer therapy. Bioconj Chem 2004; 15:1174-1181.
93. Shukla R, Thomas TP, Peters JL et al. HER2 specific tumor targeting with dendrimer conjugated anti-HER2 mAb. Bioconj Chem 2006; 17(5):1109-1115.
94. Kolhe P, Khandare J, Pillai O et al. Preparation, cellular transport, and activity of polyamidoamine-based dendritic nanodevices with a high drug payload. Biomaterials 2006; 27(4):660-669.
95. Gurdag S, Khandare J, Sarah Stapels S et al. Activity of dendrimer-methotrexate conjugates on methotrexate-sensitive and-resistant cell lines. Bioconjugate Chem 2006; 17:275-283.
96. Reuter JD, Myc A, Hayes MM et al. Inhibition of viral adhesion and infection by sialic-acid-conjugated dendritic polymers. Bioconj Chem 1999; 10:271-278.
97. Majoros I, Myc A, Thomas T et al. PAMAM dendrimer-based multifunctional conjugate for cancer therapy: Synthesis, characterization, and functionality. Biomacromolecules 2006; 7:572-579.
98. Zhao M, Sun L, Crooks RM. Preparation of Cu nanoclusters within dendrimer templates. J Am Chem Soc 1998; 120:4877-4878.
99. Esumi K, Suzuki A, Aihara N et al. Preparation of gold colloids with UV irradiation using dendrimers as stabilizer. Langmuir 1998; 14:3157-3159.
100. Lesniak W, Bielinska AU, Balogh LP. Silver/Dendrimer nanocomposites as biomarkers: Fabrication, characterization, in vitro toxicity, and intracellular detection. Nano Letters 2005; 5(11):2123-2130.
101. Balogh LP, Khan MK. Dendrimer nanocomposites for cancer therapy chapter 28. In: Mansour Amiji M, ed. Nanotechnology in Cancer Therapy. Boca Raton, London, New York: CRC Press (Taylor and Francis Group), 2006.
102. Maiti PK, Cagın T, Lin ST et al. Effect of solvent and pH on the structure of PAMAM dendrimers. Macromolecules 2005; 38:979-991.

CHAPTER 12

Quantum Dots and Other Fluorescent Nanoparticles: Quo Vadis in the Cell?

Dusica Maysinger* and Jasmina Lovrić

Abstract

An exponentially growing number of nanotechnology-based products are providing new platforms for research in different scientific disciplines (e.g., life sciences and medicine). Biocompatible nanoparticles are expected to significantly impact the development of new approaches in medical diagnoses and drug delivery; however, very little is known about the effects of long-term exposure of different nanoparticles in different cell types and tissues. The first objective of this chapter is to provide a brief account of the current status of fluorescent nanoparticles (i.e., quantum dots, fluorescently-labeled micelles, and FloDots) that serve as tools for bioimaging and therapeutics. The second objective of this chapter is to describe the modes and mechanisms of nanoparticle-cell interactions and the "potential" toxic consequences thereof.

Introduction

Recent breakthroughs of nanotechnology in biology and medicine have provided new approaches to address and solve problems either in basic research or clinics. Many consider that the applications of various nanostructures and nanodevices could significantly contribute to fighting cancer and neurodegenerative diseases.[1,2] Depending on the properties of the nanostructures used, their function could be due to their intrinsic properties[3,4] or due to the functionalization of nanomaterials that render them biologically active.[5,6]

Unfortunately some nanoparticles can constitute threats to living cells, animals or humans. Understanding the mechanisms of their negative effects is critical to preventing or exploiting their use for diagnostic or therapeutic purposes such as selectively destroying tumor cells. Currently, metal-containing nanoparticles, including quantum dots, are mainly part of the fluorescent toolbox for biological applications[7] but they may not be a clinical solution for improving diagnosis or therapy. Other types of nanoparticles such as micelles and FloDots are mostly used for delivering different agents in a controlled manner and/or as a detection system for environmental pollutants when applied to medical or detection applications.

Fluorescent Nanoparticles

An Overview

Most fluorescent nanoparticles reported either fluoresce or luminesce because of their intrinsic properties (e.g., semiconductor quantum dots) or having covalently or noncovalently

*Corresponding Author: Dusica Maysinger—Department of Pharmacology and Therapeutics, McGill University, Montreal, Canada. Email: dusica.maysinger@mcgill.ca

Bio-Applications of Nanoparticles, edited by Warren C.W. Chan. ©2007 Landes Bioscience and Springer Science+Business Media.

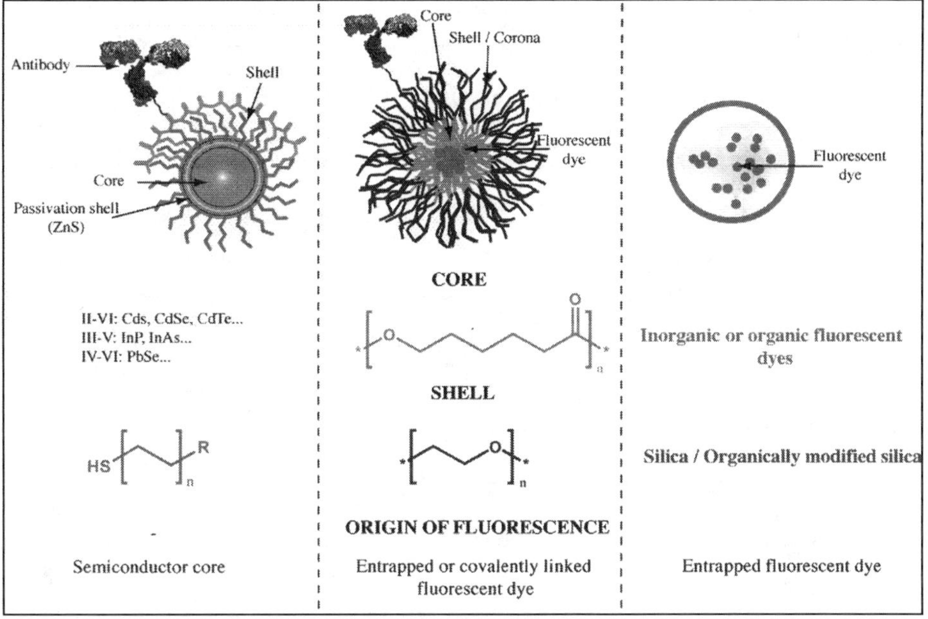

Figure 1. Some fluorescent/luminescent nanoparticles and their characteristics.

linked fluorescent or fluorogenic organic molecules (e.g., polymeric micelles). This section will focus on semi-metallic nanoparticles called quantum dots (Q-dots), fluorescent nanoparticles made of organic molecules called micelles and inorganic materials with attached organic or inorganic dyes called FloDots. Each of these classes have rather unique photo-physical and optical characteristics which can be exploited for long-term and multicolor imaging,[8-10] studies exploring nanoparticle fate[11,12] or photosensitization.[3] The main properties of selected nanoparticles discussed in this chapter are summarized in Figure 1.

Differences Between Q-Dots, Fluorescent Micelles and FloDots

Q-dots are fragments of semiconductors made of hundreds of thousands of atoms with optical and electrical properties strongly dependent on the size of these fragments.[13] Q-dots are unique because their emission maxima shift to higher wavelengths when their size increases. No matter what their size, Q-dots can be excited within a relatively large range of wavelengths. The emission spectra are rather narrow allowing for the simultaneous detection of multiple colored Q-dots upon illumination with a single-wavelength light source. Their resistance to photobleaching, which permits prolonged visualization of the target, is also a valuable addition to their properties.[14] Lastly, due to the electron dense Q-dot core, images can be acquired with electron microscopy whose high resolution complements information obtained by imaging live cells using confocal microscopy.[15]

A primary task of polymeric micelles is the delivery of poorly soluble drugs in the diseased tissue and cells. To achieve a controlled release of a drug to a specific population of cells or subcellular organelles multiple factors play a role, with micelle-cell interactions being the crucial and the most complex. Polymeric micelles are not intrinsically fluorescent but they can be made fluorescent by solubilizing hydrophobic dyes in their cores[16,17] or by covalently linking dyes to a single polymer chain.[12] Rendering them fluorescent enables the investigation of their interaction at the cellular and subcellular level. Optical properties of fluorescently labeled micelles largely depend on the properties of the dye used. However, physicochemical characteristics of block copolymers can modify the fluorescence signal intensity (both by quenching or enhancing it) and intracellular distribution of dye molecules.[12,18] For example, a fluorescent drug, doxorubicin, at high concentrations can exert an enhanced pharmacological effect (cell killing) and at the same time serve as a fluorescent agent indicating the extracellular and intracellular localization of free and polymer-bound agent.[19] In combination with photoexcitation, enhanced permeability and retention[20,21] micelles seem to provide an alternative therapeutic approach in oncology.

FloDots are silica nanoparticles doped with fluorescent dyes.[10] Dye molecules can be either luminescent organic or inorganic molecules dispersed within the silica matrix. These novel nanoparticles are extremely luminescent and are highly photostable compared to Q-dots.[22] Silica provides dispersion of nanoparticles in water, and its surface can be modified to contain functional groups. Examples of thousand fold increases in fluorescence intensities of FloDots over that of pure dyes have been reported.[10] Similar to Q-dots, FloDots can be analyzed both by fluorescent and electron microscopy.

Surface Modifications of Nanoparticles

While the inorganic core of Q-dots is responsible for their optical properties, the organic shell around the core has multiple functions. First, it enables the dispersion and stability of Q-dots in an aqueous environment. Secondly, it serves as a substrate for linking biomolecules; this gives Q-dots their precise biological functions including targeting to specific cells or subcellular organelles. Stabilization is frequently achieved by hetero-bifunctional ligands containing a mercapto group on one side and carboxylic, hydroxyl or amino group on the other side.[23,24] Mercapto groups are linked to the surface of Q-dots while carboxylic, hydroxyl or amino groups provide solubilization and further biofunctionalization.[25] In addition to short bifunctional ligands such as cysteamine and mercaptoacetic acid, more complex ones such as mercaptopropylsilane[26] and amine-containing dendrimers have been previously employed. Moreover, ligands such as thiolated peptides can stabilize and functionalize Q-dots at the same time.[27] Q-dots can also be encapsulated with diblock or triblock copolymers and phospholipids.[28]

Even though Q-dots can nonspecifically label cells in culture[9] (Fig. 2) or accumulate at site of tumors in vivo,[29] attachment of biomolecules to their surface is used to actively target them and increase their internalization in specific cells. Well-known biomolecules such as folic acid[30] or different antibodies against tumor antigens, such as prostate-specific membrane antigen[29] or over expressed HER2 receptor,[31] were exploited to achieve active targeting. Destination to distinct subcellular targets can be achieved by attaching translocation peptides such as nuclear or mitochondrial localization signal peptide.[32] In an elegant work by Dahan et al functionalized Q-dots were used to track the lateral diffusion of glycine receptors in primary spinal cord neurons. Individual receptors were seen within tens of minutes at spatial resolutions of 5-10 nm.[14] In studies employing nerve growth factor linked to Q-dots, the dynamics of ligand internalization were followed in rat pheochromocytoma cells.[33] Similar approaches can be taken to explore many other receptors in a variety of cells. Such studies could provide new insights into receptor location, distribution and trafficking in cells under normal and pathological conditions. New conjugated Q-dots with ligands for many receptors need to be developed. However, a number of versatile conjugation procedures have been described in detail and are being developed.[34] This source of information could be very useful in creating new ligand-specific imaging tools for addressing numerous unanswered questions in neuroscience and other research fields.[2]

In the case of polymeric micelles, common conjugation approaches were used as indicated above and frequently applied to functionalize liposomes.[35] An attractive feature of block copolymer micelles is their ability to accommodate a large number of drug molecules inside their cores (particularly the large micelles) and at the same time bind to the specific ligand(s) linked to the block constituting the micelle corona; enabling both drug delivery and imaging of the

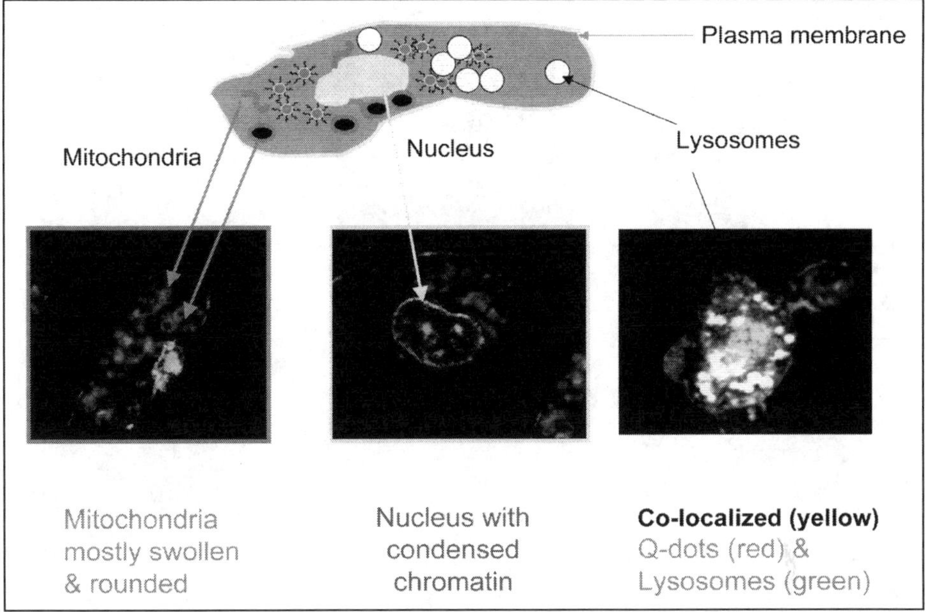

Figure 2. Multicolor imaging of QDs in live cells. Schematic of cell organelles stained with Hoechst 33342 (nucleus), mitotracker red (mitochondria), and colocalization (yellow) of internalized QDs (red) with lysotracker (green). Photomicrograph of PC12 cells treated with QDs and stained with organelle-specific fluorescent dyes show mitochondrial damage, chromatin condensation and colocalization of Q-dots with lysosomes. Color version available at Eurekah.com.

drug carrier. Chemically attached ligands to the reactive groups in the hydrophilic corona of polymeric micelles do not seem to be detrimental to the micelle formation or stability.[36]

Stability and Fate of Fluorescent Nanoparticles (Q-Dots, Micelles, FloDots)

Relatively little is known about the stability and fate of fluorescent nanoparticles in different cell types and in the whole animal. This area is relatively new and many fluorescent dyes and nanoparticles do not yet have sufficient stability and are not resistant to photobleaching upon multiple exposures to lasers to allow long term monitoring of these particles. In some instances, the synthesis of fluorescent polymers is not a trivial matter, and some dyes simply cannot be easily conjugated to the polymer. Additional problems include the autofluorescence of the tissue and a limited resolution of available instrumentation for in vivo imaging. An additional factor complicating the determination of the fate of biodegradable block copolymer micelles and Q-dots in vivo is the interference associated with components from blood and other biological molecules. One of the first studies with fluorogenic dyes incorporated into micelles show how the increasing complexity of the biological environment impacts micellar disintegration.[37]

Q-dots were mostly used for short-term imaging in live cells, fixed cells and in whole animals (in vivo).[8,23,38,39] Upon short-term exposure to Q-dots, particularly those well protected with shells such as ZnS, there is no immediate damage to the tissue, which can be readily detected. However, it is currently unclear if some tissues accumulate the semi-metallic nanoparticles after long-term exposures and if they have a limited capacity to dispose of them. These studies are ongoing and will provide important information, which will help overcome the hurdles associated with these diverse and attractive bioimaging tools.

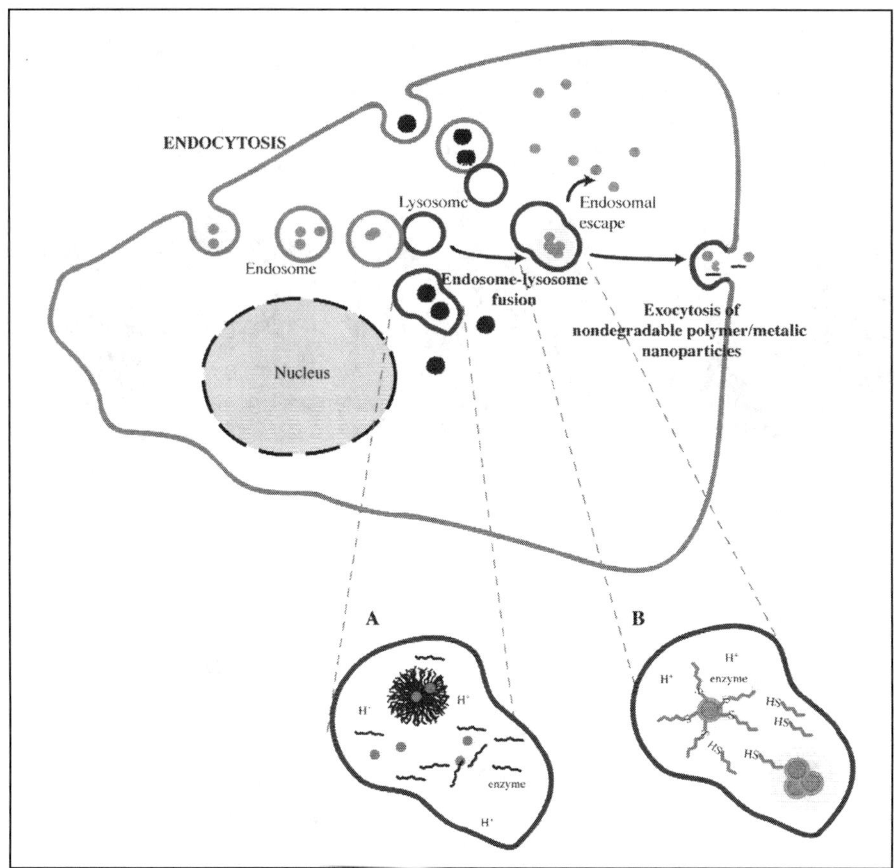

Figure 3. Stability and fate of fluorescent nanoparticles in the cell. Nanoparticles are endocytosed by the cell and can be degraded within the lysosomes: A) micelle degradation due to low pH; B) Q-dot destabilization and aggregation within the lysosomes.

For Q-dots to efficiently fluoresce, the protection of the core has proven to be an essential factor. The shell protects its core, and loss of the ligands from the shell can occur with changes in pH or due to the oxidation from prolonged illumination[40,41] (Fig. 3). Q-dots whose organic shells are built of short bifunctional ligands do not have sufficient stability in cellular environments, which is required for long-term labeling and imaging. The protection of the core with polymers of multilayered densely packed shells enhances the stability of the entire system, enabling Q-dots to strongly fluoresce for days and months after exposure to complex in vivo environments. The development of new surface coatings will allow Q-dots to demonstrate their full potential as superior imaging tools.

The problem of stability with organic nanoparticles such as micelles can be partly circumvented by cross-linking the micelle shell. However, complete control of drug release from such micelles becomes an issue to be resolved for each individual drug. One way to enhance the micelle stability is by using shell-cross-linking[42] and formation of shell cross-linked knedel-like (SCKs) nanoparticles. SCKs can be labeled with fluorescent, radioligand or dense core particles and decorated with different ligands to allow for cell or tissue targeting.[43-46]

Live Cell and Tissue Imaging with Q-Dots and FloDots

Q-dots' wide variety of emissions allows the use of numerous combinations of fluorescent dyes and Q-dots for the tracking of nanoparticles within cells[9,31] (Fig. 2). Moreover, Q-dots have an efficient multiphoton absorption cross-section which makes them brighter probes under photon-limited in vivo conditions.[29] Q-dots that can emit infrared or near infrared light are particularly suitable for deep tissue imaging because autofluorescence of hair and tissues in this range is minimal.[47-49] Q-dots emitting within 650-800 nm conjugated with polyethylene glycol (PEG) and specific ligands recognizing receptors could be particularly useful. If administered intravenously, the liver does not immediately eliminate PEG-Q-dots and their protracted circulation time allows their fate to be followed using different imaging set-ups. Injected Q-dots with PEG coatings or functionalized Q-dots can be detected with standard in vivo imaging systems for several days when administered subcutaneously or intravenously.[8,15,50] In vivo monitoring of Q-dots is appealing because it can provide needed information on time-dependent Q-dot distribution and accumulation in tissues, which is important in the evaluation of potential therapeutic applications. Data for the biodistribution and pharmacokinetics of Q-dots are emerging[51] and more systematic studies are needed to demonstrate how rapidly these particles can be eliminated from the body, where they accumulate, and what nonspecific tissue damage they may eventually cause.

Bioluminescent Q-dot conjugates for in vivo imaging were recently developed by Rao's group.[49] Instead of excitation from an external illumination source, these Q-dot nanocrystals are excited by bioluminescence resonance energy transfer and consequently tissue autofluorescence is significantly reduced;[49] a lucrative approach using bioluminescent Q-dot conjugates, particularly in small animal imaging.

Significant advances have been made in the field of biosensors based on Q-dots. Other nanoparticle types that can respond to various pathological or physiological stimuli are of particular interest for applied cell biology and medical investigations.[52-54]

Intracellular trafficking and fluorescence imaging of micelles has been recently reviewed.37,55 Ultrasenstive and high-resolution microscopic techniques are beginning to provide insight into the real-time dynamics of cellular components and macromolecular pharmacological agents as they are delivered into and travel within single cells. By combining genetic manipulations of cells with fluorescent markers and labeling of individual cellular organelles, the journey of nanoparticles and the intricacies of their interactions with cellular components within the single cells are being elucidated.

FloDots have also been employed for imaging of different cell types, e.g., leukemia cells[56] and in E. coli.[57] Owing to their greatly amplified fluorescence intensities in comparison with single fluorescent molecules, FloDots can be used as tools for detection of trace amounts of bacteria in high throughput assays.

Nanoparticles as Potential Hazards

Problems in Nanotoxicology

Is there a real reason to fear nanoparticles? The lay community and scientists are divided into two extreme and opposing camps. One of the reasons for this wide disagreement is the lack of sufficient knowledge regarding different nanoparticles. We have only begun to collect the necessary information about what is undesirable in the chemical composition of different nanoparticles; how poorly stable Q-dot surfaces can induce unwanted side effects in live cells after many hours of continuous exposure and how different cell types deal with the short and long-term exposure. Unless we have enough scientific evidence for their direct and indirect damaging effects, all the excitement and uncertainties are just speculations.[58] The novel and unique properties of nanoparticles, which have been enthusiastically explored for their advancement in therapeutics and diagnostics, could also be the source of undesirable effects on biological systems.[58] The factors that play important roles include nanoparticle size, chemical composition, surface structure, shape, solubility and aggregation.[59] The vast array of nanoparticles with diverse properties is currently being explored and each type could specifically interfere with cellular systems. It is of great importance to identify the key factors that can be used to predict their toxicity,[58] and we should use our knowledge of the toxicity of environmental pollutants in this size range to complement our understanding of functionalized nanoparticles. The objective of this section is to point out some of the potential sites of cell injury caused by some types of nanoparticles.

Nanotoxicity

Recently, a significant number of investigations have focused on the potential toxic effects of nanoparticulate systems.[59] While completely benign in certain paradigms,[9,29] nanoparticles including Q-dots,[60-62] fullerenes[63] and nanotubes[64] have induced negative effects in others.

The toxicity of Q-dots when tested in vitro, is dependent on numerous factors associated with Q-dot properties and cell status. Q-dot properties, which are critical for biocompatibility are: size, surface charge, type of surface ligand, stability, and concentration. The core, when not protected, can gradually disintegrate and cause the release of toxic ions (such as cadmium ions) into the medium.[60] Also, the diffusion of oxygen is facilitated by the loose shell around the core; this can result in electron or energy transfer resulting in the formation of toxic reactive oxygen species (ROS).[61,65-67] The protective coating of Q-dots can also breakdown (e.g., by irradiation, low lysosomal pH or possibly metabolic degradation), leading to the cell damage and death.[68] A great deal of toxicological data is available on individual metals, such as cadmium, selenium, zinc, lead and copper. Much less is known about tellurium, indium and gallium, which are also components of different types of Q-dots. Almost twenty years ago tellurium was shown to induce significant damage in the peripheral nervous system.[69]

The cellular environment plays a critical role in influencing the physiological status of cells and has an impact on their response to Q-dots. For example, cells exposed to Q-dots are more vulnerable when deprived of growth factors, whereas cells cultured in the presence of serum will be more resistant to Q-dot induced cell death. Moreover, certain types of Q-dots can interact with serum and other biological fluid components reducing or increasing their damaging effects on cells. One of the best documented components in cell culture medium which can modify nanoparticle entry and modify the extent of nanoparticle-induced cytotoxicity is serum albumin. This biomolecule can reduce the internalization of Q-dots and micelles and also delay or reduce cell death induced by these nanoparticles.

Current status and limitations of different nanoparticles were recently reviewed.70 The source of polymeric micelle-induced harmful effects in living cells is often the result of residual unpurified catalyst. These trace catalytic metals and unstable (decomposed) fluorescent dyes can be toxic. Micelles have been used as nonviral gene carriers, and some of them have certain advantages, such as safety, simplicity of preparation and large-scale production over viruses. However, DNA-polymer complexes (polyplexes) may often not be suitable for in vivo gene delivery because of their considerable toxicity, e.g., polyethylene imines proposed for gene delivery are often toxic.[70] Recent applications of micelles for gene delivery, photosenzitizers and small interfering RNA have been facilitated by employing newly developed biocompatible block copolymers by Kataoka's group.[71] Micelles made of modified polyaspartamides linked to PEG and poly-lysines, in contrast to many other positively charged materials can be well tolerated by live cells in relatively high concentrations. The attraction of these new PEG-b-(Asp) block copolymer particles is their remarkable loading capacity for siRNA (close to 90%) and appreciable gene silencing in a sequence-specific manner. It is anticipated that calcium phosphate core-shell type nanoparticles incorporating siRNA will prove sufficiently biocompatible as nanocarriers for other siRNAs.[71]

Mechanisms Involved in Nanoparticle-Induced Cell Death

A common property of nanoparticles seems to be their ability to induce the generation of ROS in the cellular environment. The exact mechanisms for the production of these reactive oxygen species are not well understood; through high levels of free metal ions[68] leading to the formation of ROS and oxidative damage are detected in live cells exposed to the nanomaterials. Some types of nanoparticles, such as Q-dots, due to their electronic configuration have the ability to spontaneously induce ROS. At the same time, it seems that cells themselves can be the source of ROS in contact with nanoparticles lacking this intrinsic property.[72]

The disturbance of cellular ROS equilibria has an important impact on cellular physiology. ROS, aside from being toxic mediators, are also important signaling molecules. Cells are responsive to minute changes in the intracellular redox state; which is sufficient to inhibit proliferation, induce differentiation[73] or cell death.[74] The effect of these disequilibria will depend on the cell type or organelle in contact with the nanoparticles and the redox status of cells and their organelles. A number of fluorescent probes have been used to reveal ROS in live cells,[75] but there is a need for new, more species specific probes for such studies. A complex interplay between nanoparticles, cells, and reactive oxygen species could be decisive factors, which can affect the performance, behavior and fate of both nanoparticles and cells.

In summary, cell death by nanoparticles is a relatively unexplored area of research. Further studies are required to understand the mechanisms involved in nanoparticle cell death, and to discover the ways to prevent it. We should also keep in mind that nanoparticles could have an

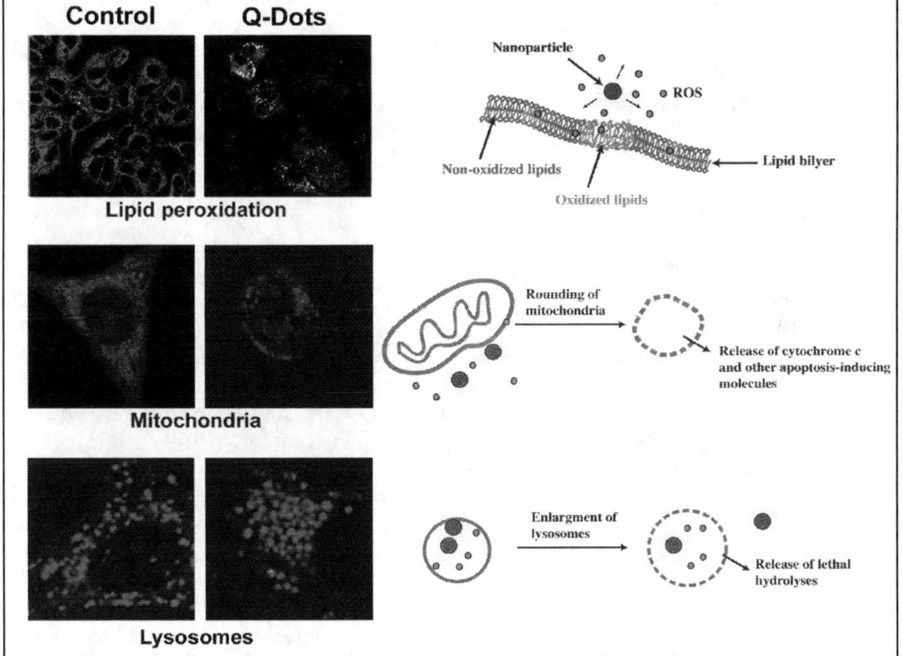

Figure 4. Multiple organelle damage induced by long term treatment with Q-dots. Long-term treatments of cells with certain types of Q-dots lead to lipid peroxidation and to cell membrane damage. Note the change in fluorescent dye from mostly red (control) to mixed green, red and yellow (Q-dots treated). Mitochondrial damage due to the oxidative stress can be revealed by fluorescent mitotrackers. Note the rounding and agglomeration of mitochondria. Lysosomes can become enlarged and leaky if nanoparticles induce damage to these organelles. Color version available at Eurekah.com.

impact on cellular function even if non toxic. Studying all these effects will be important to the safe preparation, handling and utilization of nanoparticles in science and beyond.

Cell Organelle Responses to Nanoparticles

Several cell organelles are implicated in different types of cell deaths (Fig. 4). In this section we will briefly discuss some of the key roles of these organelles, both individually and in combination.

Mitochondria play a central role in energy metabolism; they are responsive to even small stresses in multiple ways.[76] The intermembrane space of mitochondria harbor proteins that are lethal to cells once released into the cytosol, which gives mitochondria a critical role in cell death. When cells are exposed to nanoparticles (e.g., Q-dots), which can induce the generation of ROS, mitochondria are among the first and most sensitive organelles affected. If nanoparticles are not capable of producing ROS themselves, dysfunctional mitochondria in stressed cells could be the major source of ROS produced in cells.[77] In Q-dots injured cells, reduction of mitochondrial membrane potential and swelling of mitochondria can be detected.[61]

Lysosomes are organelles commonly associated with cell death.[78,79] Lysosomes and lysosomal hydrolyses play a role in the engulfment and digestion of dead and dying cells and in cellular/tissue autolysis during necrosis. Studies showing the involvement of lysosomes in cell death, induced by either pharmacological or genetic manipulations, were recently reviewed.79 Lysosomes are also very sensitive to oxidative stress, and this property is exploited in photodynamic therapy. Lysosomal membranes can potentially be damaged by nanoparticle-produced ROS internalized by endocytosis. This can lead to the release of lethal hydrolyses and consequent cell death.

Endoplasmic reticulum (ER) is another organelle at risk in cells exposed to certain types of nanoparticles. The ER is a sensor for oxidative stress (reviewed in refs. 80,81,82). If the ER damage is extensive, programmed cell death is initiated by the unfolded protein response or by the release of calcium. The Bcl-2 family of proteins and cytoplasmic calcium coordinate the cross talk between the mitochondria and the ER.[83,84]

Nanoparticles; Prospects

Developments of new nanotechnologies are gaining momentum and claims have been made that they will dominate the economy of the 21st century. The application of these technologies already extends to solar technology, new means of transportation, telecommunications, medical diagnostics and devices. However, the evident and tremendous potential benefits of nanotechnology in our society will only be achieved if we apply it sensibly and learn how to reduce the potential health risks associated with nanoparticle contamination in our environment. In the context of biological systems we must especially understand the mechanisms underlying the interactions between nanoparticles with different physical and chemical properties in conjunction with biological structures and functions.

Multidisciplinary efforts to understand the chemistry, physics, and biology of nanoparticle-cell interactions will provide a foundation to better address numerous questions, including 'Quo vadis, nanoparticle?' and how to use them to our benefits.

Acknowledgements

We acknowledge the partial financial support of the Canadian Institutes of Health Research, the National Sciences and Engineering Research Council (Canada) and the Juvenile Diabetes Research Foundation. The authors acknowledge the editorial assistance by F. Manganaro and M. Jain.

References

1. Ferrari M. Cancer nanotechnology: Opportunities and challenges. Nat Rev Cancer 2005; 5(3):161-171.
2. Silva GA. Neuroscience nanotechnology: Progress, opportunities and challenges. Nat Rev Neurosci 2006; 7(1):65-74.

3. Bakalova R, Ohba H, Zhelev Z et al. Quantum dots as photosensitizers? Nat Biotechnol 2004; 22(11):1360-1361.
4. Kam NW, O'Connell M, Wisdom JA et al. Carbon nanotubes as multifunctional biological transporters and near-infrared agents for selective cancer cell destruction. Proc Natl Acad Sci USA 2005; 102(33):11600-11605.
5. Jain KK. Nanoparticles as targeting ligands. Trends Biotechnol 2006; 24(4):143-145.
6. Weissleder R, Kelly K, Sun EY et al. Cell-specific targeting of nanoparticles by multivalent attachment of small molecules. Nat Biotechnol 2005; 23(11):1418-1423.
7. Giepmans BN, Adams SR, Ellisman MH et al. The fluorescent toolbox for assessing protein location and function. Science 2006; 312(5771):217-224.
8. Medintz IL, Uyeda HT, Goldman ER et al. Quantum dot bioconjugates for imaging, labelling and sensing. Nat Mater 2005; 4(6):435-446.
9. Jaiswal JK, Mattoussi H, Mauro JM et al. Long-term multiple color imaging of live cells using quantum dot bioconjugates. Nat Biotechnol 2003; 21(1):47-51.
10. Yao G, Wang L, Wu Y et al. FloDots: Luminescent nanoparticles. Anal Bioanal Chem 2006; 385(3):518-524.
11. Savic R, Eisenberg A, Maysinger D. Block copolymer micelles as delivery vehicles of hydrophobic drugs: Micelle-cell interactions. J Drug Target 2006; 14(6):343-355.
12. Savic R, Luo L, Eisenberg A et al. Micellar nanocontainers distribute to defined cytoplasmic organelles. Science 2003; 300(5619):615-618.
13. Alivisatos AP. Semiconductor Clusters, Nanocrystals, and Quantum Dots. Science 1996; 271:933-937.
14. Dahan M, Levi S, Luccardini C et al. Diffusion dynamics of glycine receptors revealed by single-quantum dot tracking. Science 2003; 302(5644):442-445.
15. Larson DR, Zipfel WR, Williams RM et al. Water-soluble quantum dots for multiphoton fluorescence imaging in vivo. Science 2003; 300(5624):1434-1436.
16. Maysinger D, Berezovska O, Savic R et al. Block copolymers modify the internalization of micelle-incorporated probes into neural cells. Biochim Biophys Acta 2001; 1539(3):205-217.
17. Torchilin VP. Fluorescence microscopy to follow the targeting of liposomes and micelles to cells and their intracellular fate. Adv Drug Deliv Rev 2005; 57(1):95-109.
18. Allen C, Yu Y, Eisenberg A et al. Cellular internalization of PCL(20)-b-PEO(44) block copolymer micelles. Biochim Biophys Acta 1999; 1421(1):32-38.
19. Philipp Seib F, Jones AT, Duncan R. Establishment of subcellular fractionation techniques to monitor the intracellular fate of polymer therapeutics I: Differential centrifugation fractionation B16F10 cells and use to study the intracellular fate of HPMA copolymer-doxorubicin. J Drug Target 2006; 14(6):375-390.
20. Duncan R. Polymer conjugates as anticancer nanomedicines. Nat Rev Cancer 2006; 6(9):688-701.
21. Maeda H, Wu J, Sawa T et al. Tumor vascular permeability and the EPR effect in macromolecular therapeutics: A review. J Control Release 2000; 65(1-2):271-284.
22. Wang L, Yang C, Tan W. Dual-luminophore-doped silica nanoparticles for multiplexed signaling. Nano Lett 2005; 5(1):37-43.
23. Pathak S, Cao E, Davidson MC et al. Quantum dot applications to neuroscience: New tools for probing neurons and glia. J Neurosci 2006; 26(7):1893-1895.
24. Gaponik N, Talapin DV, Rogach AL et al. Thiol-capping of CdTe nanocrystals: An alternative to organometallic synthetic routes. J Phys Chem B 2002; 106:7177-7185.
25. Chan WC, Nie S. Quantum dot bioconjugates for ultrasensitive nonisotopic detection. Science 1998; 281(5385):2016-2018.
26. Pavlovic E, Quist AP, Gelius U et al. Surface functionalization of silicon oxide at room temperature and atmospheric pressure. J Colloid Interface Sci 2002; 254(1):200-203.
27. Pinaud F, King D, Moore HP et al. Bioactivation and cell targeting of semiconductor CdSe/ZnS nanocrystals with phytochelatin-related peptides. J Am Chem Soc 2004; 126(19):6115-6123.
28. Dubertret B, Skourides P, Norris DJ et al. In vivo imaging of quantum dots encapsulated in phospholipid micelles. Science 2002; 298(5599):1759-1762.
29. Gao X, Cui Y, Levenson RM et al. In vivo cancer targeting and imaging with semiconductor quantum dots. Nat Biotechnol 2004; 22(8):969-976.
30. Bharali DJ, Lucey DW, Jayakumar H et al. Folate-receptor-mediated delivery of InP quantum dots for bioimaging using confocal and two-photon microscopy. J Am Chem Soc 2005; 127(32):11364-11371.
31. Wu X, Liu H, Liu J et al. Immunofluorescent labeling of cancer marker Her2 and other cellular targets with semiconductor quantum dots. Nat Biotechnol 2003; 21(1):41-46.

32. Hoshino A, Fujioka K, Oku T et al. Quantum dots targeted to the assigned organelle in living cells. Microbiol Immunol 2004; 48(12):985-994.
33. Vu TQ, Maddipati R, Blute TA et al. Peptide-conjugated quantum dots activate neuronal receptors and initiate downstream signaling of neurite growth. Nano Lett 2005; 5(4):603-607.
34. Hermanson GT. Bioconjugate Techniques. San Diego: Academic Press, 1996.
35. Torchilin VP, Narula J, Halpern E et al. Poly(ethylene glycol)-coated anti-cardiac myosin immunoliposomes: Factors influencing targeted accumulation in the infarcted myocardium. Biochim Biophys Acta 1996; 1279(1):75-83.
36. Torchilin VP, Lukyanov AN, Gao Z et al. Immunomicelles: Targeted pharmaceutical carriers for poorly soluble drugs. Proc Natl Acad Sci USA 2003; 100(10):6039-6044.
37. Savic R, Azzam T, Eisenberg A et al. Assessment of the integrity of poly(caprolactone)-b-poly(ethylene oxide) micelles under biological conditions: A fluorogenic-based approach. Langmuir 2006; 22(8):3570-3578.
38. Michalet X, Pinaud FF, Bentolila LA et al. Quantum dots for live cells, in vivo imaging, and diagnostics. Science 2005; 307(5709):538-544.
39. Jiang W, Papa E, Fischer H et al. Semiconductor quantum dots as contrast agents for whole animal imaging. Trends Biotechnol 2004; 22(12):607-609.
40. Aldana J, Lavelle N, Wang Y et al. Size-dependent dissociation pH of thiolate ligands from cadmium chalcogenide nanocrystals. J Am Chem Soc 2005; 127(8):2496-2504.
41. Aldana J, Wang YA, Peng X. Photochemical instability of CdSe nanocrystals coated by hydrophilic thiols. J Am Chem Soc 2001; 123(36):8844-8850.
42. Rossin R, Pan D, Qi K et al. 64Cu-labeled folate-conjugated shell cross-linked nanoparticles for tumor imaging and radiotherapy: Synthesis, radiolabeling, and biologic evaluation. J Nucl Med 2005; 46(7):1210-1218.
43. Thurmond KB, Kowalewski T, Wooley KL. Shell cross-linked knedels: A synthetic study of the factors affecting the dimensions and properties of amphiphilic core-shell nanospheres. J Am Chem Soc 1997; 119(28):6656-6665.
44. Thurmond KB, Kowalewski T, Wooley KL. Water-soluble knedel-like structures: The preparation of shell-cross-linked small particles. J Am Chem Soc 1996; 118(30):7239-7240.
45. Cheng C, Qi K, Khoshdel E et al. Tandem synthesis of core-shell brush copolymers and their transformation to peripherally cross-linked and hollowed nanostructures. J Am Chem Soc 2006; 128(21):6808-6809.
46. Joralemon MJ, O'Reilly RK, Hawker CJ et al. Shell click-crosslinked (SCC) nanoparticles: A new methodology for synthesis and orthogonal functionalization. J Am Chem Soc 2005; 127(48):16892-16899.
47. Kim S, Lim YT, Soltesz EG et al. Near-infrared fluorescent type II quantum dots for sentinel lymph node mapping. Nat Biotechnol 2004; 22(1):93-97.
48. Voura EB, Jaiswal JK, Mattoussi H et al. Tracking metastatic tumor cell extravasation with quantum dot nanocrystals and fluorescence emission-scanning microscopy. Nat Med 2004; 10(9):993-998.
49. So MK, Xu C, Loening AM et al. Self-illuminating quantum dot conjugates for in vivo imaging. Nat Biotechnol 2006; 24(3):339-343.
50. Akerman ME, Chan WC, Laakkonen P et al. Nanocrystal targeting in vivo. Proc Natl Acad Sci USA 2002; 99(20):12617-12621.
51. Fisher HC, Liu L, Pang SK et al. Pharmacokinetics of nanoscale quantum dots: In vivo distribution, Sequestration, and clearance in the rat. Adv Function Mater 2006; 16(10):1299-1305.
52. Murphy L. Biosensors and bioelectrochemistry. Curr Opin Chem Biol 2006; 10(2):177-184.
53. Zhang Y, Lim CT, Ramakrishna S et al. Recent development of polymer nanofibers for biomedical and biotechnological applications. J Mater Sci Mater Med 2005; 16(10):933-946.
54. Gruner G. Carbon nanotube transistors for biosensing applications. Anal Bioanal Chem 2006; 384(2):322-335.
55. Watson P, Jones AT, Stephens DJ. Intracellular trafficking pathways and drug delivery: Fluorescence imaging of living and fixed cells. Adv Drug Deliv Rev 2005; 57(1):43-61.
56. Santra S, Zhang P, Wang K et al. Conjugation of biomolecules with luminophore-doped silica nanoparticles for photostable biomarkers. Anal Chem 2001; 73(20):4988-4993.
57. Zhao X, Hilliard LR, Mechery SJ et al. A rapid bioassay for single bacterial cell quantitation using bioconjugated nanoparticles. Proc Natl Acad Sci USA 2004; 101(42):15027-15032.
58. Stone V, Donaldson K. Nanotoxicology: Signs of stress. Nature Nanotechnology 2006; 1(1):23-24.
59. Nel A, Xia T, Madler L et al. Toxic potential of materials at the nanolevel. Science 2006; 311(5761):622-627.
60. Derfus AM, Chen WCW, Bhatia SN. Probing the cytotoxicity of semiconductor quantum dots. Nano Letters 2004; 4:11-18.

61. Lovric J, Cho SJ, Winnik FM et al. Unmodified cadmium telluride quantum dots induce reactive oxygen species formation leading to multiple organelle damage and cell death. Chem Biol 2005; 12(11):1227-1234.
62. Kirchner C, Liedl T, Kudera S et al. Cytotoxicity of colloidal CdSe and CdSe/ZnS nanoparticles. Nano Lett 2005; 5(2):331-338.
63. Oberdorster E. Manufactured nanomaterials (fullerenes, C60) induce oxidative stress in the brain of juvenile largemouth bass. Environ Health Perspect 2004; 112(10):1058-1062.
64. Lam CW, James JT, McCluskey R et al. Pulmonary toxicity of single-wall carbon nanotubes in mice 7 and 90 days after intratracheal instillation. Toxicol Sci 2004; 77(1):126-134.
65. Ipe BI, Lehnig M, Niemeyer CM. On the generation of free radical species from quantum dots. Small 2005; 1(7):706-709.
66. Samia AC, Chen X, Burda C. Semiconductor quantum dots for photodynamic therapy. J Am Chem Soc 2003; 125(51):15736-15737.
67. Green M, Howman E. Semiconductor quantum dots and free radical induced DNA nicking. Chem Commun (Camb) 2005; (1):121-123.
68. Cho S, Maysinger D, Jain M et al. Long-term exposure to CdTe quantum dots causes functional impairments in live cells. Langmuir 2007; 23(4):1974-1980.
69. Toews AD, Lee SY, Popko B et al. Tellurium-induced neuropathy: A model for reversible reductions in myelin protein gene expression. J Neurosci Res 1990; 26(4):501-507.
70. Moghimi SM, Hunter AC, Murray JC. Nanomedicine: Current status and future prospects. FASEB J 2005; 19(3):311-330.
71. Nishiyama N, Kataoka K. Current state, achievements, and future prospects of polymeric micelles as nanocarriers for drug and gene delivery. Pharmacol Ther 2006; 112(3):630-648.
72. Xia T, Kovochich M, Brant J et al. Comparison of the abilities of ambient and manufactured nanoparticles to induce cellular toxicity according to an oxidative stress paradigm. Nano Lett 2006; 6(8):1794-1807.
73. Noble M, Mayer-Proschel M, Proschel C. Redox regulation of precursor cell function: Insights and paradoxes. Antioxid Redox Signal 2005; 7(11-12):1456-1467.
74. Finkel T, Holbrook NJ. Oxidants, oxidative stress and the biology of ageing. Nature 2000; 408(6809):239-247.
75. Clarke SJ, Hollmann CA, Zhang Z et al. Photophysics of dopamine-modified quantum dots and effects on biological systems. Nat Mater 2006; 5(5):409-417.
76. Hansen JM, Go YM, Jones DP. Nuclear and mitochondrial compartmentation of oxidative stress and redox signaling. Annu Rev Pharmacol Toxicol 2006; 46:215-234.
77. Li N, Sioutas C, Cho A et al. Ultrafine particulate pollutants induce oxidative stress and mitochondrial damage. Environ Health Perspect 2003; 111(4):455-460.
78. Broker LE, Kruyt FA, Giaccone G. Cell death independent of caspases: A review. Clin Cancer Res 2005; 11(9):3155-3162.
79. Kroemer G, Jaattela M. Lysosomes and autophagy in cell death control. Nat Rev Cancer 2005; 5(11):886-897.
80. Xu C, Bailly-Maitre B, Reed JC. Endoplasmic reticulum stress: Cell life and death decisions. J Clin Invest 2005; 115(10):2656-2664.
81. Boyce M, Yuan J. Cellular response to endoplasmic reticulum stress: A matter of life or death. Cell Death Differ 2006; 13(3):363-373.
82. Zhang K, Kaufman RJ. The unfolded protein response: A stress signaling pathway critical for health and disease. Neurology 2006; 66(2 Suppl 1):S102-109.
83. Oakes SA, Lin SS, Bassik MC. The control of endoplasmic reticulum-initiated apoptosis by the BCL-2 family of proteins. Curr Mol Med 2006; 6(1):99-109.
84. Ferri KF, Kroemer G. Organelle-specific initiation of cell death pathways. Nat Cell Biol 2001; 3(11):E255-263.

CHAPTER 13

Toxicity Studies of Fullerenes and Derivatives

Jelena Kolosnjaj, Henri Szwarc and Fathi Moussa*

Abstract

Due to their unique properties, fullerenes, a model of carbon-based nanoparticles, have attracted considerable interest in many fields of research including material science and biomedical applications. The potential and the growing use of fullerenes and their mass production have raised several questions about their safety and environmental impact. Available data clearly shows that pristine C_{60} has no acute or sub-acute toxicity in a large variety of living organisms, from bacteria and fungal to human leukocytes, and also in drosophila, mice, rats and guinea pigs. In contrast to chemically—either covalently or noncovalently—modified fullerenes, some C_{60} derivatives can be highly toxic. Furthermore, under light exposure, C_{60} is an efficient singlet oxygen sensitizer. Therefore, if pristine C_{60} is absolutely nontoxic under dark conditions, this is not the case under UV-Visible irradiation and in the presence of O_2 where *fullerene solutions* can be highly toxic through $^1O_2$ formation.

This chapter offers a general review of the studies on the toxicity of [60]fullerene or C_{60}, the most abundant fullerene, and its derivatives.

Introduction

[60]fullerene was discovered in 1985 by H. Kroto, R. Smalley and Curl.[1] The molecule -buckminsterfullerene or simply fullerene- was named after the American architect Richard Buckminster Fuller (1895-1985), since it had the shape of a truncated icosahedron that looked like his geodesic domes. Since their discovery fullerenes have attracted considerable interest in many fields of research including material science and biomedical applications. Fullerenes were proved useful in the chemical and material science applications, such as semiconductors and microscopic engineering and polymers.[2-5] Biomedical applications include antioxidant, antiviral, antibiotic and anti-cancerous activities, enzyme inhibition, cell signalling, DNA cleavage as well as imaging and nuclear medicine.[6-8]

As underivatized and derivatized fullerenes are becoming increasingly available, it is of great importance to assess their safety and environmental impact. Indeed, since 1993, different groups of researchers performed toxicity studies in vitro as well as in vivo in various model systems. Available data clearly shows that pristine C_{60} has no acute or sub-acute toxicity in a large variety of living organisms. However, it is obvious that chemical modifications can change the general properties of pristine fullerene. As a matter of fact, in contrast to pristine C_{60}, (Fig. 1A) several of its derivatives can be highly toxic. Besides, some modifications, sometimes even noncovalent interaction with a solubilizing agent (like for example polyvinylpyrrolidone, PVP, which forms a charge transfer complex with fullerene), can have a considerable impact on its toxicity. Several authors cleverly avoided the problem of chemical modification

*Corresponding Author: Fathi Moussa—UMR CNRS 8612, Faculté de Pharmacie, Université Paris-Sud 11, 5, Rue J.-B. Clément, 92296, Châtenay-Malabry, France. E-mail: fathi.moussa@u-psud.fr

Bio-Applications of Nanoparticles, edited by Warren C.W. Chan. ©2007 Landes Bioscience and Springer Science+Business Media.

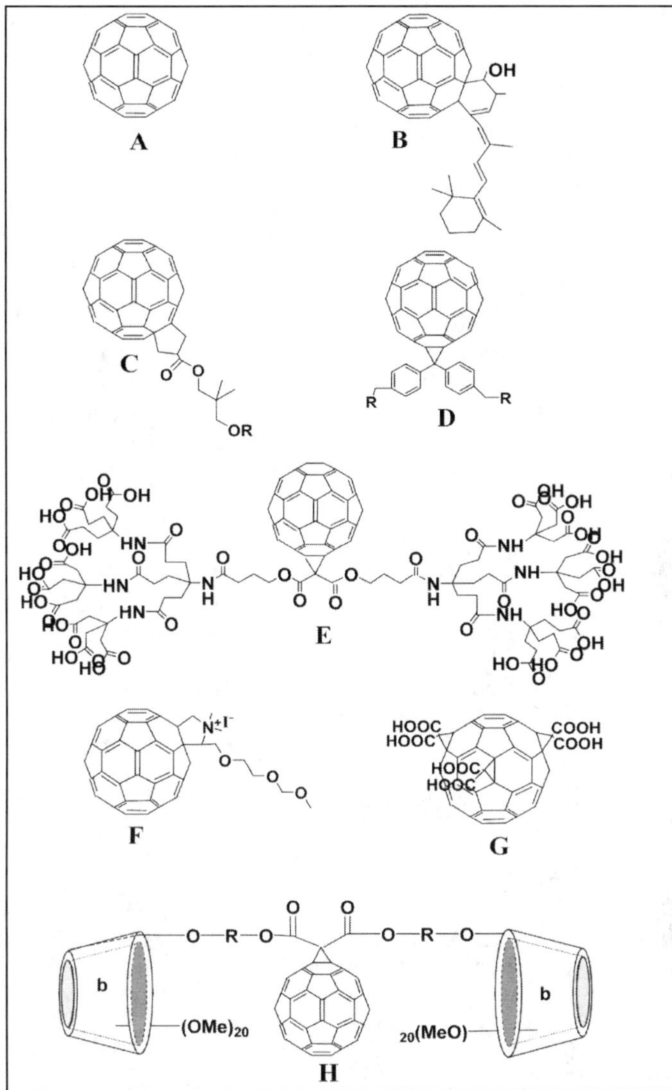

Figure 1. Schematic of [60]fullerenes with different surface modifications.

by forming highly stable aqueous suspensions or molecular-colloid systems of C_{60} in water. Similar molecular-colloid systems can be prepared by using intermediate organic solvents, notably THF. However, the molecular-colloid resulting complex can lead to surface modification and hence a different toxicological profile. On another hand, C_{60} is an efficient singlet oxygen sensitizer under light exposure.[9] Then, in the presence of O_2, this fullerene and some of its derivatives can be highly toxic through $^1O_2$ formation. The latter is then able to damage crucial biological molecules such as DNA, lipids and proteins. This chapter presents a general review of toxicity studies of C_{60} and derivatives, the most abundant and most studied fullerene. For a better understanding, the following section will be divided in three subsections including toxicity studies performed with pristine C_{60}, noncovalently modified C_{60} and covalently modified C_{60}. However, to understand the reactivity and the behaviour of the fullerene, it is useful to review some of its general physical properties.

Physical Properties

C_{60} is a granular, dark-brown powder which is not soluble in polar solvents. However, this molecule is soluble in some organic solvents, like chloronaphtalene, dichlorobenzene, toluene etc.[10] Other ways of solubilization might involve the encapsulation or micro-encapsulation in carriers (such as γ-cyclodextrines, calixarenes or liposomes) or the addition of PVP, octanol, Triton X or lecithin, Tween and phospholipids. The latter methods have a great advantage for using C_{60} and its derivatives for medicinal purposes.[6,11] C_{60} is a symmetric molecule which consists of 60 carbon atoms that form 12 pentagons and 20 hexagons, like a ball of the European football. As a consequence, it has been called by several other names, notably buckyball, soccerballene or, in Europe, footballene. C_{60} is the most perfectly symmetrical molecule in nature (symmetry group I_h) and even though its molecular weight seems large, the van der Waals diameter of the fullerene molecule is 1 nm.[1] This corresponds approximately to the length of a C_8H_{18} molecule. The carbon atoms are connected by sp^2 bonds and the remaining 60 π electrons distribution gives rise to an aromatic character. The reactivity of C_{60} is that of a strained, electron deficient poly-alkene with rather localized double bonds.[12] The fullerene can be reversibly reduced up to six electrons. It readily forms adducts with nucleophiles and carbenes, participates as an electron deficient dienophile component in a variety of thermal cycloaddition reactions. The additions are exothermic and are presumably driven by the relief of strain in the C_{60} cage that largely results from the pyramidalization of its sp^2 C atoms. In adducts, the functionalized fullerene C atoms change their hybridization from trigonal sp^2 to the less strained tetrahedral sp^3 state.[10] Due to its spherical molecules with 30 carbon double bonds, [60]fullerene or C_{60} can easily react with free radicals: it is a very efficient free radical scavenger, which labels this molecule as a "radical sponge".[9] Indeed, C_{60} and derivatives has been widely used in vitro as well as in vivo in different model systems to protect living organisms against free-radical damage.[13-16]

C_{60} strongly absorbs light, especially in the UV domain of the spectrum. Singlet excited state of C_{60} ($^1C_{60}$) is initially formed upon light excitation. The main decay mode of fullerene singlets consists in intersystem crossing to triplet states ($^3C_{60}$).[17] Additionally, fullerene triplets can be formed indirectly using triplet sensitizers such as acridine and anthracene, and are quenched by triplet state quenchers such as rubrene, tetracene and ground state triplet oxygen.[17] It is well established that under UV irradiation and in the presence of O_2, $^3C_{60}$ can efficiently (almost 100 %) sensitize formation of $^1O_2$, a highly reactive form of O_2 that can damage bio-molecules.[18] Several in vitro studies showed that C_{60} and derivatives can be highly toxic under UV irradiation, including DNA cleavage.[19-20]

Toxicity Studies of Pristine C_{60}

Available data clearly show that pristine C_{60} has no acute or sub-acute toxicity in a large variety of living organisms, from bacteria and fungal to human leukocytes, and also in drosophilae, mice, rats and guinea pigs. The first toxicity study of pristine C_{60} has been conducted in the United States (Tucson, Arizona).[21] The authors used a C_{60}-benzene solution to determine the effects of acute and sub-chronic exposure of topically applied fullerene extracts on mouse skin. This study has not revealed any acute toxic effects on mouse skin epidermis.[21] As benzene is highly toxic, it cannot represent the appropriate solvent for studying the toxicity of C_{60}. Nevertheless, this experiment was relevant for the scientists working in the area of fullerenes as they were routinely exposed to such C_{60}-solutions. One year later, another American team synthesized ^{14}C labelled-C_{60} and suspended it in water in order to study its uptake by cultured human keratinocytes.[22] Although the authors have not succeeded in visualizing the fullerenes inside cells, they observed that it was rapidly up-taken by the cells without any sign of acute toxicity. At the same time, a Russian team used the somatic mosaicism model to demonstrate that this fullerene has no genetic toxicity.[23] In 1995, a French team used a simply hand-grinded C_{60} to study its effects on cultured human leucocytes.[24] The authors of this study demonstrated for the first time that C_{60} can be internalised by normal human phagocytes (Fig. 2) without any sign of acute toxicity, which confirm the results obtained previously in the United States on human keratinocytes.[22] One year later, a German group confirmed the absence of cytotoxic effects on cultured bovine alveolar macrophages and HL-60-macrophages.[25] The authors also measured the effects of C_{60} on the production by these cells of tumour-necrosis-factor and interleukin-8 as well as on superoxide anion release. The same year, the French group prepared a highly concentrated aqueous suspension of micronized C_{60} (100 mg/ml), in order to study for the first time the in vivo toxicity of this fullerene after

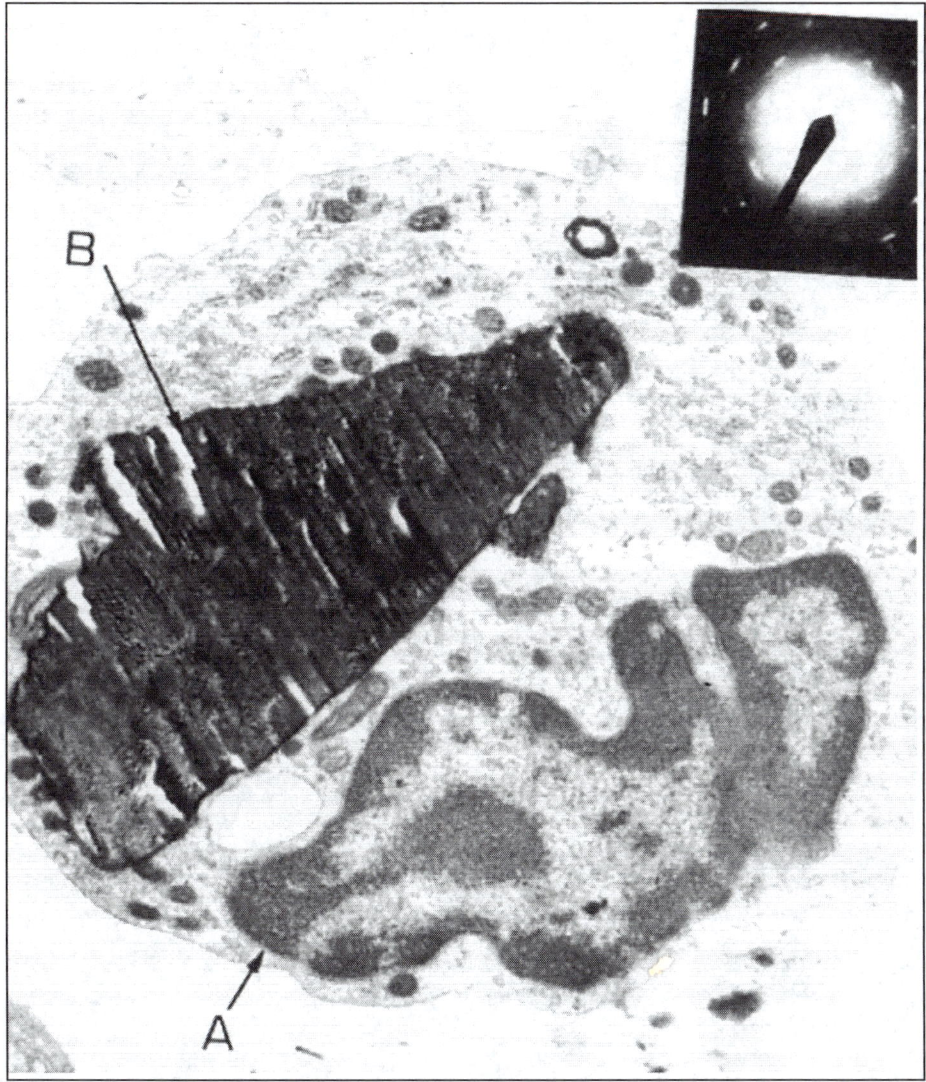

Figure 2. Micrograph of a 5μm long C_{60} crystal (B) inside a human leukocyte (A: nucleus); insert: electron diffraction spectrum of (B) (Moussa et al 1995[24]).

intra-peritoneal injection to rodents.[26-27] Despite the large amounts injected (2.5 to 5.0 g/kg of body weight), C_{60} showed neither lethal nor acute[26] or sub-acute[27] toxicity with respect to this animal species. This work clearly demonstrated that C_{60} can cross the membrane barriers in vivo (Fig. 3).

In early stages, C_{60} induced hypertrophy and hyperplasia of hepatic stellate cells (HSC: liver resident nonparenchymal cells also referred to as fat-storing or perisinusoidal cells, lipocytes and Ito cells) where it mainly accumulates (Fig. 3). HSC play a central role in the production of extracellular matrix both in normal and fibrotic liver.[16] The phenomena of hypertrophy and hyperplasia of HSC, showing the activation of these cells, usually occurs under different pathological conditions leading to liver fibrosis.[28] The activation of HSC irremediably leads to their

Figure 3. C_{60} crystals (arrows) inside an Ito cell (Moussa et al 1996[26]).

transformation into myofibroblast-like cells (MFC).[28] This phenomenon is now recognised as the main event in hepatic fibrogenesis, which occurs through the amplification of extracellular matrix production by these cells.[28] Furthermore, the transformation of HSC into MFC can be modulated by oxidative stress-related products[29] as reflected in rat intoxication with CCl_4, a well-known in vivo free radical initiator.[30,31] In the case of C_{60} however, in spite of HSC activation, the mouse liver structure remained normal and no fibrosis either sinusoidal or portal developed.[26] Fifty-six days after C_{60}-treatment, HSC decreased in number without any transformation into MFC.[16] This surprising phenomenon of inhibition of HSC transformation into MFC may be attributed to the free radicals-scavenging properties of C_{60}.[16] Not much later, the Russian group confirmed the absence of mutagenic effect of C_{60} in a prokaryotic in vitro test using the Escherichia coli strain PQ37 and in a eukaryotic in vivo system on somatic wing cells of Drosophila melanogaster larvae.[32] In another way, an English group showed during the same period that fungal can grow on C_{60}.[33] At the same time, the French group observed that C_{60} did not induce interleukin1-β secretion in cultured human leukocytes[34] and demonstrated for the first time that this fullerene can react in vivo inside liver cells with vitamin A, according to a Diels-Alder-like reaction (Fig. 1B).[35] Two years later, in France, a study on the microbial growth of 22 collection strains including *E. coli* (5) ; *P. aeruginosa* (2); *S. typhimurium* (6); *S. aureus* (2); *L. monocytogenes* (2) *E. hirae* (1) ; *B. cereus* (1) ; *B. subtilis* (1) ; *B. pumilis* (1) and *C. albicans* (1) showed that C_{60} has no effect on the normal growth of these micro-organisms.[36] Very recently, through collaboration with an American team, the French group confirmed the absence of acute and sub-acute toxicity in another species of rodents

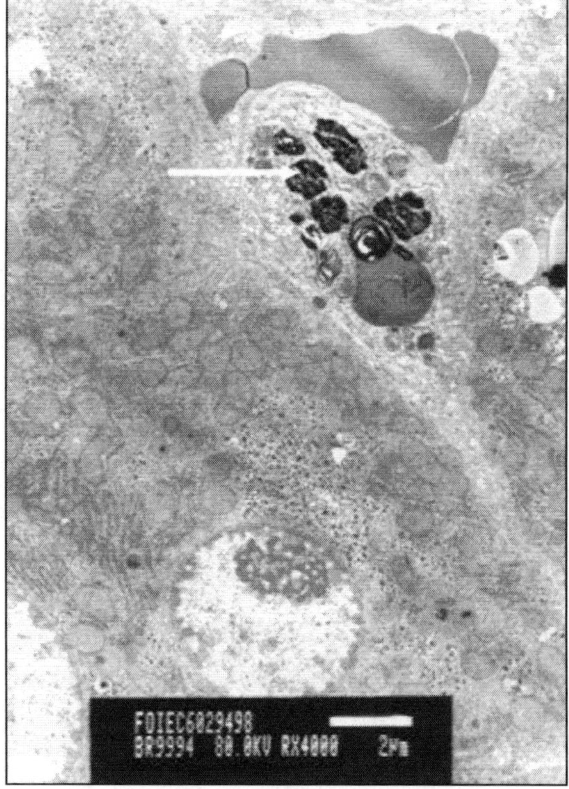

Figure 4. C_{60} crystal (arrow) inside a rat Kuppfer cell (Gharbi et al 2005[16]).

                                                         Bio-Applications of Nanoparticles

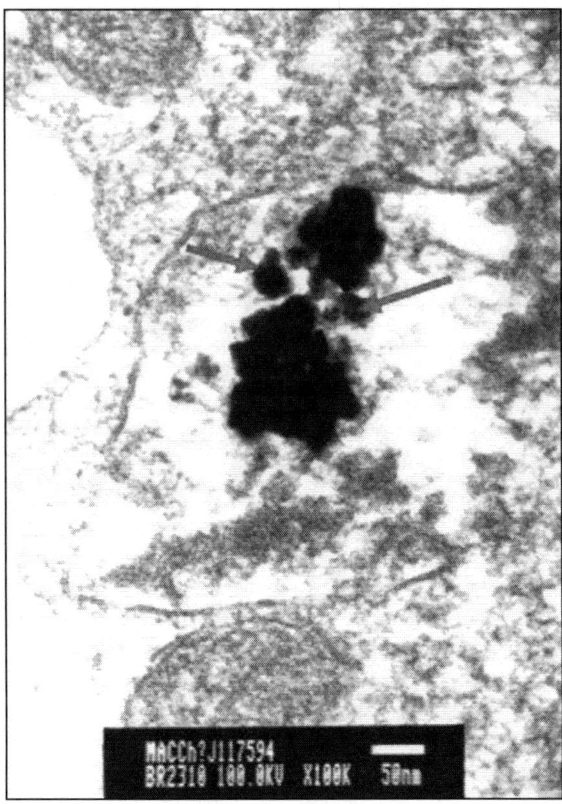

Figure 5. TEM micrograph of C_{60} clusters (arrows) inside a rat hepatocyte (Gharbi et al 2005[16]).

(Figs. 4 and 5).[16] The latter work also showed that C_{60} can be partly eliminated through the bile ducts and transformed into C_{60}-retinyl adducts like that which occurs in mice.[35] Recently, the Japonese group confirmed the result obtained by the French scientists in 1996[26] on the absence of toxicity in rodents[37] upon exposure. The authors used the same C_{60} aqueous suspension than that described by Moussa et al[16,26,27] and administered it orally as a single dose of 2g/kg of bodyweight to male and female Sprague-Dawley rats. No deaths were observed in treated rats and the growth was normal in all instances. The absence of genotoxicity was also assessed by the same group in a bacterial reverse mutation assay (Ames test) and the chromosomal aberration test in cultured Chinese hamster lung (CHL/IU) cells. The authors concluded that fullerenes are not of high toxicological significance.[37] In conclusion these studies performed on various experimental models by different groups from different countries showed that pristine C_{60} has no acute or sub-acute toxicity.

Toxicity Studies of Noncovalently Modified C_{60}

In contrast to what occurs with pristine C_{60} directly suspended in water or by means of biocompatible additives, some aqueous C_{60} preparations can be highly toxics. These preparations include C_{60}-solutions prepared with solubilizing agents that could form charge-transfer complexes C_{60} such as PVP, as well as aqueous C_{60}-suspensions using polar organic solvents like THF. The C_{60}/PVP complexes have the highest equilibrium constant K found so far for organic charge-transfer complexes.[38] Such charge-transfer complexes formed in aqueous medium generally exhibit either one or multiple coordinating nitrogens, depending on the concentration of the ligand and its ionization potential.[38] In biological media, a C_{60}/PVP complex could then bind to different receptors in comparison to C_{60} alone; this could lead to different effects. As a matter of fact, the first study pointing out a possible negative effect of C_{60} was observed as early as 1995 in Japan by Satoh et al.[39] The authors studied the effects of acute and short term repeated application of a PVP-solution of fullerene C_{60} (4 µM) on agonist-induced responses in various tissues of Guinea pigs and rats. This study has not revealed any direct effect on all tissues or antagonistic properties towards drug receptors, however, they reported that sub chronic exposure decreased responsiveness.[39]

Actually, the first harmful effect of C_{60}-PVP solutions was observed in 1996 by Tsuchiya et al on mouse embryos in vitro as well as in vivo.[40] In vitro, the authors used a mouse midbrain cell differentiation system and observed that the incubation of C_{60} with various PVP concentrations inhibited cell differentiation and proliferation but the effect was weaker than the vehicle controls. In vivo, they observed that C_{60}-PVP solutions strongly alter embryogenesis while PVP alone has no effect.[40] In 1996 however, the interactions between C_{60} and PVP were still unknown. Ignoring the strong interactions between C_{60} and PVP, Tsuchiya et al attributed this harmful effect to C_{60}.[40] We can now give a better interpretation of these results. As both C_{60}-PVP and PVP are toxic in vitro while only C_{60}-PVP exhibits harmful effects in mouse embryos, the obvious conclusion is that C_{60}-PVP complexes can cross the placental barrier while PVP cannot. Surprisingly, at that time, the media ignored this "serious harmful effect of C_{60}" according to the authors, perhaps because fullerenes were not yet mass-produced. This was not the case in 2004 when a study performed in the Southern Methodist University in Dallas, Texas linking a fullerene aqueous suspension (nano-C_{60} aggregated water-soluble fullerenes or nC_{60}) to brain damage in fish.[41] This article made international headlines with coverage in a number of newspapers.[42,43] According to Rittner M. N., provocative headlines turned this single small study into a poster child for the alleged dangers of fullerenes and nanotechnologies.[43] Nevertheless, this study raised a serious doubt about the safety of C_{60} and its possible ecotoxic effects. Other reports emanating from a limited group of authors using the same nC_{60} preparation continued to maintain that pristine C_{60} might be very toxic in living systems such as bacteria, algae and fishes by inducing oxidative stress.[44-48] According to these authors, C_{60} derivatives are less toxic than pristine C_{60}, which is contrary to all previous studies.[49-54,56] Let us examine how the nC_{60} suspensions used in these alarming studies are obtained. To get them, the authors first dissolve fullerene in THF. The resulting solution is then mixed with water and the organic solvent is removed through an evaporative step, leaving clusters of nC_{60} as a stable colloidal suspension in water.[57] According to the authors, "This suspension consisted of stable 30- to 100 nm aggregates in which the fullerenes facing the water were most likely partially modified but the central core of the aggregate contained unmodified fullerenes".[41,57] Since 1995, a number of papers published by several research groups clearly demonstrated that the toxic effects observed with such nC_{60} suspensions[47,58] can be attributed to THF/C_{60} complexes. The first criticism came from Andrievsky, the Ukrainian pioneer of biological applications of nC_{60}.[59] Andrievsky et al. also showed by FTIR analysis that organic groups remained in the nC_{60} structure prepared using THF as organic solvent.[60] This finding was soon confirmed by NMR imaging by a Chinese group.[61] In the same way, a Serbian group demonstrated that γ-irradiation of THF/nC_{60} modifies the physico-chemical properties of THF-prepared nanocrystalline C_{60}, resulting in a complete loss of its toxic effects.[62] Furthermore, Deguchi et al., a Japanese group at the origin of the preparation of nC_{60} suspensions, demonstrated that avoiding the use of THF during the preparation of nC_{60} suspensions prevents the toxic effects observed with THF/C_{60}.[63] In another way, Jia et al[64] a Chinese group, did not observe any significant toxicity for C_{60} up to a dose of 226 microg/cm^2 on alveolar macrophage cell lines after 6h exposure. At the same time in France, Gharbi et al. confirmed the absence of acute and sub acute toxicity in rodents.[16] One year later, an Italian-French collaborative study explored the inflammatory response of murine and human macrophage cells in vitro.[65] The authors showed that C_{60} does not stimulate the release of NO by murine macrophages in culture and that this fullerene has no toxicity against human macrophages. Last summer in Japan, Mori et al. confirmed the absence of toxicity of pristine C_{60} in rats.[37] Finally, a recent study[58] highlighted the impact of the method of preparation on nC_{60} properties. The authors examined the physical and chemical characteristics of colloidal dispersions of fullerene materials produced through several solvent exchange processes and through extended mixing in water only. According to these authors, the nC_{60} produced via the different methods differs from one another with respect to size, morphology, charge, and hydrophobicity. They found that the greatest dissimilarities were observed between the nC_{60} produced by extended mixing in water alone and the nC_{60} produced by solvent exchange processes. The role of the respective solvents in determining the characteristics of the various nC_{60} were attributed to differences in solvent-C_{60} interactions and the presence of residual solvent within the nC_{60} structure; this emphasizes the importance of the solvent properties in determining the ultimate characteristics of colloidal fullerene. Thus, the authors concluded that fullerene C_{60} that may become mobilized through natural processes (agitation in water) may behave in dramatically different ways than those produced through more artificial means.[58] Taken together; all these reports definitively show that pristine C_{60} is not of high toxicological significance. However, the authors[41,44,45,57] who announced the toxicity of fullerene C_{60} induced confusion in the mass media[43] but have to be granted merit. They attracted considerable interest on the safety and the environmental impact of nanotechnologies as reflected in the importance of the funds which are now allowed to research in the emerging area of nano-toxicology.

Toxicity of Covalently Modified C_{60}

Despite the large number of papers devoted to the potential of medicinal applications of fullerenes derivatives,[6,8,13-15,50,57,66-79] little is known about their toxicity and their in vivo behaviour. Few studies were performed in vitro[8,15,53,67,70-78] as well as in vivo.[19,20,49,51,69,79-82] They show that in contrast with pristine C_{60}, some derivatives can be highly toxic (Fig. 1). Furthermore, their effects in vitro generally do not reflect exactly their in vivo behaviour. The first study on the toxicity of C_{60} derivatives was conducted by Yamago et al.[19] After a single intra-peritoneal injection of 0.5 g/ kg of bodyweight of a water-soluble methanofullerene (Fig. 1C), the mice survived for one week. One year later Rajagopalan et al. studied the pharmacokinetics of another water-soluble methanofullerene (Fig. 1D).[49] Although the in vitro toxicity of this derivative was quite low,[50] it appeared highly toxic in vivo.[49] At a dose of 15 mg/kg, this derivative seemed well tolerated. However, a dose of 25 mg/kg resulted in shortness of breath and violent movement of the rats, followed by death within 5 min of dosing.[49] We also observed in our laboratory a similar behaviour in mice treated with a cationic C_{60}-derivative for a dose as small as 2 mg/kg of bodyweight (unpublished work). In 2000 we studied the in vivo behaviour and the toxicity of two C_{60} derivatives, a highly water-soluble dendrofullerene (DF, Fig. 1E) and a less soluble pyridinium salt (Fig. 1F).[51] After intra-peritoneal injection to mice, the two compounds both exhibited very low toxicity with a LD_{50} about 0.7 g/kg and a LD_{50} higher than 1.2 g/kg, respectively. Compound E (Fig. 1) was readily eliminated through the kidneys while Compound F (Fig. 1) was not. This work showed for the first time that some C_{60}-derivatives can be readily eliminated by a living organism.[51] Later, Dugan et al.[52] reported a LD_{50} higher than 0.4 g/kg in mice for another C_{60}-derivative (Fig. 1G). According to these authors, Compound G can also be eliminated through the kidneys in this rodent species.[52] Very recently, we tested the toxicity of a highly water-soluble cyclodextrine-C_{60} derivative (Fig. 1H), synthesized by Zhang's group.[55] After intra-peritoneal injection up to 1g/kg, the treated mice survived for 2 weeks without any sign of toxicity (unpublished work).

Taken together, these preliminary results show that in contrast with pristine C_{60}, some of its derivatives can be highly toxic and highlight the difficulties in generalizing the in vivo behaviour and the toxicity of chemically modified C_{60} (Table 1).

Table 1. In vivo toxicity of some C_{60} derivatives

Compound (Fig. 1)	Toxicity in Rodents
A	> 5 g/Kg
C	> 0.5 g/kg
D	0.015 - 0.025 g/kg
E	0.7 ± 0.2 g/kg
F	> 1.2 g/kg
G	> 0.4 g/kg
H	> 1 g/Kg

Conclusion

It has been considered for a long time that "the dose makes the poison". In the case of fullerenes and nanoparticles in general however, evidence proves that we should expand this term. The dose by itself is no longer sufficient, as numerous parameters can have a crucial impact. Some of these parameters include crystal structure, chemical composition, size distribution and surface area, surface chemistry and surface charge, shape, porosity and agglomeration state. Most of these parameters obviously depend of the method of preparation of the suspension. To sum up, this review shows that if pristine C_{60} is not toxic, the toxicity of each C_{60} preparation or each C_{60}-derivative must be determined before use. For the promising medicinal applications of fullerenes, sufficient data regarding in vivo behaviour of fullerenes should be accumulated. As to environment impact, new experimental models of C_{60} release have to be developed.

References

1. Kroto HW, Heath JR, O'Brien SC et al. C60: Buckminsterfullerene. Nature 1985; 318:162-163.
2. Smalley RE, Yakobson BI. The future of the fullerenes. Solid State Communications 1998; 107(11):597-606.
3. Granja F, Dorantes-Dávila J, Morán-López JL et al. Electronic structure of some semiconductor fullerenes. Nanostructured Materials 1993; 3(1-6):469-477.
4. Spence JCH. The future of atomic resolution electron microscopy for materials science. Materials Science and Engineering Reports 1999; 26(1-2):1-49.
5. Kuzmany H, Winter J, Burger B. Polymeric fullerenes. Synthetic Metals 1997; 85(1-3):1173-1177.
6. Jensen AW, Wilson SR, Schuster DI. Biological applications of fullerenes. Bioorg Med Chem 1996; 4(6):767-79.
7. Wilson LJ, Cagle DW, Thrash TP et al. Metallofullerene drug design. Coordination Chemistry Reviews 1999; 190-192:199-207.
8. Tsao N, Kanakamma P, Luh TY et al. Inhibition of escherichia coli-induced meningites by carboxyfullerene. Antimicrobial Agents and Chemotherapy 1999; 43(9):2273-7.
9. Krusic PJ, Waserman E, Keizer PN et al. Radical reaction of C60. Science 1991; 254:1183-5.
10. Hirsh A. The chemistry of fullerenes: George Thieme Verlang. 1st ed. NY: 1994.
11. Yamakoshi YN, Yagami T, Fukuhara K et al. Solubilization of fullerenes into water with polyvinylpyrrolidone applicable to biological tests. J Chem Soc Chem Commun 1994; 4:517-18.
12. Diederich F, Thilgen C. Covalent fullerene chemistry. Science 1996; 271(5247):317-323.
13. Dugan LL, Gabrielsen JK, Yu SP. Buckminsterfullerenol free radical scavengers reduce excitotoxic and apoptotic death of cultured cortical neurons. Neurobiology of Disease 1996; 3:129-35.
14. Dugan LL, Lovett EG, Quick J et al. Fullerene based antioxidants and neurodegenerative disorders. Parkinsonism and Related Disorders 2001; 7:243-46.
15. Dugan LL, Turetsky DM, Du C et al. Carboxyfullerenes as neuroprotective agents. Proc Natl Acad Sci USA 1997; 94:9434-9.
16. Gharbi N, Pressac M, Hadchouel M et al. [60]fullerene is a powerful antioxidant in vivo with no acute or subacute toxicity. Nano Lett 2005; 5(12):2578-85.

17. Arbogast J, Darmanyan A, Foote C et al. Photophysical properties of C60. J Phys Chem 1991; 95(1):11-12.
18. Sera N, Tokiwa H, Miyata N. Mutagenicity of the fullerene C60-generated singlet oxygen dependent formation of lipid peroxides. Carcinogenosis 1996; 17(10):2163-9.
19. Yamago S, Tokuyama H, Nakamura E et al. In vivo biological behavior of a water-miscible fullerene: ^{14}C labeling, absorption, distribution, excretion and acute toxicity. Chem Biol 1995; 2:385-9.
20. Boutorine AS, Tokuyama H, Takasugi M et al. Fullerene-oligonucleotide conjugates: Photoinduced sequence-specific DNA cleavage. Angew Chem Int ed Engl 1994; 33(23-24):2462-5.
21. Nelson MA, Domann F, Bowden GT et al. Acute and subchronic exposure of topically applied fullerene extracts on the mouse skin. Toxicol Ind Health 1993; 9:623-30.
22. Scrivens WA, Tour JM, Kreek KE et al. Synthesis of ^{14}C labeled C60, its suspension in water, and its uptake by human keratinocytes. J Am Chem Soc 1994; 116:4517-8.
23. Zakharenko LP, Zakharov IK, Lunegov SN et al. Somatic mosaicism demonstrates that fullerene C60 has no genetic toxicity. Biological Sciences 1994; 335:153-154.
24. Moussa F, Chrétien P, Dubois P et al. The influence of C60 powders on cultured human leukocytes. Fullerene Science and Technology 1995; 3:333-42.
25. Baierl T, Seidel A. In vitro effects of fullerenes C60 and fullerenes black on immunofunctions of macrophages. Fullerenes Science and Technology 1996; 5:1073.
26. Moussa F, Trivin F, Céolin R et al. Early effects of C60 administration in Swiss Mice: A preliminary account for in vivo C60 toxicity". Fullerenes Science and Technology 1996; 4:21-29.
27. Moussa F, Pressac M, Hadchouel M et al. C60 fullerene toxicity: Preliminary account of an in vivo study. In: Pennington NJ, Kadish K, Ruoff R, eds. Fullerenes. Proceedings Series of the 191st Meeting of the Electrochem Soc 1997; 97-42:332-336.
28. Geerts A, De Bleser P, Hautekecte M et al. In: Arias IM, Boyer JL, Fausto N, Jakoby WB, Schater DA, Shafritz DA, eds.The Liver: Biology and Pathobiology. 3rd ed. New York: Raven Press Ltd., 1994:819-839.
29. Poli G. Pathogenesis of liver fibrosis: Role of oxidative stress. Mol Aspects Med 2000; 21:49-98.
30. Slater TF. Necrogenic action of carbon tetrachloride in the rat: A speculative mechanism based on activation. Nature 1966; 209:36-40.
31. Slater TF, Cheesman KH, Ingold KU et al. Carbon tetrachloride toxicity as a model for studying free-radical mediated liver injury. Trans R Soc London 1985; B311:633-645.
32. Zakharenko LP, Zakharov IK, Vasiunina EA et al. Determination of the genotoxicity of fullerene C60 and fullerol using the method of somatic mosaics on cells of Drosophila melanogaster wing and SOS-chromotest. Genetika 1997; 33(3):405-9.
33. Wainwright M, Falih AM. Fungal growth on buckminsterfullerene. Microbiology 1997; 143:2097-2098.
34. Moussa F, Chrétien P, Pressac M et al. Preliminary study of the influence of cubic C60 on cultured human monocytes: Lack of interleukin-1β secretion. Fullerenes Science and Technology 1997; 5:503-510.
35. Moussa F, Roux S, Pressac M et al. In vivo reaction between [60]fullerenes and vitamin A in mouse liver. N J Chem 1998; 32:989-92.
36. Chiron JP, Lamande J, Moussa F et al. Effect of «micronized» C60 fullerene on the microbial growth in vitro. Ann Pharm Fr 2000; 58(3):170-5.
37. Mori T, Takada H, Ito S et al. Preclinical studies on safety of fullerene upon acute oral administration and evaluation for no mutagenesis. Toxicology 2006; 225(1):48-54.
38. Ungurenasu C, Airinei A. Highly stable C60/poly(vinylpyrrolidone) charge-transfer complexes afford new predictions for biological applications of underivatized fullerenes. J Med Chem 2000; 43(16):3186-8.
39. Satoh M, Matsuo K, Kiriya H et al. Effects of acute and short term repeated application of fullerene C60 on agonist-induced responses in various tissues of Guinea pig and rats. Gen Pharmac 1995; 26(7):1533-8.
40. Tsuchiya T, Oguri I, Yamakoshi YN et al. Novel harmful effects of (60)fullerene on mouse embryos in vitro and in vivo. FEBS Lett 1996; 393:139-145.
41. Oberdorster E. Manufactured nanomaterials (fullerenes, C60) induce oxidative stress in the brain of juvenile largemouth bass. Environ Health Perspect 2004; 112(10):1058-62.
42. Barnaby JF. New York Times, 2004.
43. Rittner MN. Confusion in the mass media. Nanoparticles News, 2004.
44. Sayes CM, Fortner JD, Guo W et al. The differential cytotoxicity of water-soluble fullerenes. Nano Lett 2004; 4(10):1881-7.
45. Sayes CM, Gobin AM, Ausman KD et al. Nano-C60 cytotoxicity is due to lipid peroxidation. Biomaterials 2005; 26(36):7587-95.

46. Isakovic A, Markovic Z, Todorovic-Markovic B et al. Distinct cytotoxic mechanisms of pristine versus hydroxylated fullerene. Toxicol Sci 2006; 91(1):173-83.

47. Oberdorster E, Zhu S, Blickley M et al. Ecotoxicology of carbon-based engineered nanoparticles: Effects of fullerene (C60) on aquatic organisms. Carbon 2006; 44(6):1112-1120.

48. Lovern SB, Klaper R. Daphnia magna mortality when exposed to titanium dioxide and fullerene (C60) nanoparticles. Environ Toxicol Chem 2006; 25(4):1132-7.

49. Rajagopalan P, Wudl F, Schinazi RF et al. Pharmacokinetics of a water-soluble fullerene in rats. Antimicrob Agents Chemother 1996; 40(10):2262-5.

50. Schuster DI, Wilson SR, Schinagi RF. Anti-human immunodeficiency virus activity and cytotoxicity of derivatized buckministerfullerenes. Bioorg Med Chem Lett 1996; 6:1253-1256.

51. Gharbi N, Pressac M, Tomberli V et al. In vivo behaviour of two C60 derivatives. Electrochemical Society Proceedings 2000; 240-43.

52. Dugan LL, Lovett EG, Quick KL et al. United States Patent Application Publication, 2003, (Pub no US 2003; 2003/0162837 A1, Pub. Date: Aug. 28).

53. Rancan F, Rosan S, Boehm F et al. Cytotoxicity and photocytotoxicity of a dendritic C(60) mono-adduct and a malonic acid C(60) tris-adduct on Jurkat cells. J Photochem Photobiol B 2002; 67:157-162.

54. Bosi S, Feruglio L, Da Ros T et al. Hemolytic effects of water-soluble fullerene derivatives. J Med Chem 2004; 47(27):6711-5.

55. Yang J, Wang Y, Rassat A et al. Synthesis of novel highly water-soluble 2:1 cyclodextrin/fullerene conjugates involving the secondary rim of b-cyclodextrin. Tetrahedron 2004; (60):12163-8.

56. Yamawaki H, Iwai N. Cytotoxicity of water-soluble fullerene in vascular endothelial cells. Am J Physiol Cell Physiol 2006; 290(6):C1495-502.

57. Colvin V, Sayes CM, Ausman KD et al. Environmental chemistry and effects of engineered nanostructures [Abstract]. In: Anaheimn CA, ed. Proceedings of the 227th ACS National Meeting. IEC 18. Washington, DC: American Chemical Society, 2004, (Available: http://oasys2.confex.com/acs/227nm/techprogram/ P721792.HTM [accessed 20 May 2004]).

58. Brant JA, Labille J, Bottero JY et al. Characterizing the impact of preparation method on fullerene cluster structure and chemistry. Langmuir 2006; 22(8):3878-85.

59. Andrievsky G, Klochkov V, Derevyanchenko L. Is C60 fullerene molecule toxic?! Fullerenes Nanotubes and Carbon Nanostructures 2005; (13):1-14.

60. Avdeev MV, Khokhryakov AA, Tropin TV et al. Structural features of molecular-colloidal solutions of C60 fullerenes in water by small-angle neutron scattering. Langmuir 2004; 20(11):4363-8.

61. Zha QQ, Wei XW, Sun J et al. Synthesis and properties of [60]fullerene based nanocomposites. Proceedings of the 7th Biennial international Workshop on fullerenes and atomic clusters, St. Petersburg, Russia, June 27-July 1, 2005. Abingdon, UK: Taylor and Francis, 2006.

62. Isakovic A, Markovic Z, Nikolic N et al. Inactivation of nanocrystalline C60 cytotoxicity by gamma-irradiation. Biomaterials 2006; 27(29):5049-58.

63. Deguchi S, Mukai S, Tsudome M et al. Facile generation of fullerene nanoparticles by hand-grinding. Advanced Materials 2006; 18(6):733-737.

64. Jia G, Wang H, Yan L et al. Cytotoxicity of carbon nanomaterials: Single-wall nanotube, multi-wall nanotube, and fullerene. Environ Sci Technol 2005; 39(5):1378-83.

65. Fiorito S, Serafino A, Andreola F et al. Effects of fullerenes and single-wall carbon nanotubes on murine and human macrophages. Carbon 2006; 44(6):1100-1105.

66. Lin YL, Lei HY, Wen YY et al. Light-dependent inactivation of Dengue-2 virus by carboxyfullerene C60 isomer. Virology 2000; 275:258-62.

67. Straface E, Natalini B, Monti D et al. C3-Fullero-tris-methanodicarboxylic acid protects epithelial cells from radiation-induced anoikia by influencing cell adhesion ability. FEBS Lett 1999; 454:335-40.

68. Monti D, Moretti L, Salvioli S et al. C60 carboxyfullerene exerts a protective activity against oxidative stress-induced apoptosis in human peripheral blood mononuclear cells. Biochemical and Biophysical Research Communications 2000; 277:711-7.

69. Lin AM, Chyi BY, Wang SD et al. Carboxyfullerene prevents iron-induced oxidative stress in rat brain. Journal of Neurochemistry 1999; 72(4):1634-40.

70. Tsao N, Luh TY, Chou CK et al. Inhibition of group A streptococcus infection by carboxyfulleren. Antimicrobial Agents and Chemotherapy 2001; 1788-93.

71. Tsao N, Luh TY, Chou CK et al. In vitro action of carboxyfullerene. Journal of Antimicrobial Chemotherapy 2002; 49:641-9.

72. Sijbesma R, Srdanov G, Wudl F et al. Synthesis of a fullerenes derivative for the inhibition oh HIV enzymes. J Am Chem Soc 1993; 115:6510.

73. Schinazi RF, McMillan A, Juodawlkis AS et al. Synthesis and virucidal activity of a water-soluble, configurationally stable, derivatized C60 Fullerene. Proc Electrochem Soc 1994; 94-24:689.

74. Yamakoshi Y, Yagami T, Sueyoshi S et al. Acridine adduct of [60]fullerenes with enhanced DNA-cleaving activity. The Journal of Organic Chemistry 1996; 61:7236-7.

75. Schinazi RF, Sijbesma R, Srdanov G et al. Synthesis and virucidal activity of a water-soluble, configurationally stable, derivatized C60 fullerene. Antimicrobial Agents and Chemotherapy 1993; 1707-10.

76. Huang HM, Ou HC, Hsieh SJ et al. Blockage of amyloid beta peptide-induced cytosolic free calcium by fullerenol-1, carboxylate C60 in PC12 cells. Life Science 2000; 66(16):1525-33.

77. Bisaglia M, Natalini B, Pellicciari R et al. C3-fullero-tris-methanodicarboxylic acid protects cerebellar granule cells from apoptosis. Journal of Neurochemistry 2000; 74:1197-204.

78. Rancan F, Helmreich M, Molich A et al. Fullerene-pyropheophorbide a complexes as sensitizer for photodynamic therapy: Uptake and photo-induced cytotoxicity on Jurkat cells. J Photochem Photobiol B 2005; 80(1):1-7.

79. Chueh SC, Lai MK, Chen SC et al. Chiang Fullerenols in canine renal preservation-A preliminary report. Kidney 1997; 1313-5.

80. Lai YL, Chiang LY. Water-soluble fullerene derivatives attenuate exsanguination-induced bronchoconstriction of guinea-pigs. Journal of Autonomic Pharmacology 1997; 17:229-35.

81. Chen HH, Yu C, Ueng TH et al. Acute and subacute toxicity study of water-soluble polyalkylsulfonated C60 in rats. Toxicol Pathol 1998; 26(1):143-51.

82. Huang SS, Tsai SK, Chih CL et al. Neuroprotective effect of hexasulfobutylated C60 on rats subjected to focal cerebral ischemia. Free Radical Biology and Medicine 2001; 30(6):643-9.

83. Yang DY, Wang MF, Chen IL et al. Systemic administration of a water-soluble hexasulfonated C(60) (FC(4)S) reduces cérébral ischemia-induced infarct volume in gerbils. Neuroscience Letters 2001; 311:121-4.

84. Oberdorster G, Oberdorster E, Oberdorster J. Nanotoxicology: An emerging discipline evolving from studies of ultrafine particles. Environ Health Perspect 2005; 113(7):823-39.

85. Hans C, Fischer HC, Liu L et al. Pharmacokinetics of nanoscale quantum dots: In vivo distribution, sequestration, and clearance in the rat. Adv Funct Mater 2006; 16:1299-1305.

CHAPTER 14

Toxicity Studies of Carbon Nanotubes

Jelena Kolosnjaj, Henri Szwarc and Fathi Moussa*

Abstract

As for fullerenes, the potential and the growing use of CNT and their mass production have raised several questions about their safety and environmental impact. Research on the toxicity of carbon nanotubes has just begun and the data are still fragmentary and subject to criticisms. Preliminary results highlight the difficulties in evaluating the toxicity of this new and heterogeneous carbon nanoparticle family. A number of parameters including structure, size distribution and surface area, surface chemistry and surface charge, and agglomeration state as well as purity of the samples, have considerable impact on the reactivity of carbon nanotubes. However, available data clearly show that, under some conditions, nanotubes can cross the membrane barriers and suggests that if raw materials reach the organs they can induce harmful effects as inflammatory and fibrotic reactions. Therefore, many further studies on well-characterized materials are necessary to determine the safety of carbon nanotubes as well as their environmental impact.

Introduction

Along with fullerenes, carbon nanotubes (CNT) represent the third allotropic crystalline form of carbon.[1] CNT are basically rolled sheets of graphite terminated by two end caps similar to half fullerene skeletons (Fig. 1). Since their discovery in 1991, CNT with their unique structural, mechanical and electronic properties attracted considerable interest in many fields of materials science, including conductive and high-strength composites; energy storage and energy conversion devices; sensors; field emission displays and radiation sources; hydrogen storage media; and nanometer-sized semiconductor devices, probes, as well as biomedical applications.[2-15]

As for fullerenes, the potential and the growing use of CNT and their mass production have raised several questions about their safety and environmental impact. Since 2001, many researchers investigated the toxicity of CNT and derivatives. The results are highly contradictory: whereas some authors reported high toxicity, others affirm that the toxicity is low.

This chapter presents a general review of toxicity studies on CNT and their derivatives. We will initially present the physical properties of these unique nanostructures. Subsequently, this chapter will be divided in two subsections including in vitro and in vivo toxicity studies performed with pristine CNT and covalently modified or functionalized CNT (*f*-CNT).

General Properties

CNT are composed entirely of sp^2 bonds, similar to graphite. Stronger than the sp^3 bonds found in diamond, CNT bonding structure provides them with their unique strength.[2]

*Corresponding Author: Fathi Moussa—UMR CNRS 8612, Faculté de Pharmacie, Université Paris-Sud 11, 5, Rue J.-B. Clément, 92296, Châtenay-Malabry, France. Email: fathi.moussa@u-psud.fr

Bio-Applications of Nanoparticles, edited by Warren C.W. Chan. ©2007 Landes Bioscience and Springer Science+Business Media.

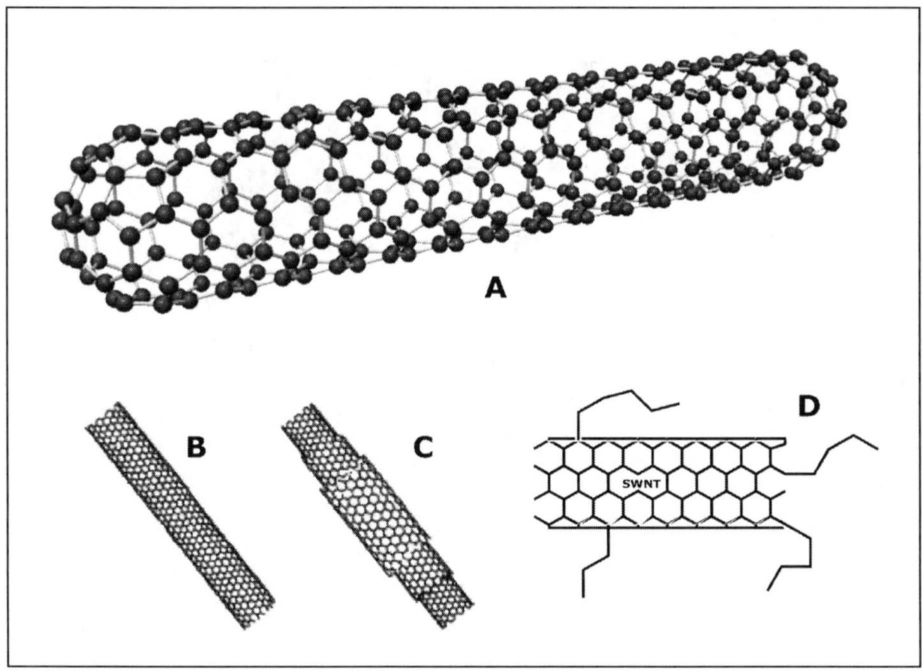

Figure 1. Structures of carbon nanotubes. A) Full length; B,C) Short-cut SWNT and MWNT. D) Functionalized SWNT.

Nanotubes are usually classified according to two headings: single-walled carbon nanotubes (SWNT), which are composed of rolled monolayered graphene sheet and multi-walled carbon nanotubes (SWNT), which are composed of several graphitic concentric layers (Fig. 1). The distance between each layer of a MWNT is about 0.34 nm. The diameter varies from 0.7 to 20 nm for SWNT and from 1.4 to 100 nm for MWNT, while the length usually reaches several micrometers. CNT naturally align themselves into "ropes" held together by van der Waals forces.[2]

Depending on their structure CNT can have different thermal properties as well as metallic or semi-conductive characteristics. MWNT also strongly absorb and emit light in the near-infrared (NIR) emitting region (800-1600 nm) with extremely photostable fluorescence, showing no blinking or photo-bleaching after prolonged exposure to excitation at high fluence.[16]

The chemical reactivity of CNT depends on their structure. Semiconducting nanotubes are analogous to aromatic [4n+2]annulenes, whereas metallic nanotubes are analogous to antiaromatic [4n]annulenes.[17] Such reasoning explains the high reactivity of metallic nanotubes over their semiconducting counterparts. However, CNT are materials practically insoluble and hardly dispersed in any kind of solvent: this is a major practical difficulty for their functionalization. Successful approaches toward chemically functionalized SWNT reported so far can be generally divided into three categories, namely, defect functionalization, noncovalent (supramolecular) functionalization, and covalent functionalization of the sidewalls (Fig. 1).[17-19]

Three processes are being used for producing CNT: carbon arc-discharge, laser-ablation and chemical vapour deposition (CVD). Among the CVD processes, high-pressure carbon monoxide process (HiPCO) can be easily scaled-up to industrial production. This affords the production of controllable diameter and length of the CNT as well as acceptable purity.[20]

To integrate the nanotube technology with biological environment, the dispersion of the nanotubes have been improved by addition of anionic and cationic surfactants including sodium dodecyl sulfate, Triton X-100, sodium dodecyl benzene sulfonate, etc.[21,22] Although surfactants may be efficient in the dispersion of CNT, they are known to destabilize plasma membranes and some of them exhibit intrinsic toxicity. Dispersion of CNT with biocompatible polymers has been proposed as an alternative to surfactants although they do not have better dispersion efficiency.[22,23]

Furthermore, CNT samples typically contain up to 30% metal catalyst (mainly iron and nickel particles), some amorphous carbon, and nanoparticles residual from the production

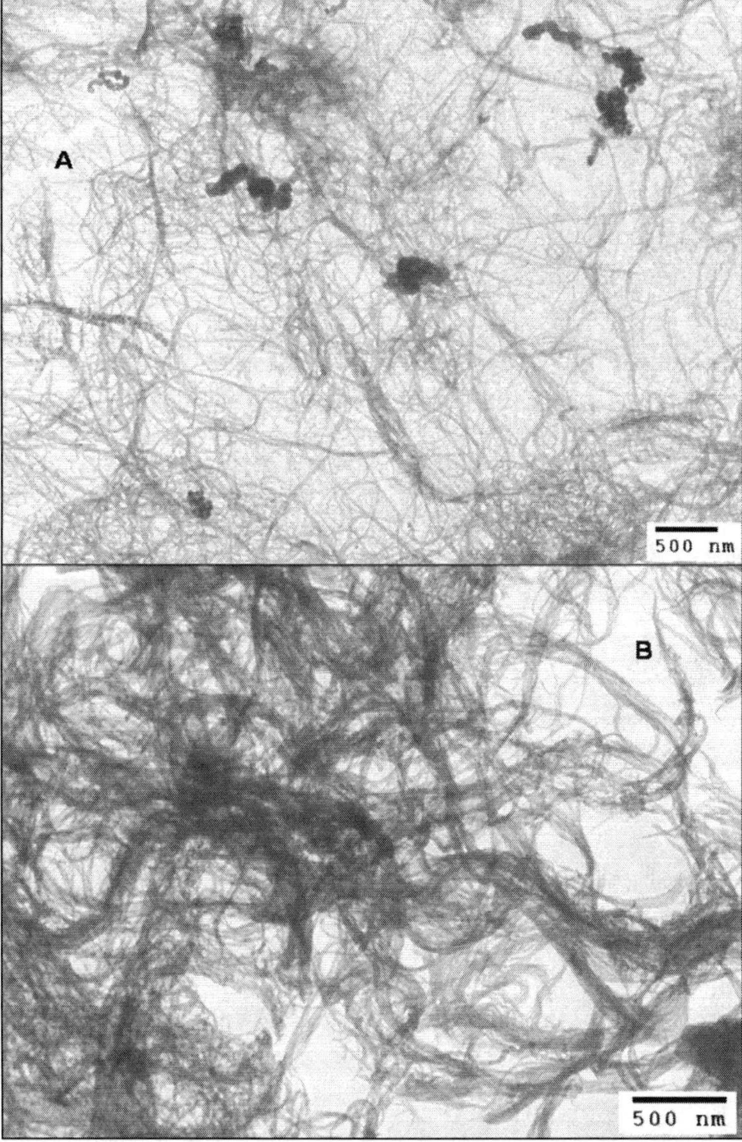

Figure 2. TEM images showing a SWNT sample with iron impurities before (A) and after (B) purification (unpublished work).

procedure (Fig. 2). Several methodologies have been described for CNT purification.[20,24] Strong acidic treatments are commonly used for removing metallic impurities. However, this methodology has a deep impact on the structure of the nanotubes. Strong acidic treatments irremediably lead to oxidation of the nanotubes and this results in cutting them into short pieces (Fig. 1) and generating carboxylic acid and hydroxyl groups at the tips and around the sidewalls where the curvatures present a higher strain.[25]

In Vitro Toxicity Studies on Pristine CNT

A relatively large number of in vitro studies on CNT toxicity have been performed since 2003. Their results vary as a function of the origin of the sample (method of production), the method of purification, the methodology of dispersion in the biological medium as well as the assays' reagents used to evaluate the viability of the cells (Tables 1, 2).

The first study was conducted by Shvedova et al from the National Institute of Occupational Safety and Health (Morgantown, West Virginia, USA).[26,27] The authors studied the effects of unrefined SWNT produced by the high-pressure carbon monoxide process (HiPCO) on human epidermal keratinocyte cell line (HaCaT) and human bronchial epithelial cells (BEAS-2B). The results obtained from this study showed that exposure of the cells to SWNT (0.24 mg/mL for 18 h) induced oxidative stress with ultra-structural and morphological changes as well as loss of cell viability. Adding desferrioxamine, a classical metal chelator, in the medium dramatically decreased the electron spin resonance signal intensity, indicating a key role of iron in the radical generation. The authors therefore concluded that cytotoxicity of the SWNT is associated with the well-known iron catalytic effects.[26,27] Indeed, the nanotubes used in these studies contained 30% of iron catalyst. It is worth noting here that free as well as *chelated* ferrous iron can catalyse the reduction of oxygen to superoxyde radical.[26]

Two years later, several studies have been performed by other authors. Cui et al evaluated the effects of SWNT on human embryonic kidney cells (HEK293).[28] The authors observed that SWNT can inhibit the proliferation of these cells with cell apoptosis, decrease of cell adhesive ability, and secretion of some 20-30 kd proteins, which wrap SWNT into nodular structures and isolate SWNT-attached cells from the main cell populations. The maximal dose tested was 200 μg/mL and the viability of the cells was determined with the MTT [3-(4, 5-dimethyl-2-thiazolyl)-2, 5-diphenyl-2H-tetrazolium bromide] test. However, the CNT samples used in this study were not characterized and no details on the preparation of the suspension were given.

Jia et al reported a comparative study of the effects of SWNT, MWNT and C_{60} on guinea pig alveolar macrophage cell line.[29] Using the MTT viability test, the authors concluded that SWNT were highly toxic under their conditions. In these studies, the iron impurities present in the CNT samples were removed by an acidic treatment. The authors did not focus on the subsequent surface modification of the nanotubes. The samples used in these studies also contained amorphous carbon.

Monteiro-Riviere et al explored the capacity of MWNT to enter human keratinocytes and to elicit a biological effect by IL-8 release.[30] MWNT (0.4 mg/ml) were suspended by sonication in the culture medium. According to the author the nanotubes were free from iron impurities due to the synthesis method and there was no evidence that the process led to breaking or defect formation in the MWNT. The authors observed that the morphologies of keratinocytes exposed to MWNT were different from those of the control. The nuclei of the cells were free of MWNT while several cytoplasmic vacuoles contained MWNT of various sizes, up to 3.6 μm in length. In addition, MWNT were found within the cytoplasm and, at times, were seen lying close to the nucleus and appeared to pierce the nuclear membrane. The nanotubes within cells did not undergo structural changes; they preserved their typical multi-walled 'bamboo' shoot structure. The concentration of IL-8 increased with time and the viability of the cells, assayed by neutral red, slightly decreased in a dose-dependent manner. After 24h under exposure of 0.4mg/mL, 75% of the cells were still viable.[30]

Table 1. In vitro toxicity of pristine carbon nanotubes

Author	Type of CNT	Concentration	Cell Line	Effects
Shvedova et al[26,27]	Raw SWNT	0.24 mg/mL	Human epidermal keratinocytes (HaCaT), Human bronchial epithelial cells (BEAS-2B)	Oxidative stress Ultrastructural/ morphological cell viability
Cui et al[28]	SWNT	200 µg/mL	Human embryonic kidney cells (HEK293)	Inhibition of proliferation Induction of apoptosis Decrease in cell adhesion Induction of protein secretion Loss of cell viability
Jia et al[29]	Purified SWNT and MWNT	Up to 226.0 µg/cm^2 of SWNT, up to 22.6 µg/cm^2 of MWNT	Guinea pig alveolar macrophages (AM)	High toxicity Loss of cell viability changes Increased IL-8 secretion slightly
Monteiro-Riviere et al[30]		MWNT	0.4 mg/mL	Human keratinocytes (HEK) morphological decreased viability
Murr et al[31]	Raw SWNT, MWNT in DMSO	Up to 2.5 µg/mL	Murine lung macrophages	No cytokine production Loss in cell viability
Sato et al[32]	Purified MWNT of 220 and 825nm of length	Up to 0.5 µg/mL	Human acute monocytic leukemia cells	Slight induction of TNF-α
Ghibelli et al[33]	MWNT	Up to 125 µg/mL	Phagocytes	No apoptosis and no necrosis
Manna et al[34]	SWNT in dimethilformamide	Up to 10 µg/mL (HaCaT) epithelial	Human keratinocytes	Oxidative stress Activation of NF-κB
Bottini et al[36]	Raw and purified MWNT suspended in water	Up to 400 µg/mL	Human T lymphocytes isolated from healthy human blood donors Jurkat T cells	No toxicity at 40 µg/mL Toxicity at 400 µg/mL
Sayes et al[37]	SWNT in 1% Pluronic F108	0.2 mg/mL	Human dermal fibroblasts	Loss in cell viability

Continued on next page

Table 1.   **Continued**

Author	Type of CNT	Concentration	Cell Line	Effects
Fiorito et al[41]	Purified SWNT	Up to 60 µg/mL	Murine macrophage cell line Human monocytes-derived macrophages (MDMs)	Low toxicity, Low inflammation No morphological changes
Worle-Knirsch et al[43]	Raw and purified SWNT in SDS	Up to 200 µg/mL	Human alveolar epithelial cell line A549 Endothelial cells derived from umbilical cord Rat macrophage cell line	Decrease of cell viability apparently lower in some viability assays (MTT) but not (MTT) but not (WST-1, LDH)
Magrez et al[44]	Raw and purified	Up to 0.2 µg/mL	Human lung-tumor cell lines: H596, H446, and Calu-1	Cell proliferation inhibition Loss of cell viability
Kagan et al[45]	Raw and purified SWNT	Up to 0.5 mg/mL	RAW264.7 macrophages	No intracellular production of superoxide radicals or NO
Simon et al[46]	MWNT	Not specified	Human alveolar epithelial cell line A549	Significant toxicity
Witzmann et al[47]	MWNT in cell growth medium	0.4 mg/mL	Human keratinocytes (HEK)	Increase in IL-8 and IL-1β production Decrease in IL-6 production No difference in TNF-α secretion Altered expression of some signaling, cytoskeleton and membrane trafficking proteins
Soto et al[48]	MWNT in DMSO	Up to 5 mg/mL	Murine macrophage RAW 267.9	Loss in cell viability

Continued on next page

Table 1. *Continued*

Author	Type of CNT	Concentration	Cell Line	Effects
Tian et al[49]	Raw and purified SWNT, MWNT	Up to 100 µg/mL	Human fibroblasts	Cell detachment from the substrate Morphological changes Altered protein expression Loss in cell viability
Davoren et al[50]	SWNT with and without 5% FBS	Up to 800 µg/mL	Human alveolar epithelial cell line A549	Very low acute toxicity Morphological changes Minor loss in cell viability
Meng and al.[53]	Non-woven SWNT scaffolds	-	3T3-L1 mouse fibroblasts	Enhanced long-term proliferation
Pulskamp et al[54]	Raw SWNT and purified SWNT in 1%SDS	Up to 100 µg/mL	Human epithelial cell line A549 Rat alveolar macrophage cell line	Intracellular ROS due to contaminants No acute toxicity No NO and IL-8 production No significant viability loss

Table 2. In vitro toxicity studies of functionalized carbon nanotubes

Author	Type of CNT	Concentration	Cell Line	Effects
Pantarotto et al[56]		FITC-labeled	Up to 10 μM	Human 3T6 90% viability at
	SWNT Peptide-SWNT conjugate		fibroblasts Murine 3T3 fibroblasts Human keratinocytes	5 μM 20% viability at 10 μM
Shi Kam et al[57]	Carboxylic group charged SWNT Fluorescein-functionalized SWNT Biotin-functionalized SWNT SWNT-biotin-streptavidin	0.05 mg/mL	Human promyelocytic leukemia HL60 cells Jurkat-cells Chinese hamster ovary Human 3T3 fibroblast	Toxicity observed only with the SWNT-biotin-streptavidin conjugate
Sayes et al[37]	SWNT in 1% Pluronic F108 SWNT-phenyl-SO$_3$H SWNT-phenyl-SO$_3$Na SWNT-phenyl-(COOH)$_2$	Up to 30 mg/mL	Human dermal fibroblast	Cytotoxic response is dependant on the degree of functionalization of SWNT
Nimmaga et al[58]	Raw and purified SWNT SWNT-glucosamine conjugate	Up to 1.0% (wt/vol)	3T3 mouse fibroblasts	No differences in metabolic activities in cells treated with 0.25% glucosamine-functionalised SWNT Raw and purified tubes cause cell viability decrease in smaller concentrations than the functionalized nanotubes
Dumortier et al[59]	Lymphocytes and macrophages of BALB/c mice	Up to 10 μg/mL	Mouse spleen cells B and T mouse lymphocytes Mouse macrophages	No activation of lymphocytes and macrophages, except for the pegilated compound that activated the macrophages No disturbances of cell functions No loss in cell viability

Murr et al[31] performed a comparative toxicity study on murine lung macrophage cell line between carbon nanotubes (SWNT, two different types of MWNT from different sources and with different size characteristics), carbon black and chrysotile asbestos nanotubes, the latter two were used as toxicity standards. Their conclusion was that the cytotoxicity of carbon nanoparticles is dose dependent.[31] Iron impurities (roughly 5 to 10%) were present only in the SWNT samples. All materials were suspended (5 μg/mL) in a stock solution in dimethyl sulfoxide (DMSO) and concentrations up to 2.5 μg/mL were tested on the cell line. The cell viability was assessed with the MTT viability test. Murr et al also assessed the cytokine production of murine macrophages after exposure to raw SWNT or MWNT.[31] The concentrations above 2 μg/mL caused cell death, however no IL-10 or IL-12 response was observed.

Sato et al[32] studied the effect of length of MWNT on cytotoxicity by investigating the activation of the human acute monocytic leukemia cell line (THP-1). Since the purity of the nanotubes was about 80% (the impurities being amorphous carbon, Fe, Mo, Cr, Al), they further purified the nanotubes with an acidic treatment (HCl) followed by a neutralization by NaOH.[32] The nanotubes were then cut with a mixture of sulphuric and nitric acid (95% H_2SO_4-60% HNO_3), ultrasonic irradiation and heating. They finally separated the nanotubes in two length fractions, 220 and 825 nm, respectively. These authors used human acute monocytic leukemia cell line because it provides a macrophage system that not only eliminates foreign bodies but also produces cytokines that work as signalling mediators serving as initial triggers of antibody production. As a positive control diacylated lipopeptide (FSL-1) was used, since FSL-1 is known to induce macrophages production of TNF-α. Both length fractions induced the production of TNF-α in a dose dependent manner, but the level of induction by the MWNT was much lower than that of the microbial peptide. FSL-1 at a concentration of 0.5 μg/mL induced a 30-time bigger production of TNF-α(1780 pg/mL) than the nanotubes (59 pg/mL and 48 pg/mL for 220 nm and 825 nm, respectively). The lowest dose of CNT (5 ng/μL) did not elicit TNF-α production. The inflammatory potential of the nanotubes used in this study can be considered low because because the nanotube stimulated the macrophages more than the positive control. At the same time, Ghibelli et al found no direct cytotoxicity of MWNT of 10 to 50 nm of average size. The nanotubes were internalized in phagocytes, but no apoptosis or necrosis was observed at doses up to 125 μg/mL.[33]

Manna et al also reported the induction of oxidative stress by SWNT in human keratinocytes through the activation of the nuclear transcription factor-kappaB (NF-κB).[34] These authors used dimethylformamide to disperse the particles and MTT to determine the viability of the cells. Unfortunately, they did not mention the amount of residual catalysts or an eventual purification of the nanoparticles. Purification is crucial to the interpretation of the data since iron itself can induce NF-κB.[35] Finally, Bottini et al compared the effects of pristine and acid-treated MWNT on human T lymphocytes isolated from healthy human blood donors and Jurkat T leukaemia cells.[36] Incubation of cells was performed by adding 0.5 mL of nanomaterial (carbon black or MWNT, both pristine and oxidized) dispersed in water, or 0.5 mL water alone, to 4.5 mL of cell suspension. The incubation lasted up to 5 days. Pristine MWNT showed no toxicity to the lymphocytes at doses up to 40 μg/mL or approximately 10^6 individual MWNT per cell. However, oxidized MWNT were toxic at the same dose, and both types were toxic at a dose of 400μg/mL.

One year later, many studies were performed. Their results lead to more confusion. While some studies reported harmful effects, other observed either no toxicity or beneficial effects (Table 2).

Sayes et al reported that pristine SWNT are more toxic than their covalently functionalized counterparts to human dermal fibroblasts.[37] The nanotubes used in this study were suspended in 1% Pluronic F108. The surfactant itself exhibited a decrease in cell viability of 10%, as most of the surfactants usually do by destabilizing cell membranes. Anyway, at a concentration of 0.2 μg/mL the surfactant-SWNT assembly decreased the viability of the cells up to 50% after 48 h of incubation. Unfortunately, the authors did not compare the toxicity of SWNT with that of the SWNT-Pluronic F108 assembly, which could correlate the relationship of the integrity of the membrane and toxicity. The end result would be a better understanding of the mechanism of toxicity.

As several authors suggested that CNT radical generation activities observed so far can be attributed to metal impurities, Fenoglio et al tried to examine the possible radical generation by purified CNT.[38] They used MWNT suspended in 5% SDS after purification by subsequent treatment with sodium hydroxide leading to a final metal content of less than 1%. Surprisingly, their results showed that MWNT exhibit strong radical scavenging activity towards hydroxyl and superoxide radicals, under their experimental conditions. Basically they compared the radical generation/scavenging activity of normal and ground nanotubes immediately after grinding and after ageing. Grinding a covalent solid may cause exposure of free radicals generation sites at the surface of the particles as previously described for silica.[38] To exclude the eventual involvement of particle sizes, they also compared the effect of non-ground, amorphous silica, with a specific surface that equalled or was smaller than that of the nanotubes. Two radical generating mechanisms have been investigated: 1- the generation of hydroxyl radicals in the presence of H_2O_2: this reaction mimics the contact of particles with physiological fluids during their phagocytosis by alveolar macrophages and recruited granulocytes; and 2- the generation of CO_2^- radicals following cleavage of the C-H bond of formate ions. Such reaction may be triggered directly by active sites at the particle surface, on contact with formate ions, or by a short-lived radical (e.g., HO· radicals) which would react with formate ions as primary traps. The abstraction of a hydrogen atom may occur with several endogenous molecules, such as proteins, nucleic acids, and lipids, yielding reactive products which may in turn be involved in oxidative damage. In this study, the hydroxyl radicals were generated following the Fenton reaction ($H_2O_2 + Fe^{2+} \rightarrow HO. + OH^- + Fe^{3+}$) as well as by photolysis of hydrogen peroxide by irradiation with an UV lamp ($H_2O_2 \rightarrow 2HO·$). Under all conditions, CNT were able to scavenge high amounts of hydroxyl radicals. When the scavenging reaction toward hydroxyl radicals generated from H_2O_2 UV irradiation was conducted in a non-buffered solution, the pH drastically decreased (from 6.6 to 2.0) only in the presence of nanotubes. Acidity may arise both from reactions taking place in the supernatant and from surface modifications of oxidized CNT. The authors also investigated the effects of CNT on the oxidation of cytochrome c by superoxide radicals (O_2^-), generated by the xanthine/xanthine oxidase system. Under their conditions, the authors observed that CNT drastically decreased, or completely abolished, the oxidation of cytochrome c by superoxide radicals.[38] Due to their high electron affinity, MWNT can directly react with cytochrome c. Moreover, the high hydrophobic surface of CNT may adsorb and possibly inactivate both cytochrome c and xanthine oxidase. To exclude these hypotheses, superoxide radicals were also generated by irradiating with a UV lamp a phosphate buffered solution (pH 7.4) of riboflavine and 5,5-dimethyl-1-pyrroline-N-oxide (DMPO) but this method led to the same results as the former one.[38]

Although the idea of radical quenching is not new, three years ago Watts et al[39] have already described CNT as polymer antioxidants, the mechanism has not been elucidated yet. However, radical addition reactions might occur as it happens with fullerenes.[40]

In another way, Fiorito et al recently tried to assess the toxicity and the capacity of eliciting an inflammatory reaction of highly purified SWNT on murine and human macrophages.[41] The induction of an inflammatory response was evaluated through the release of nitric oxide (NO) by murine macrophages and the stimulation of phagocytic activity of human macrophages.[41] The nanotubes were purified by a thermal acidic treatment and were administered to cell cultures in a dose up to 60 µg/mL. The nanotubes did not induce apoptosis and cell death compared to that induced by graphite particles. This suggests that the nanotubes are not able to elicit a reactive-inflammatory response by mammalian cells in vitro and were not cytotoxic under their conditions. The nanotubes were internalized in a smaller extent than graphite particles used a positive control and, in contrast to the control, did not stimulate the release of NO by murine macrophage cells in culture. In this study cytotoxicity of graphite nanoparticles and nanotubes on cell cultures was evaluated by observing the nuclear morphology and counting the number of apoptotic or necrotic cells counter-stained with propidium iodide. The cytotoxicity was low (~ 4% apoptotic/necrotic cells after 48 h treatment, whereas the positive control led to 25% apoptotic/necrotic cells). On top of that, no morphological changes due to surface damage were noticed in cells treated with the nanotubes.[41]

In the same way, Tsien et al reported redox cycling and DNA damage induced by iron-containing carbon nanomaterials.[42] Highly pure and iron-doped carbon nanofibers were synthesized and characterized and their toxicity compared to positive and negative control (with non-toxic particles and asbestos fibers) in an in vitro system using iron mobilization, plasmid DNA breaks, and cell death as endpoints. Iron mobilization and DNA breaks were increased in the presence of ascorbate. Addition of ferrozine, an iron chelator, prevented DNA breaks induced by asbestos fibers or iron-doped nanomaterials. The investigators concluded that iron-catalyzed generation of reactive oxygen species could play an important role in the toxicity of particulate nanomaterial if the iron residues are not fully encapsulated.

At the same time, a major breakthrough in in vitro studies of CNT was made by Worle-Knirsch et al.[43] These authors showed that CNT toxicity studies can be heavily perverted by using the MTT cell viability test. They demonstrated that MTT-formazan crystals can clump with CNT. The resulting agglomerates could not be dissolved with SDS, 2-propanol/HCl, not even when heating (60°C, 10 minutes). Measuring the viability of cells after incubation with chemicals is a routinely made method in toxicological laboratories. The MTT salt is reduced by mitochondrial dehydrogenases to the water-insoluble MTT-formazan, then extracted and photometrically quantified at 550 nm. When applying the MTT assay to cells incubated with SWNT, MTT-formazan-SWNT agglomerates cannot be extracted leading to an apparent loss in viability.[43] Indeed, the authors tested different viability assays after human A549 lung epithelial cells, endothelial ECV304 cells and rat NR8383 macrophages incubation with SWNT (with mainly cobalt and nickel impurities ranging from 8 to 2%) were suspended in 1% SDS. Except for MTT, other tests like the tetrazolium salt 2-(4-iodophenyl)-3-(4-nitrophenyl)-5-(2,4-disulfophenyl)-2H-tetrazolium and lactate dehydrogenase showed no decrease in viability. According to the authors these findings strongly suggest verifying in vitro cytotoxicity data of this new class of materials with at least two or more independent test systems.[43] Magrez et al also compared the toxicity of MWNT, carbon fibers and flake-like-shaped carbon black nanoparticles on three different human lung-tumor cell lines, H596, H446, and Calu-1.[44] All particles were suspended in a strongly diluted gelatine solution and the cell proliferation and cytotoxicity of the materials were evaluated by the MTT assay. The results showed a clear particle morphological dependence on the cell toxicity: carbon black particles exhibited the highest cytotoxicity evidenced by the lowest number of viable cells at all concentrations and time points tested. In particular, at low nanoparticle concentrations (0.002 and 0.02 µg/mL), the number of viable cells decreased in the following sequence: carbon black > carbon nanofibers > MWNT. At higher concentrations (0.2 µg/mL), differences diminished especially for carbon nanofibers and carbon black. However, MWNT always appeared to be less toxic than the other two materials. A possible explanation for the observed aspect-ratio dependence of the carbon nanoparticle toxicity is the presence of dangling bonds, which are highly reactive sites. In general, they are present in carbon black with a high density, whereas in carbon nanotubes they preferentially occur at lattice defects or end caps.

To explore the effects of surface chemical properties of SWNT, these authors performed another set of experiments in which they modified the surface of the nanotubes and the fibres. The surface of MWNT and carbon fibres has been modified by acidic treatment resulting in the addition of carbonyl (C=O), carboxyl (COOH), and hydroxyl (OH) groups onto the nanotube and nanofiber surfaces. The surface treatment increased the toxicity of the particles, especially in the case of MWNT, where surface modification led to cell proliferation inhibition and cell death.[44]

In a second study, Kagan et al recently compared purified (0.23 wt. % iron) and non purified SWNT (26 wt. % iron) on RAW264.7 macrophages.[45] Neither purified nor non-purified SWNT were able to generate intracellular production of superoxide radicals or nitric oxide in RAW 264.7 macrophages as documented by flow-cytometry and fluorescence microscopy.

SWNT with different iron contents displayed different redox activity in a cell-free model system as revealed by EPR-detectable formation of ascorbate radicals resulting from ascorbate oxidation. In the presence of zymosan-stimulated RAW 264.7 macrophages, non-purified SWNT were more effective in generating hydroxyl radicals than purified SWNT. Similarly, EPR spin-trapping experiments in the presence of zymosan-stimulated RAW 264.7 macrophages showed that non-purified SWNT more effectively converted superoxide radicals generated by xanthine oxidase/xanthine into hydroxyl radicals as compared to purified SWNT. Adding catalase in the medium partially protected macrophages against SWNT-induced elevation of biomarkers of oxidative stress, thus confirming the effects of iron impurities on macrophages response.[45]

Anyway, a comparative toxicity study has been performed very recently by Simon et al.[46] According to the authors, MWNT were much more toxic to A549 lung carcinoma human cells than other nanoparticles (TiO2, Al2O3 and ZnO). However, the authors did not give any specification about the eventual impurities present in the samples or the surface state of the tested materials.

In a second study, Witzmann et al examined the cytokine release and the proteomics of human keratinocytes (HEK) after 24 and 48 h exposure to MWNT.[47] As previously,[30] they used 0.4 mg/mL of MWNT dispersed in KGM by sonication. HEKs showed a significant increase in IL-8 at both time points compared to controls. IL-1β concentrations showed a significant increase only at 48 hours. IL-6 release was lower than that of the controls. TNF-α levels were not increased, although this cytokine is considered as one of the most common initiators of keratinocyte activation. MWNT exposure for 24 hours resulted in the altered expression of 36 HEK protein spots, measured by two-dimensional electrophoresis and peptide mass fingerprinting, whereas 48 hours of MWNT exposure altered 106 protein spots. In the control group, on the other hand, only 48 protein spots were significantly different. Changes in proteins expression involved signalling, cytoskeleton, and membrane trafficking proteins. The proteins that have been altered by CNT exposure were identified, yet their impact on the keratinocyte specifically and the skin more generally remains difficult to define.

Soto et al also performed a comparative study of the toxicity of Ag, NiO, TiO$_2$, MWNT, and chrysotile asbestos nanoparticles on murine macrophage cell line (RAW 267.9).[48] They used dimethyl sulfoxide (DMSO) to suspend the nanoparticles at concentrations up to 5 μg/mL and the viability of the cells was determined by the MTT test. According to these authors, cytotoxicity did not appear to be associated with specific particle morphologies. Fibers, platelets, and spherules corresponding to MWNT, NiO, and Ag, respectively have all demonstrated recognizable cytotoxic response for the murine macrophage cell line. Whereas there was no toxicity observed with TiO$_2$ treated cells, MWNT had almost the same toxicity as asbestos (the cell viability was smaller than 50%).[48]

An ancillary comparative study between refined and unrefined SWNT, MWNT, active carbon, carbon black, and graphite was performed by Tian et al on human fibroblast cells.[49] These authors purified all the nanomaterials by acidic treatment and dispersed them in water by sonication. The cells were treated with concentrations up to 100 μg/mL of SWNT from 1 to 5 days. The other carbon nanoparticles were tested only at a concentration of 25 μg/mL. At the end of the experiment, the survival rate of cells as compared for the 25 μg/mL concentration was 84% for graphite, 78% for MWNT and 58% for SWNT. Cell survival was assessed with the MTT test and cell death, on the other hand, was assessed by Bio-rad Model 680 using Cellular DNA Fragmentation ELISA kit. After this experiment, the authors found out that a significant difference in cell death occurs after treatment of the cells with 25 μg/mL of purified SWNT, compared to the control. When they compared the non-purified SWNT and the control, no significant difference between the cells was observed.[49] Cell adhesion and cell morphology were evaluated and results showed that cells treated with purified SWNT detached from the substrate and showed ruffles on their cell membranes and the cell shape appeared rounded in comparison to the normal cell. Cells treated with the nanotubes had their nuclei moved towards the regions where the nanotubes were attached. As their membranes and nuclei moved closer, the adhesion-related proteins exhibited a punctuate distribution along the cell periphery. Normal fibroblast cells showed a rather organized radial distribution of actin network, which turned random and irregular when treated with SWNT. While the concentration of SWNT increased the western blot results of proteins, fibronectin, laminin and collagen IV did exhibit a strong decrease of expression levels. However, focal adhesion and cell-cell adhesion protein (FAK and P-cadherin) expressions showed a less severe decreasing tendency. Cell cycle related protein cyclin D3 expression also decreased while the β-actin protein expression remained unchanged.[49]

Very recently, Davoren et al described the in vitro cytotoxicity assessment of SWNT on a human lung cell line (A549 cells).[50] In their experiments, the authors used HiPCO derived SWNT with 10%wt iron suspended in water without or with 5% foetal bovine serum (FBS) in which nanotubes appeared to disaggregate more readily. Concentrations up to 800 µg/mL were tested and cellular viability was determined using the Alamar blue, Neutral red and MTT assays, which evaluated metabolic, lysosomal and mitochondrial activity, respectively. In addition, the total protein content of the cells was measured using the coomassie brilliant blue assay. Supernatants were also assayed for adenylate kinase release and IL-8 used as a marker of loss of cell membrane integrity and of inflammation response, respectively. Results from the cytotoxicity tests revealed that SWNT have very low acute toxicity to the A549 cells. Cytotoxicity was recorded at 400 and 800 µg/mL of SWNT tested in presence of serum for the Alamar blue assay, with approximately 33% and 53% inhibition in comparison to untreated controls. For CNT exposures in presence of serum, the EC_{50} value was about 800 µg/mL. In serum free media, cytotoxicity was recorded at 400 and 800 µg/mL of SWNT where 42% and 51% of inhibition were observed in comparison to unexposed controls. In the absence of serum, the EC_{50} was found to be greater than 800 µg/mL SWNT. The cytotoxicity of the positive control-quartz particles in the FBS was determined after 18% and 23% cell inhibition corresponding to 200 to 400 µg/mL suspension. On the other hand, the same particles without the FBS proteins exhibited cytotoxicity already from a concentration of 25 µg/mL and reached 90% inhibition at the concentration of 800 µg/mL. No EC_{50} values were determined for the coomassie blue assay for both serum and serum free exposures, as maximum cytotoxicity determined in both instances was less than 50%. For the MTT assay, exposure in serum containing media resulted in significant cytotoxicity from much lower doses-12.5 µg/mL of SWNT upwards, and a maximum of 32% inhibition was recorded at 800 µg/mL of SWNT. For exposures in the absence of serum, significant cytotoxicity was recorded from 3.125 µg/mL of SWNT upwards and 800 µg/mL of SWNT resulted in approximately 45% inhibition. Since the inhibition assessed by MTT never reached 50%, EC_{50} values could not be determined for the MTT assay. Of the multiple cytotoxicity assays used, the Alamar blue assay was found to be the most sensitive and reproducible.[50]

Transmission electron microscopy studies confirmed that there was no intracellular localization of SWNT in A549 cells following 24 h of exposure; however, morphological features such as lamellar body structures, microvilli, tonofilaments, desmosome junctions, characteristic of alveolar epithelial type II cells were observed in TEM images of control cell cultures.[50] Following exposure to 400 µg/mL of SWNT, an increase in lamellar bodies was recorded. Multivesicular bodies were also observed and the presence of extracellular (excreted) lamellar bodies was noted. Following exposure to the highest test concentration of SWNT, substantially more lamellar bodies were observed when compared to untreated controls. A reduction in microvilli and an increase in lipid droplet numbers were also evident compared to control cultures. Multivesicular bodies were also observed at the cell surface. Since transformed cell lines are typically less resistant to toxic effects than other cells derived from normal tissues, Davoren et al are currently using normal lung cell lines e.g., HFL and BEAS-2B in their on-going toxicity studies.[50] While conducting the cytotoxicity assays it was observed that the SWNT were interacting with some of the colorimetric and fluorescent dyes used in the toxicity assessment, resulting in unexpected absorption/fluorescence data. The most obvious interference was found with the Neutral red assay as reflected in the widely scattered results obtained with this viability test.[50]

Casey et al[51] conducted spectroscopic studies in order to elucidate the interactions of SWNT with cell culture medium and its components, prepared both with and without FBS. Although interactions between the nanotubes and the cell medium and the FBS growth supplement are most likely interpreted as a physisorption through van der Waals forces, this phenomenon implies that the availability of the constituents to the cells will be reduced, potentially resulting in a secondary rather than primary toxicity.[51]

Monteiro-Riviere and Inman[52] have also reported that adsorbing properties of carbon interfere with viability markers in the assay systems resulting in false cytotoxicity interpretation. The adsorbing properties of CNT were also suspected to interfere with the adenylate kinase and cytokine assays in the study of Davoren et al.[50] Therefore no results for Neutral red test, adenylate kinase or IL-8 assay for either SWNT or the positive control quartz were presented.[50]

As it has been proved by Worle-Knirsh et al,[43] when cells are grown in sub-confluent culture and are challenged with SWNT, the nanotubes adhere to the cells and the cells start to form agglomerates around the site of attachment. Human epithelial A549 cells grown to confluence 24 h after exposure with SWNT start to enclose the nanotubes. Nanotube preparations that were used in this study were stored in a 1% SDS solution and before use in the experiment, and then SWNT were acetone precipitated and resuspended in bi-distilled water, centrifuged and finally taken up in growth media and diluted to final concentrations. Enclosing the nanotubes seems to be an active process because strong condensation of actin filaments and FAK could be observed in the overlay image of SWNT, actin, and FAK. The CNT bundles are actively anchored to the cells and kept at this location; the cells start to detach from the culture dish and grow out of the plain. This tendency to form agglomerates may explain the formation of granuloma in animals but it could also be used as a substrate for mammalian cells' growth.[43] Meng et al[53] recently showed that non-woven SWNT scaffolds (Macroscopic membranes made with carbon nanotubes) have promotional influence on the cells growing upon the cells cultivated on the other substrates through cell-cell communication by using a modified trans-well device.[53] The non-woven SWNT were composed of thousands of highly entangled SWNT bundles with 20-30 nm diameters and several hundred micrometers in length. The average size of pores formed by the entangled SWNT bundles ranged from 50 to 200 nm. The thickness of the non-woven SWNT observed was about several micrometers. Experimental results indicated that non-woven SWNT exhibited excellent functions of enhancing long-term 3T3-L1 mouse fibroblasts proliferation in vitro, which could have implications in tissue regeneration and repairing.[53] The cell adhesion and proliferation were assessed by the CellTiter 96 AQueous One Solution Cell Proliferation Assay (MTS assay) and cell morphology was observed by fluorescence and laser scanning confocal microscopy. 3T3 L1 fibroblasts were spreading and proliferating on non-woven SWNT exhibiting highly organised skeletal systems. Their survival persisted for at least 3 weeks, indicating a more desirable growth environment for the cells compared to the other substrates including carbon fibers with microscopic structure features, and the polyurethane films.[53]

Finally, a very recent study reported that CNT show no sign of acute toxicity but induce intracellular reactive oxygen species resulting from contaminants. In this study, Pulskamp et al[54] used particle suspensions prepared freshly before the experiments in the complete culture medium and sonicated to break up agglomerates. The authors observed no acute toxicity on cell viability of rat (NR8383) and human (A549) lung cells upon incubation with all CNT products up to 100 µg/mL with exposure times up to 72h. None of the CNT induced the inflammatory mediators NO and IL-8. Cells that were priory stimulated with LPS and afterwards with the nanotubes showed an increase of TNF-α production, which was not statistically significant compared to the only LPS-treated cells. A rising tendency of TNF-α release from LPS-primed cells due to CNT treatment could be observed. They detected, however, a dose- and time-dependent increase of intracellular reactive oxygen species and a decrease of the mitochondrial membrane potential with the commercial CNT in both cell types after particle treatment whereas incubation with the purified nanotubes had no effect: this leads to the conclusion that metal traces associated with the commercial CNT are responsible for their biological effects.[54]

Functionalized CNT

A great deal of interest in functionalized carbon nanotubes was evoked after Mattson et al reported the first application of carbon nanotube technology to neuroscience research.[55] They developed a method for growing embryonic rat-brain neurons on MWNT. Although the study was not really a toxicological one, we can extrapolate that no toxicity was observed against rat neurons, since neurons grown on nanotubes coated with the bioactive molecule 4-hydroxynonenal elaborated multiple neurites, which exhibited extensive branching.[55]

In the end of 2003 Pantarotto et al[56] reported the effects of fluorescein isothiocyanate (FITC)-labeled SWNT and the peptide-SWNT conjugate on Human 3T6, murine 3T3 fibroblasts and human keratinocytes. The peptide in cause was taken from the α subunit of the G_s protein. The capacity of the two compounds to penetrate into the cells was studied by epifluorescence and confocal microscopy. The cells were stained with DAPI (4A,6-diamidino-2-phenylindole). The cytotoxicity caused by a peptide-functionalised CNT was studied by flow cytometry (FACS) in the range 1 to 10 μM concentration. Annexin V-APC (allophycocyanine) and propidium iodide were used as apoptotic and necrotic fluorescent probes, respectively. When the fibroblasts were incubated with 5 μM of peptide-CNT derivative, 90% of the cell population remained alive. In contrast, increasing twice the concentration of nanotubes induced 80% of cell death.[56]

The toxicity of *f*-CNT were also evaluated by Shi Kam et al[57] during their study of the uptake of functionalized SWNT (carboxylic group charged SWNT, fluorescein-functionalized SWNT, biotin-functionalized SWNT and SWNT-biotin-streptavidin conjugates) into human promyelocytic leukemia (HL60). Stable aqueous suspensions of purified, shortened, and functionalized nanotubes were obtained by oxidation and sonication. Zeta potential confirmed the existence of numerous negatively charged acidic groups at the sidewalls of the nanotubes.[57] The toxicity of the four derivatives was examined after 1h incubation of the cells with 0.05 mg/mL isolated by centrifugation and observed after 24 and 48h. No appreciable cell death was observed for all tested *f*-SWNT, however, the SWNT-biotin-streptavidin conjugate was found to cause extensive cell death when examined after 48 h of incubation. The authors attributed the toxicity to the ability of CNT to make the streptavidin enter the cells. The same results were obtained with other cells as well, including Jurkat-cells and Chinese hamster ovary (CHO) and 3T3 fibroblast cell lines.

As mentioned above, Sayes et al[37] compared the cytotoxicity of non functionalized SWNT suspended with a surfactant and some covalently functionalized SWNT on human dermal fibroblast cell line. SWNT samples used in this exposure include SWNT-phenyl-SO_3H and SWNT-phenyl-SO_3Na (six samples with carbon/-phenyl-SO_3X ratios of 18, 41, and 80), SWNT-phenyl-$(COOH)_2$ (one sample with carbon/-phenyl-$(COOH)_2$ ratio of 23), and underivatized SWNT stabilized in 1% Pluronic F108. The authors concluded that sidewall functionalized SWNT are substantially less cytotoxic than surfactant stabilized SWNT and that the cytotoxic response of cells in culture is dependent on the degree of functionalization of SWNT.[37]

Nimmaga et al[58] also performed a comparative in vitro study of raw SWNT (with nickel and yttrium impurities), acid purified SWNT, SWNT functionalized with glucosamine over concentrations of 0.001-1.0% (wt/vol) in order to improve the nanotube solubility.

As many debates followed fullerene-THF associated toxicity, it is worth noting that the nanotubes reacted with glucosamine dissolved in THF for 72h under reflux (66-68°C).

The authors investigated the effects of various nanotubes on 3T3 mouse fibroblasts.[58] In contrast to raw nanotubes, the purified and glucosamine functionalised ones did not cause significant cell viability decrease at concentrations between 0.001% and 0.016% (wt/vol). At concentrations up to 0.25% glucosamine-functionalised SWNT showed no significant difference in metabolic activity from control values.[58]

Dumortier et al[59] reported that SWNT functionalized via the 1,3-dipolar cycloaddition reaction are not cytotoxic to mouse spleen cells, B and T lymphocytes and macrophages of BALB/c mice, since they induced neither cell death nor activation of lymphocytes and macrophages and since they do not disturb cell functions. In contrast PEG-functionalized CNT, obtained using an oxidation/amidation procedure, can activate macrophages and modify their subsequent capacity to respond to a physiological stimulus.

In Vivo Toxicity Studies

Pristine CNT

Although still fragmentary, in vivo toxicity studies performed hitherto in order to assess the dangers of respiratory and skin exposure to these nanoparticles, especially concerning people working in the carbon nanotubes production industry, showed some harmful effects (Tables 3, 4).

In 2001 Huczko and Lange[60] evaluated the potential of CNT to induce skin irritation by conducting two routine dermatological tests. Initially, 40 volunteers with allergy susceptibilities were exposed for 96 h to a patch test consisting in a filter paper saturated with a water suspension of unrefined CNT synthesized via the arc discharge process. Secondly, a modified Draize rabbit eye test using a water suspension of unrefined CNT was conducted with four albino rabbits monitored for 72 h after exposure. Both tests showed no irritation in comparison to a CNT-free soot control and it was concluded that "no special precautions have to be taken while handling these carbon nanostructures".[60]

In a two-part study, preliminary investigations have been carried out by Maynard et al[61] into the potential exposure routes and toxicity of SWNT. The study was undertaken to evaluate the physical nature of the aerosol formed from SWNT during mechanical agitation. This was complemented by a field study in which airborne and dermal exposure to SWNT was investigated while handling unrefined material. Although laboratory studies indicated that unrefined SWNT material can release fine particles into air under sufficient agitation, concentrations generated while handling material in the field were very low. Estimates of the airborne concentrations of nanotube materials generated during handling suggest that concentrations were lower than 53 $\mu g/m^3$ in all instances. In another way, glove deposits of SWNT during handling were estimated at between 0.2 mg and 6 mg per hand.[61]

However, unprocessed nanotubes are very light and could become airborne and potentially reach the lungs; therefore their pulmonary toxicity was investigated.

Lam et al[62] studied three different groups of SWNT: raw and purified HiPCO and CarboLex nanotubes. The first was rich in iron impurities and the last contained nickel and yttrium impurities. The particles were dispersed by brief shearing (2 min in a small glass homogenizing tube) and subsequent sonication (0.5 min) in heat-inactivated mouse serum. Mice were then intratracheally instilled with 0, 0.1, or 0.5 mg of CNT or carbon black or quartz particles used as negative and positive control, respectively. Seven and 90 days after this single treatment, the animals were sacrificed for histopathological examination of the lungs. All CNT treatments induced dose-dependent epithelioid granulomas and, in some cases, interstitial inflammation in the animals euthanized after 7 days. These lesions persisted and were more pronounced in the group euthanized after 90 days; the lungs of some animals also revealed peribronchial inflammation and necrosis that had extended into the alveolar septa. The lungs of mice treated with carbon black were normal, whereas those treated with high-dose quartz revealed mild to moderate inflammation. These results show that, under these conditions and on an equal-weight basis, if carbon nanotubes reach the lungs, they are much more toxic than carbon black and can be more toxic than quartz, which is considered a serious occupational health hazard in chronic inhalation exposures.[62]

Warheit et al[63] performed another pulmonary toxicity assessment of pristine SWNT. The aim of this study was to evaluate the acute lung toxicity of intratracheally instilled SWNT in rats. The nanotubes used in this study were produced by laser ablation and contained about 30 to 40% amorphous carbon (by weight) and 5% each of nickel and cobalt.

The lungs of rats were instilled either with 1 or 5 mg/kg of the following control or particle types: SWNT, quartz particles (positive control), carbonyl iron particles (negative control), and the vehicle-phosphate buffered saline (PBS) and 1% Tween 80, or graphite particles. Following exposures, the lungs of treated rats were assessed using bronchoalveolar fluid biomarkers and cell proliferation methods, as well as by histopathological examination of lung tissue at 24 h, 1 week, 1 month, and 3 months post-instillation. Exposures to high-dose (5 mg/kg) of SWNT produced mortality in approximately 15% of the instilled rats within 24 h

Table 3. *In vivo toxicity studies of pristine carbon nanotubes*

Author	Type of CNT	Dose	Species	Route of Administration	Effects
Huczko and Lange[60]	Filter saturated with raw SWNT Water suspension		Man Rabbit	Dermal Ocular	No irritation
Lam et al[62]	Raw and purified	0,5 mg/animal	Mouse	Intra-tracheal instillation	Epitheloid granulomas Interstitial and peribronchial inflammation Necrosis
Warheit et al[63]	Raw SWNT	Up to 5 mg/kg	Rat	Intra-tracheal instillation	Temporary inflammatory and cell injury effects Granulomas, non-progressive beyond 1 month of exposure
Sato et al[32]	Purified MWNT of 220 and 825nm of lenght	0,1 mg clusters	Rat	Subcutaneous implants	Shorter tubes slighter inflammatory response then longer ones
Muller et al[65]	MWNT and ground MWNT, suspended in 0,9% saline and 1% Tween 80	Up to 5 mg/animal	Rat	Intratracheal instillation	Inflammation. Fibrosis, granulomas
Shvedova et al[66]	Purified SWNT	Up to 40 µg/animal	Mouse	Pharingeal aspiration	Inflammation. Fibrosis, granulomas Functional respiratory deficiencies Decreased bacterial clearence
Zhu et al[67]	MWNT	Up to 200 µg/mL	*Stylonychia mytilus*	Oral	Growth inhibition
Koyama et al[68]	SWNT, MWNT (20nm average diameter), MWNT (80 nm average diameter), CSNT	2 mg/animal	Mouse	Subcutaneous implants	Activation of MHC I and MHC II Granulomas

Table 4. In vivo toxicity studies of functionalized carbon nanotubes

Author	Type of CNT	Dose	Species	Route of Administration	Effects
Singh et al[70]	SWNT-DTPA and MWNT-DTPA radiolabelled with [111]In	400 µg/animal	Mouse	Intravenous	No acute toxicity
Carrero-Sanchez et al[71]	MWNT and N-MWNT	Up to 5 mg/kg	Mouse	Nasal Oral Intratracheal Intraperitoneal	N-MWNT cause granuloma at some intratracheally instilled animals but no distress when other routes of administration were used

post-instillation. This mortality resulted from mechanical blockage of the upper airways by the instilled particulate SWNT. In the surviving animals, SWNT produced temporary inflammatory and cell injury effects. Results from the lung histopathology indicated that pulmonary exposures to SWNT in rats produced a non-dose-dependent series of multifocal granulomas, which were evidence of a foreign tissue body reaction. However, they were non-uniform in distribution and not progressive beyond one month of post-exposure.[63]

The observation of SWNT-induced multifocal granulomas was inconsistent with the following: lack of lung toxicity by assessing lavage parameters, lack of lung toxicity by measuring cell proliferation parameters, apparent lack of a dose response relationship, non-uniform distribution of lesions, the paradigm of dust-related lung toxicity effects, and possible regression of effects over time. The observation of granulomas, in the absence of adverse effects measured by pulmonary endpoints was surprising, and did not follow the normal inflammogenic/fibrotic pattern produced by fibrogenic dusts, such as quartz, asbestos, and silicon carbide whiskers.[63]

While Lam et al concluded that SWNT exposures were more toxic than similar exposures to quartz and crystalline silica particles;[62] in contrast Warheit's results indicated a transient pulmonary inflammation and granuloma formation after SWNT exposure, but sustained lung inflammation, cytotoxicity, enhanced lung cell proliferation, foamy macrophage accumulation and lung fibrosis after exposure to quartz particles.[63] The differences between these findings may be related in part to species differences (mouse vs. rat), but are more likely due to the differences in the experimental designs of the two studies.[64]

Respiratory toxicity of MWNT has been assessed as well by Muller et al.[65] They administered intratracheally MWNT or ground MWNT suspended and sonicated in sterile 0.9% saline containing 1% of Tween 80, at doses of 0.5, 2 or 5mg per animal corresponding to approximately 2.2 mg/kg, 8.9 mg/kg and 22.2 mg/kg to Sprague-Dawley rats. The applied nanotubes were still present in the lung after 60 days (80% and 40% of the lowest dose) and both induced inflammatory and fibrotic reactions. At 2 months, pulmonary lesions induced by MWNT were characterized by the formation of collagen-rich granulomas protruding in the bronchial lumen, in association with alveolitis in the surrounding tissues. These lesions were caused by the accumulation of large MWNT agglomerates in the airways. Ground nanotubes were better dispersed in the lung parenchyma and also induced inflammatory and fibrotic responses. Both MWNT and ground MWNT stimulated the production of TNF-α in the lung of treated animals.[65]

The physiological relevance of the intratracheally instilled CNT is questionable since particles that are inspired encounter several barriers before reaching the trachea and lungs. Nevertheless, even if administered by pharyngeal aspiration, as it has been shown by Shvedova et al,[66] purified SWNT elicited inflammation, fibrosis and granulomas formation in C57BL/6 mice. The nanotubes used in this study were produced by HiPCO and where further purified by acidic treatment. The analysis also proved that CNT accounted for more than 99% of carbon. The animals were treated with either SWNT (0, 10, 20, 40 μg/mouse) or two reference materials (ultrafine carbon black or SiO_2 at 40 μg/mouse). The animals were sacrificed at 1, 3, 7, 28, and 60 days following exposures. A rapid progressive fibrosis found in mice exhibited two distinct morphologies: 1- SWNT-induced granulomas mainly associated with hypertrophied epithelial cells surrounding dense micrometer-scale SWNT aggregates and 2- diffuse interstitial fibrosis and alveolar wall thickening likely associated with dispersed SWNT. These differences in fibrosis morphology were attributed to the distinct particle morphologies of compact aggregates and dispersed SWNT structures. Importantly, deposition of collagen and elastin was also observed in both granulomatous regions as well as in the areas distant from granulomas. Increased numbers of alveolar type II (AT-II) cells, the progenitor cells that replicate following alveolar type I (AT-I) cell death, were also observed as a response to SWNT administration. Moreover, functional respiratory deficiencies and decreased bacterial clearance (Listeria monocytogenes) were found in mice treated with SWNT.[66]

In another way, Sato et al[32] evaluated the response of subcutaneous tissue in rats to MWNT of different lengths. They used 220 nm and 825 nm-long MWNT. Two pockets were made in the subcutaneous tissue and 0.1 mg of clusters of the MWNT samples were implanted bilaterally in each rat. The degree of inflammatory response around the 220nm long MWNT was slight in comparison with that around the 825nm long ones, indicating that macrophages could envelop the shorter nanotubes more readily than the longer ones. However, no severe inflammatory response such as necrosis, degeneration or neutrophiles infiltration was observed around both MWNT.[32]

The first ecotoxicological study of MWNT was performed by Zhu et al.[67] They examined the interaction of CNT with living unicellular protozoan *Stylonychia mytilus*. They found that MWNT were largely ingested by this kind of organisms. They also examined by optical microscopy the distribution of MWNT in the cells, the redistribution during dividing process as well as the excretion from the cells. The dependence of viability of *Stylonychia mytilus* was determined on the concentration of MWNT, ranging from 0.1 μg/mL to 200 μg/mL. It was found that exposure of *Stylonychia mytilus* to the MWNT with concentration higher than 1.0 μg/mL induced a dose-dependent growth inhibition to the cells and the damage occurred on the macronucleus and external membrane of the cells. The ultra-structural changes observed by electron microscopy revealed that MWNT exclusively localized on the mitochondria. Therefore it was proposed that the damage of macronucleus, micronucleus, and membrane of the cells, as well as growth inhibition of the cells might be a result of the damage of mitochondria.[67]

Very recently Koyama et al[68] evaluated the biological responses to SWNT, two different types of MWNT (20 and 80 nm of average diameter) and cup- stacked carbon nanotubes (CSNT made with stacked truncated cones) subcutaneously implanted in mice (2 milligrams per animal) for up to 3 months. The nanotubes used in this study were purified by thermal treatment process during which the metal particles evaporate. After 1, 2, 3 weeks, 1 month, 2 months and 3 months post-implantation the animals were sacrificed, blood was collected for CD4+ and CD8+ T-cells counting by flow-cytometry and tissue of skin including muscle layers was collected for histopathological examination. All mice survived, and no large changes in their weights were observed during the experimental period. After one week of implantation, only SWNT activated the major histocompatibility complex (MHC) class I pathway of antigen-antibody response system (higher CD4+/CD8+ value), leading to the appearance of an oedematous aspect. After two weeks, significantly high values of CD4+ without changes in CD8+ signified the activation of MHC class II for all samples. The authors noted that antigenic mismatch becomes less evident with time, notably one month post-implantation, indicating an establishment of a granuloma formation. Furthermore, the toxicological response of CNT was absolutely lower than that of asbestos.[68]

Functionalized CNT (f-CNT)

Although it was not for a toxicological study per se, the first on the in vivo behaviour of f-CNT was conducted by Pantarotto et al.[69] The purpose of their study was to produce such a peptide-functionalized SWNT that could immunize mice and serve as synthetic vaccine for foot-and-mouth disease virus (FMDV). They administered an i.p. dose of 100 µg of SWNT-peptide conjugate (the peptide was a neutralizing B cell epitope from FMDV) together with 100 µg of ovalbumin in a 1:1 emulsion in complete Freund's adjuvant and 3 weeks later they repeated the injection in incomplete Freund's adjuvant. The mice used in this study succeeded in producing antibodies to the FMDV and the authors did not report any mortality.[69]

Singh et al[70] functionalized SWNT and MWNT with a chelating agent, the diethylene-triamine-penta-acetic acid (DTPA) in order to label them with indium (^{111}In). After administering a single dose of 400 µg intra-venously to BALB/c mice, the authors followed the radioactivity tracing using gamma scintigraphy. No animal exhibited signs of acute toxicity. The nanotubes were rapidly cleared from systemic blood circulation through the renal excretion route. Urine samples of the treated animals revealed unmodified excreted nanotubes.[70]

A very recent study conducted by Carrero-Sanchez et al[71] compared the toxicological effects between pure and nitrogen doped MWNT (N-MWNT). Different doses of nanotubes suspended in PBS (up to 5 mg/kg) were administered in various ways to mice: nasal, oral, intratracheal, and intraperitoneal. When MWNT were injected into the mice's trachea, the mice could die by dyspnea depending on the MWNT doses. However, N-MWNT nanotubes never caused any death to animals. They found that N-MWNT were far more tolerated by the mice when compared with pristine MWNT. Extremely high concentrations of nitrogen-doped nanotubes administrated directly into the mice's trachea only induced granulomatous inflammatory responses. The other routes of administration did not induce signs of distress or tissue changes on any treated mouse. N-MWNT could therefore be less harmful than MWNT or SWNT.[71]

Conclusion

The data presented in this chapter are still fragmentary and subject to criticisms because of the lack of characterisation of the administered material as well as the non-physiological modes of administration used. Taken together, these preliminary results highlight the difficulties in evaluating the toxicity of this new and so heterogeneous carbon nanoparticles family. Indeed, a number of parameters have considerable impact on the reactivity of CNT. These parameters include CNT structure, size distribution and surface area, surface chemistry and surface charge and agglomeration state as well as purity of the sample. Most of these parameters obviously depend of the method of production of the sample, the method of purification and the method of preparation of the tested suspension. As the reactivity and the general behaviour of CNT in biological media are not completely understood, assessing the safety of these carbon nanoparticles should include a careful selection of appropriate experimental methods.

The concern about the nanotubes in the industrial context was somewhat reduced in 2005 by the relative risk analysis of several manufactured nanomaterials.[72] The results from the analysis performed by Robichaud et al[72] determined that relative environmental risk from manufacturing SWNT, bucky balls, one variety of quantum dots, alumoxane nanoparticles, and nano-titanium dioxide, was comparatively low in relation to other common industrial manufacturing processes.[72]

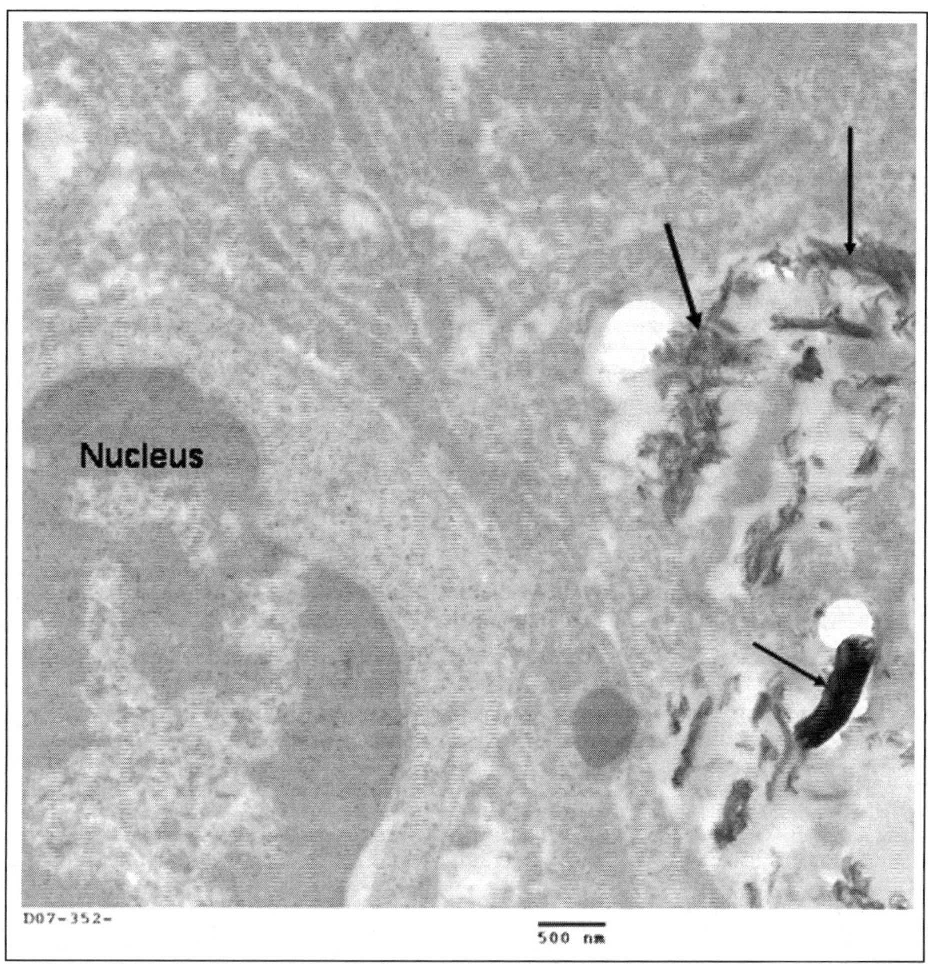

Figure 3. TEM image showing SWNT ropes and bundles (arrows) inside a rat macrophage (Unpublished work).

Nevertheless, available data clearly show that, under some conditions, nanotubes can cross the membrane barriers (Fig. 3) and suggest that if CNT raw materials reach the organs they can induce harmful effects as inflammatory and fibrotic reactions. Thus, many more studies are necessary in order to determine the safety of CNT and their environmental impact.

Acknowledgements

We are thankful to Jéril Degrouard and Danielle Jaillard from the Common Microscopy Center of Orsay.

References

1. Saito R, Dresselhaus G, Dresselhaus MS. Physical properties of carbon nanotubes. London: Imperial College Press, 1998.
2. Popov VN. Carbon nanotubes: Properties and application. Mater Sci Eng 2004; R43:61-102.
3. Baughman RH, Zakhidov AA, De Heer WA. Carbon nanotubes-The route toward applications. Science 2002; 297:787-792.

4. Special issue on carbon nanotubes. Acc Chem Res 2002; 35:997-1113.
5. Balavoine F, Richard C, Mioskowski C et al. Helical crystallization of proteins on carbon nanotubes: A first step towards the development of new biosensors. Angew Chem Int Ed 1999; 38:1912-1915.
6. Bekyarova E, Ni Y, Malarkey EB et al. Applications of carbon nanotubes in biotechnology and biomedicine. J Biomed Nanotechnol 2005; 1:3-17.
7. Lin Y, Taylor S, Li H et al. Advances toward bioapplications of carbon nanotubes. J Mater Chem 2004; 14:527-541.
8. Richard C, Balavoine F, Mioskowski C et al. Supramolecutar self-assembly of lipid derivatives on carbon nanotubes. Science 2003; 300:775-778.
9. Wang J, Musameh M, Lin Y. Solubilization of carbon nanotubes by nafion toward the preparation of amperometric biosensors. J Am Chem Soc 2003; 125:2408-2409.
10. Wang S, Delduco DF, Lustig SR et al. Peptides with selective affinity for carbon nanotubes. Nat Mater 2003; 2:196-200.
11. Mattson MP, Haddon RC, Rao AM. Molecular functionalization of carbon nanotubes and use as substrates for neuronal growth. J Mol Neurosci 2000; 14:175-82.
12. Chen X, Lee GS, Zettl A et al. Biomimetic engineering of carbon nanotubes by using cell surface mucin mimics. Angew Chem Int Ed 2004; 43:6112-6116.
13. Park KH, Chhowalla M, Iqbal Z et al. Single-walled carbon nanotubes are a new class of ion channel blockers. J Biol Chem 2003; 278:50212-50216.
14. Bianco A, Kostarelos K, Partidos CD et al. Biomedical applications of functionalised carbon nanotubes. Chem Commun 2005; 571-577.
15. Bianco A, Prato M. Can carbon nanotubes be considered useful tools for biological applications? Adv Mater 2003; 15:1765-1768.
16. Barone PW, Baik S, Heller D et al. Near-infrared optical sensors based on single-walled carbon nanotubes. Nature Materials 2005; 4:86-92.
17. Lu X, Chen Z. Curved pi-conjugation, aromaticity, and the related chemistry of small fullerenes (< C60) and single-walled carbon nanotubes. Chem Rev 2005; 105(10):3643-96.
18. Chen RJ, Zhang Y, Wang D et al. Noncovalent sidewall functionalization of single-walled carbon nanotubes for protein immobilization. J Am Chem Soc 2001; 123:3838-3839.
19. Minko T. Soluble polymer conjugates for drug delivery. Curr Drug Discov Technol 2005; 2:15-20.
20. Sinha N, Yeow JT. Carbon nanotubes for biomedical applications. IEEE Trans Nanobioscience 2005; 4(2):180-95.
21. Islam MF, Rojas E, Bergey DM et al. High weight fraction surfactant solubilization of single-wall carbon nanotubes in water. Nano Lett 2003; 3:269-273.
22. Moore VC, Strano MS, Haroz EH et al. Individually suspended single-walled carbon nanotubes in various surfactants. Nano Lett 2003; 3:1379-1382.
23. Star A, Stoddart JF, Steuerman D et al. Preparation and properties of polymer- wrapped single-walled carbon nanotubes. Angew Chem Int Ed 2001; 40:1721-1725.
24. Niyogi S, Hu H, Hamon MA et al. Chromatographic purification of soluble single-walled carbon nanotubes. J Am Chem Soc 2001; 123:733-734.
25. Ziegler KJ, Gu Z, Peng H et al. Controlled oxidative cutting of single-walled carbon nanotubes. J Am Chem Soc 2005; 127:1541-1547.
26. Shvedova AA, Castranova V, Kisin ER et al. Exposure to carbon nanotube material: Assessment of nanotube cytotoxicity using human keratinocyte cells. J Toxicol Environ Health A 2003; 66(20):1909-26.
27. Shvedova AA, Kisin ER, Murray AR et al. Exposure of human bronchial epithelial cells to carbon nanotubes causes oxidative stress and cytotoxicity. Ioannina, Greece: 2004:91-103, (Proc Soc Free Rad Research Meeting, European Section, June 26-29, 2003).
28. Cui D, Tian F, Ozkan CS et al. Effect of single wall carbon nanotubes on human HEK293 cells. Toxicol Lett 2005; 155(1):73-85.
29. Jia G, Wang H, Yan L et al. Cytotoxicity of carbon nanomaterials: Single-wall nanotube, multi-wall nanotube, and fullerene. Environ Sci Technol 2005; 39(5):1378-83.
30. Monteiro-Riviere NA, Nemanich RJ, Inman AO et al. Multi-walled carbon nanotube interactions with human epidermal keratinocytes. Toxicol Lett 2005; 155(3):377-84.
31. Murr LE, Garza KM, Soto KF et al. Cytotoxicity assessment of some carbon nanotubes and related carbon nanoparticle aggregates and the implications for anthropogenic carbon nanotube aggregates in the environment. Int J Environ Res Public Health 2005; 2(1):31-42.
32. Sato Y, Yokoyama A, Shibata K et al. Influence of length on cytotoxicity of multi-walled carbon nanotubes against human acute monocytic leukemia cell line THP-1 in vitro and subcutaneous tissue of rats in vivo. Mol Biosyst 2005; 1(2):176-82.

33. Ghibelli L, De Nicola M, Somma G et al. Lack of direct cytotoxic effect of intracellular nanotubes. G Ital Med Lav Ergon 2005; 27(3):383-4.
34. Manna SK, Sarkar S, Barr J et al. Single-walled carbon nanotube induces oxidative stress and activates nuclear transcription factor-kappaB in human keratinocytes. Nano Lett 2005; 5(9):1676-84.
35. Reelfs O, Tyrrell RM, Pourzand C. Ultraviolet A radiation-induced immediate iron release is a key modulator of the activation of NF-kappaB in human skin fibroblasts. J Invest Dermatol 2004; 122(6):1440-7.
36. Bottini M, Bruckner S, Nika K et al. Multi-walled carbon nanotubes induce T lymphocyte apoptosis. Toxicol Lett 2006; 160(2):121-6.
37. Sayes CM, Liang F, Hudson JL et al. Functionalization density dependence of single-walled carbon nanotubes cytotoxicity in vitro. Toxicol Lett 2006; 161(2):135-42.
38. Fenoglio I, Tomatis M, Lison D et al. Reactivity of carbon nanotubes: Free radical generation or scavenging activity? Free Radic Biol Med 2006; 40(7):1227-33.
39. Watts PCP, Fearon PK, Hsu WK et al. Carbon nanotubes as polymer antioxidants. J Mater Chem 2003; 13:491-495.
40. Krusic PJ, Wasserman E, Keizer PN et al. Radical reaction of C60. Science 1991; 254:1183-1185.
41. Fiorito S, Serafino A, Andreola F et al. Effects of fullerenes and single-wall carbon nanotubes on murine and human macrophages. Carbon 2006; 44(6):1100-1105.
42. Tsien M, Morris D, Petruska J et al. Redox cycling and DNA damage induced by iron-containing carbon nanomaterials Abstracts of Papers. Proceedings of the 229th ACS National Meeting, San Diego, CA, United States. IEC-075. Washington, DC: American Chemical Society, 2005:13-17.
43. Worle-Knirsch JM, Pulskamp K, Krug HF. Oops they did it again! Carbon nanotubes hoax scientists in viability assays. Nano Lett 2006; 6(6):1261-8.
44. Magrez A, Kasas S, Salicio V et al. Cellular toxicity of carbon-based nanomaterials. Nano Lett 2006; 6(6):1121-5.
45. Kagan VE, Tyurina YY, Tyurin VA et al. Direct and indirect effects of single walled carbon nanotubes on RAW 264.7 macrophages: Role of iron. Toxicol Lett 2006; 165(1):88-100.
46. Simon A, Thiebault C, Reynaud C et al. Toxicity of oxide nanoparticles and carbon nanotubes on cultured pneumocytes: Impact of size, structure and surface charge. Toxicology Letters 2006; 164(1):20, (S222).
47. Witzmann FA, Monteiro-Riviere NA. Multi-walled carbon nanotube exposure alters protein expression in human keratinocytes Nanomedicine: Nanotechnology, Biology and Medicine. 2006; 2(3):158-168.
48. Soto KF, Carrasco A, Powell TG et al. Biological effects of nanoparticulate materials: Materials Science and Engineering. 2006; 26(8):1421-1427.
49. Tian F, Cui D, Schwarz H et al. Cytotoxicity of single-wall carbon nanotubes on human fibroblasts. Toxicol In Vitro 2006; 20(7):1202-12.
50. Davoren M, Herzog E, Casey A et al. In vitro toxicity evaluation of single walled carbon nanotubes on human A549 lung cells. Toxicology in Vitro 2006, (doi:10.1016/j.tiv.2006.10.007).
51. Casey A, Davoren M, Herzog E et al. Probing the interaction of single walled carbon nanotubes within cell culture medium as a precursor to toxicity testing. Carbon doi:10.1016/j.carbon.2006.08.009.
52. Monteiro-Riviere NA, Inman AO. Challenges for assessing carbon nanomaterial toxicity to the skin. Carbon 2006; 44(6):1070-1078.
53. Meng J, Song L, Meng J et al. Using single-walled carbon nanotubes non-woven films as scaffolds to enhance long-term cell proliferation in vitro. J Biomed Mater Res A 2006; 79(2):298-306.
54. Pulskamp K, Diabaté S, Krug HF. Carbon nanotubes show no sign of acute toxicity but induce intracellular reactive oxygen species in dependence on contaminants. Toxicology Letters doi:10.1016/j.toxlet.2006.11.001.
55. Mattson MP, Haddon RC, Rao AM. Molecular functionalization of carbon nanotubes and use as substrates for neuronal growth. J Mol Neurosci 2000; 14:175-82.
56. Pantarotto D, Briand JP, Prato M et al. Translocation of bioactive peptides across cell membranes by carbon nanotubes. Chem Commun (Camb) 2004; (1):16-7.
57. Shi Kam NW, Jessop TC, Wender PA et al. Nanotube molecular transporters: Internalization of carbon nanotube-protein conjugates into Mammalian cells. J Am Chem Soc 2004; 126(22):6850-1.
58. Nimmagadda A, Thurston K, Nollert MU et al. Chemical modification of SWNT alters in vitro cell-SWNT interactions. J Biomed Mater Res A 2006; 76(3):614-25.
59. Dumortier H, Lacotte S, Pastorin G et al. Functionalized carbon nanotubes are non-cytotoxic and preserve the functionality of primary immune cells. Nano Lett 2006; 6(7):1522-8.
60. Huczko A, Lange H. Carbon nanotubes: Experimental evidence for a null risk of skin irritation and allergy. Fullerene Sci Tech 2001; 9(2):247-50.

61. Maynard AD, Baron PA, Foley M et al. Exposure to carbon nanotube material: Aerosol release during the handling of unrefined single-walled carbon nanotube material. J Toxicol Environ Health A 2004; 67(1):87-107.
62. Lam CW, James JT, McCluskey R et al. Pulmonary toxicity of single-wall carbon nanotubes in mice 7 and 90 days after intratracheal instillation. Toxicol Sci 2004; 77(1):126-34.
63. Warheit DB, Laurence BR, Reed KL et al. Comparative pulmonary toxicity assessment of single-wall carbon nanotubes in rats. Toxicol Sci 2004; 77(1):117-25.
64. Warheit DB. What is currently known about the health risks related to carbon nanotube exposures? Carbon 2006; 44(6):1064-1069.
65. Muller J, Huaux F, Moreau N et al. Respiratory toxicity of multi-wall carbon nanotubes. Toxicol Appl Pharmacol 2005; 207(3):221-31.
66. Shvedova AA, Kisin ER, Mercer R et al. Unusual inflammatory and fibrogenic pulmonary responses to single-walled carbon nanotubes in mice. Am J Physiol Lung Cell Mol Physiol 2005; 289(5):L698-708.
67. Zhu Y, Zhao Q, Li Y et al. The interaction and toxicity of multi-walled carbon nanotubes with Stylonychia mytilus. J Nanosci Nanotechnol 2006; 6(5):1357-64.
68. Koyama S, Endo M, Kim YA et al. Role of systemic T-cells and histopathological aspects after subcutaneous implantation of various carbon nanotubes in mice. Carbon 2006; 44(6):1079-1092.
69. Pantarotto D, Partidos CD, Hoebeke J et al. Immunization with peptide-functionalized carbon nanotubes enhances virus-specific neutralizing antibody responses. Chem Biol 2003; 10(10):961-6.
70. Singh R, Pantarotto D, Lacerda L et al. Tissue biodistribution and blood clearance rates of intravenously administered carbon nanotube radiotracers. Proc Natl Acad Sci USA 2006; 103(9):3357-62.
71. Carrero-Sanchez JC, Elias AL, Mancilla R et al. Biocompatibility and toxicological studies of carbon nanotubes doped with nitrogen. Nano Lett 2006; 6(8):1609-16.
72. Robichaud CO, Tanzil D, Weilenmann U et al. Relative risk analysis of several manufactured nanomaterials: An insurance industry context. Environ Sci Technol 2005; 39(22):8985-94.

Index